Volume One

ENERGY AND ENVIRONMENTAL CHEMISTRY

Fossil Fuels

Edited by
LAWRENCE H. KEITH

ANN ARBOR SCIENCE
THE BUTTERWORTH GROUP

Copyright © 1982 by Ann Arbor Science Publishers
230 Collingwood, P.O. Box 1425, Ann Arbor, Michigan 48106

Library of Congress Catalog Card Number 81-69255
ISBN 0-250-40401-X

Manufactured in the United States of America
All Rights Reserved

Butterworths, Ltd., Borough Green, Sevenoaks
Kent TN15 8PH, England

PREFACE

Energy and Environmental Chemistry—they are inextricably entwined. As we progress toward an increasingly industrialized and energy hungry world the demands of producing more and more energy are going to require greater care if we are to maintain a clean environment. There is no question that we will continue to produce more energy—progress demands it and people demand progress. The only question is how much we will let our continuous quest for energy affect our environment.

Learning about the effects of energy production—directly or indirectly—on the environment is one of the first steps to controlling adverse effects of that production. That's what these volumes are all about. They do not cover it all—do not even come close; the subject area is much too large and complex and we are only in the early stages of learning about the inter-relations of energy and environmental chemistry. But it is hoped that the information in these books will bring us one step closer to understanding some of these relationships and hence to ultimately controlling unwanted pollution from energy production.

Lawrence H. Keith

ACKNOWLEDGMENTS

This is the first of a series of volumes entitled, "Energy and Environmental Chemistry." These volumes are collected papers from distinguished authors on the North American Continent and abroad, and span several national and international meetings beginning with the Second Chemical Congress of the North American Continent in Las Vegas, Nevada (1980).

These symposia were jointly sponsored by the American Chemical Society's Committee on Environmental Improvement and the Division of Environmental Chemistry. However, it was only with the help of the following people that various sessions of these symposia were organized. Without their help and commitment, the symposia, and hence this volume, could never have been accomplished. Special thanks go to:

- José Alberto Celestinos I. and Amanda Cortés-Rubio of the Refinación y Petroquímica Instituto Mexicano del Petróleo, Mexico City, Mexico, for organizing the symposium on "Assessment of the Environmental Impact of Accidental Oil Spills in the Oceans," at the Second Chemical Congress of the North American Continent;

- Dr. Robert Meglen, University of Colorado, Denver, and Dr. Donald W. Denney, Syncrude of Canada, Ltd., Ft. McMurray, Alberta, Canada, for organizing the symposium on "Oil Shale and Tar Sands," at the Second Chemical Congress of the North American Continent;

- Dr. Donald D. Rosebrook and Dr. Robert G. Wetherold, Radian Corporation, Austin, Texas, for organizing the symposium on "Fugitive Hydrocarbon Emissions" at the 181st National ACS Meeting, Atlanta, Georgia; and

- Mr. Ronald K. Patterson, U.S. Environmental Protection Agency, Research Triangle Park, North Carolina, for organizing the symposium on "Coal Gasification" at the Second Chemical Congress of the North American Continent.

Finally, I wish to express my appreciation to the many authors who, by their hard work, dedication and commitment, provided the main substance of this work—scientific facts and evaluations on newly emerging technologies.

Lawrence H. Keith's current technical interests continue to center around analyses of organic pollutants in the environment, with emphasis on developing new methods or improving on old ones. Techniques for the safe handling of carcinogenic and/or extremely toxic materials are also an important aspect of Dr. Keith's current research efforts.

Dr. Keith was formerly involved with the selection of many of the U.S. Environmental Protection Agency's initial 129 Priority Pollutants, and he also helped to formulate some of the initial methodology for analyzing for these pollutants. He is presently involved in the selection of representative compounds and methodologies for the Appendix C Priority Pollutants and for the synfuel industry.

A member of the American Chemical Society's Division of Environmental Chemistry Executive Committee, Dr. Keith has served as secretary, alternate councilor, program chairman and chairman of the division. He is also a past chairman of the Central Texas Section of the American Chemical Society and past secretary and councilor of the Northeast Georgia Section of the American Chemical Society.

In other professional activities, Dr. Keith served as Vice-Chairman of the Gordon Research Conference on Environmental Sciences: Water, and is currently a delegate of the U.S. National Committee to the International Association of Water Pollution Research. He is also a member of the National Research Council Committee on Military Environmental Research. Dr. Keith edited the two-volume *Advances In the Identification & Analysis of Organic Pollutants In Water*, published by Ann Arbor Science.

He and his wife, Virginia, reside in Austin, Texas.

To Virginia, whose help and encouragement have made this book a thoroughly enjoyable task.

CONTENTS
VOLUME 1

Part 3: Impact of Fugitive Hydrocarbon Emissions

CONTENTS
VOLUME 2

Part 1: Point Source Effects

Part 2: Regional Effects

PART 1

IMPACT OF TAR SANDS AND OIL SHALE

CHAPTER 1

THE ALBERTA OIL SANDS
ENVIRONMENTAL RESEARCH PROGRAM

W. R. MacDonald, H. S. Sandhu, B. A. Munson and J. W. Bottenheim
Research Management Division
Alberta Department of the Environment
Edmonton, Alberta, Canada, T5K 2J6

INTRODUCTION

Energy is of major concern in industrial countries and considerable activity is directed to finding, developing and using every form of energy resource. Canada is no exception, and in this country the Province of Alberta is the major supplier of hydrocarbon energy forms to the entire nation. A considerable fraction of Canada's energy is still imported, however.

Alberta has immense supplies of various forms of energy, in comparison to the total energy resource considered to exist in Canada. Natural gas and conventional oil are the two prime resources. Coal, low in sulfur, is abundant for thermal and metallurgical uses, and its production is going through a period of rapid development. Hydroelectric potential has been developed on a couple of rivers, and is presently being rigorously assessed on two others in the north. By far the largest source of energy exists in the oil sands of Alberta, containing an estimated 960 billion barrels of bitumen. (Bitumen is the unrefined hydrocarbon obtained from the oil sands.) Located in the northern half of the province, there are four distinct deposits considered viable for energy recovery: Athabasca (720 billion), Wabasca (24 billion), Cold Lake (159 billion) and Peace River (64 billion).

Knowledge of the Athabasca deposit goes back two hundred years, when

Alexander Mackenzie reported observations of oil sands along the river. A process to extract bitumen was developed by Dr. K. Clark in 1923, and this was followed by several attempts to produce a commercial product. Today there are two surface mines in operation, Suncor (formerly Great Canadian Oil Sands) and Syncrude, both located in the Athabasca deposit. Numerous pilot plants exist, all of which attempt to extract bitumen from underground by way of in situ processes: for example, by fire-flooding, or the injection of steam and hot water. At present two new plants are being proposed; one will be a surface mine in the Athabasca deposit, the other an in situ operation in the Cold Lake deposit. It should be noted that these deposits have considerably different characteristics. The Athabasca Oil Sands contain a bitumen-clay-sand mixture which requires specific treatment to release the bitumen; the Cold Lake deposit contains a heavy crude which must be thermally induced to achieve reasonable flow for pumping to the surface. As shown in Figure 1, the bituminous sands are under glacial drift, Cretaceous shale and sandstones, and overlay a limestone formation [1]. Oil sands contain 0 to 18% by weight bitumen, averaging about 12%. Overburden depth varies; surface mining can take place where overburden is up to 60 m thick, whereas other methods will be required to extract bitumen with greater overburden depths.

Industrial interests have concentrated on developing technology for energy recovery. In order to stimulate this interest, the Alberta Government established the Alberta Oil Sands Technology and Research Authority (AOSTRA) in 1974. Originally provided with a budget of $100,000,000 for five years, the Authority now estimates it will have approximately $250,000,000 in funds to be spent over several years. The major portion of funds is spent jointly with industry in pilot operations, and other funds are spent on general research.

Environmental issues related to oil sands development became of central interest only in the early 1970s as industrial activity rapidly increased. In 1975, the governments of Canada and Alberta entered into an agreement to jointly sponsor the ten-year Alberta Oil Sands Environmental Research Program (AOSERP). Funding was anticipated to be $40,000,000 during the lifetime of AOSERP.

AOSERP is not the only sponsor of environmental research in the oil sands. Industry has undertaken its own research, primarily restricted to specific leases, and numerous companies have formed the Oil Sands Environmental Study Group (OSESG) to further their general interests.

In this chapter, the work sponsored by AOSERP from 1975 to 1980 is reviewed, and particular attention is paid to baseline studies and research of interest to chemists. Emphasis is on the processes and effects of pollutants on the ecosystem.

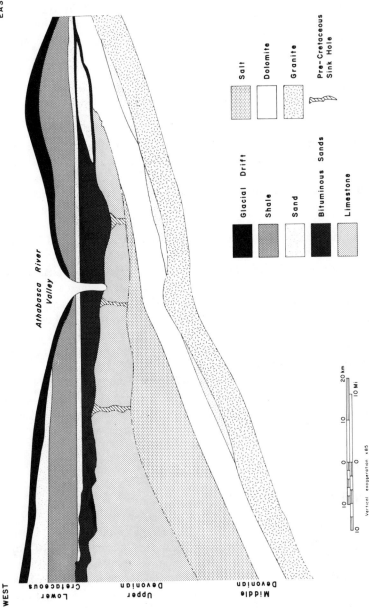

Figure 1. A simplified cross section of the Athabasca Oil Sands deposit [1].

AOSERP

As mentioned above, AOSERP was established in 1975 by the governments of Canada and Alberta. Program activities were restricted to a well-
defined region in the Athabasca Oil Sands deposit (Figure 2). The general
program objective in the formal agreement [2] stated, in part:

> To undertake environmental research relative to the renewable re-
> sources . . . directed to the solution of practical social and technical prob-
> lems resulting from oil sands development.

Under this general objective, AOSERP would not become involved in technology of extraction, plant processes or research on recovery of bitumen
from tailings ponds. Such research was considered a responsibility of industry
or AOSTRA. Moreover, AOSERP would not undertake lease-specific research, but would direct attention to regional issues in the Athabasca Oil
Sands.

The initial AOSERP structure created eight committees: air quality and
meteorology, terrestrial fauna, aquatic fauna, vegetation, hydrogeology,
hydrology, water quality and human environment. Committee membership
consisted of government representatives, an industry representative and,
infrequently, an academic representative. In 1977, AOSERP was reorganized
under a central management, and created a structure which subsumed research into the Air, Land, Water and Human Systems. Scientific advisory
committees, consisting of recognized experts in environmental sciences, were
struck to provide independent advice.

In 1979 the government of Canada withdrew from AOSERP as a result
of a fiscal restraint policy. Alberta is continuing to support research at the
level of $2,000,000 per year. As the federal-provincial agreement no longer
is in effect, research is not restricted to the Athabasca deposit.

REGIONAL CHARACTERISTICS

The AOSERP study area, in the NE quadrant of the province, lies between 56° and 59° north latitude, covering an area of about 2.86×10^4 km^2
(Figure 2). In this region, the climatic conditions resulting from latitudinal
effects and proximity to arctic air masses strongly influence meteorological
conditions. In the winter, cold high pressure areas moving from the arctic
often result in clear and cold weather with negligible precipitation and intermittent buildup of ice crystal fog. Atmospheric subsidence associated with

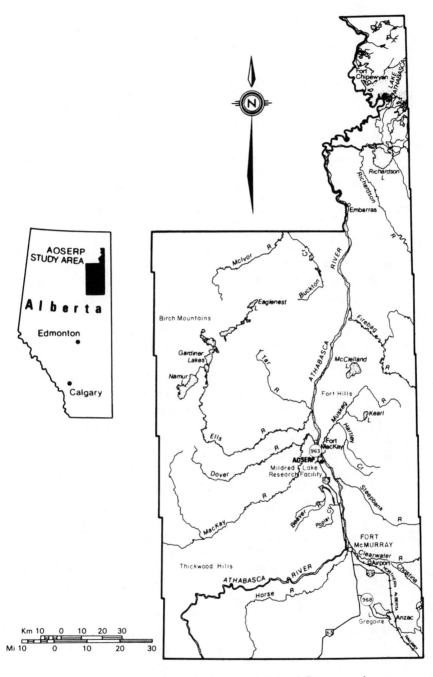

Figure 2. The Alberta Oil Sands Environmental Research Program study area.

stationary or slow-moving high pressure ridges in the summer is characterized by extended periods of clear, dry weather.

The marked variation in climate throughout the year is to a large extent due to the significant seasonal differences in solar insolation. Neglecting the influence of the earth's atmosphere, the area receives more energy than areas at the equator during the summer solstice, and during late December the daily inflow is less than 15% of that at the equator. In general, the climatic zone of the AOSERP area is classified as "cold temperate" with the added subclassification of "cool short summers". This indicates a continental type of climate with significant daily and seasonal variation in temperature (summertime maximum occasionally near 30°C, wintertime minimum near -45°C) and precipitation (435 mm annually).

The population in the region was very small until recently. Fort McMurray, located at the confluence of the Athabasca and Clearwater Rivers, had a population of only 1186 in 1961, with a regional population of 2644. By 1978, Fort McMurray's population was approximately 25,000, with about 2000 additional people in the region. In 1977, there were an additional 7000 workers, not included in annual census figures, residing in the Syncrude construction camp.

The study area lies in the Boreal Forest region, characterized by tree species such as white spruce, black spruce, jack pine, trembling aspen, balsam poplar and white birch. Little commercial timber is being harvested. Soils of the study area are mainly muskeg, which can be worked in the winter (unless previously drained), sandy soils and soils developed on glacial till.

The region lies on the Central and Mississippi flyways for migrating waterfowl. The bituminous sand region is not important for waterfowl production; however, directly to the north the Peace-Athabasca Delta is an important breeding ground as well as a staging ground during spring and fall migration.

Central to the drainage of the region is the Athabasca River, which originates in the Rocky Mountains and drains northeasterly for 1300 km before emptying into Lake Athabasca. In the region this meandering river has incised more than 150 m into the land surface. Inflow of the Athabasca River to the region averages 17.2 Gm^3 ($10^9 m^3$) annually, with additions of tributaries adding 6.9 Gm^3 annually before the river reaches Lake Athabasca. Several other much smaller lakes exist in the region, only two of which are presently impacted directly by industrial operations, while others are accessible for recreation. Most of these lakes are characterized by low mean depths, high nutrient levels, abundant macrophytic growth and moderate to high productivities.

A more detailed account of regional characteristics may be found in the AOSERP Interim Report to 1978 [1].

GENERAL BASELINE STUDIES AND RESEARCH

As AOSERP was originally planned to be a ten-year program, during the first three years emphasis was on conducting baseline studies. Research assumed a priority role in the fourth and fifth years and current work is almost exclusively in a research mode.

Air System

Air System research has been concerned with the transport, dispersion, transformation and deposition of airborne pollutants within the study area. A major emphasis has been on data collection by means of a nine-station meteorological network, a 152-m tower, minisonde profiling programs and a three-station air quality network. In addition, cooperative ventures with industry continue to collect both meteorology and air quality data in the immediate vicinity of presently operating oil sands plants. Data collection techniques such as precipitation radar, satellite imagery and Doppler acoustic sounding systems also have been investigated.

Specialized data collection through several summer and winter field studies has made use of airborne and ground-based measurements to examine plume rise, dispersion and chemical reaction rates for the existing operating plants. Baseline states of the atmosphere in the area have been documented and analysis of the extensive data bases is continuing. Additional data collection will be required to examine atmospheric processes in regions of complex terrain.

Pollutant deposition characteristics in the area have been investigated. Snowpack loading has been measured and analyzed, preliminary estimates of dry deposition rates are documented, and precipitation chemistry data for an event sampling network are being analyzed.

A substantial effort has been directed toward employing computer models to synthesize the data and atmospheric processes in the study area. The modeling work has proceeded after assessment of model suitability and perceived user needs for model predictions. The Climatological Dispersion Model, CRSTER, and Alberta Environment's Gaussian models have all been run to compute air quality characteristics for comparison with measured data. Examination of a model of the convective boundary layer and the applicability of more complex models such as LIRAC and ADPIC also have been accomplished. A plume chemistry model, discussed in a later section, has been developed for incorporation into the atmospheric dispersion models.

Land System

Extensive surveys were conducted in the region to identify and map soils, topography, geology and vegetation.

Faunal investigations were aimed at establishing baseline conditions and populations, and some research was undertaken to examine management practices. The major ungulates are moose and caribou, with an estimated population in 1976 of 5000 and 400, respectively. Trapping of furbearing animals, especially beaver and muskrat, occurs throughout the region. The black bear, estimated at between 5000 and 7000 in 1977, has begun to interact with the increased human population in a predictable manner, as nuisance complaints are becoming more common.

Reclamation research has been designed to take advantage of natural successional trends by utilizing native species and naturalized introduced species to accelerate revegetation. Responsibility for governmental reclamation research now resides in another provincial interdepartmental committee, and is not within the AOSERP mandate. An interesting discovery in reclamation trials on the Suncor tailings pond dike has been the substantial rise in small mammal populations as a result of habitat provided by heavy growth of grass. These small mammals then damage trees planted for long-term reclamation. Various control methods have been examined.

Water System

Early work in the program established networks to gather baseline data on surface water hydrology and water quality. Water quality studies have provided considerable information on major ions, suspended materials and sediments, nutrients and dissolved metals. Because the Athabasca River cuts through the bitumen deposit, and many of its tributaries flow through muskeg regions, these watercourses have a naturally high level of organics and trace metals. Groundwater discharge also contributes dissolved constituents to surface waters.

Biological studies have characterized the microbiota, macro- and micro-invertebrate communities and fish populations in the surface waters with the respective habitat conditions which are important to the maintenance of the biotic community. Fish surveys have identified a total of 27 species in the area, many of which are important for commercial or recreational purposes.

Human System

The tremendous growth in Fort McMurray, a direct result of energy development, and its effects on the local population has been the subject of several studies of the Human System. Such studies have examined the history of the region, demographic structure, economic indicators, employment and labor composition, housing, human adjustment, environmental health and environmental perceptions.

CHEMISTRY-RELATED RESEARCH

Air

There are currently two sources of industrial emissions in the oil sands: Suncor, which began operation in 1967, and Syncrude, which started in 1979. Both plants discharge SO_2 and NO_X as the pollutants of concern. The total emissions of these two species are compared with totals for other areas in Table 1.

Table 1. Emissions of SO_2 and NO_X

Region	SO_2	NO_X
	(10^4/metric ton/yr)	
Oil Sands Area	11 (1978)[a]	2 (1979)[b]
Alberta	55 (1978)[a]	37 (1979)[b]
Canada	500 (1975)[c]	190 (1975)[c]
United States	2570 (1975)[c]	2220 (1975)[c]

[a]From Ref. 3.
[b]From Ref. 4.
[c]From Ref. 5.

During the program, an extensive air quality monitoring network has been collecting relevant data using exposure cylinders and eight instrumented continuous monitoring stations [6]. Background sulfur dioxide concentrations at ground level are in the 0.001 ppm range; at sites closer to the emission source the monthly concentrations range from 0.003 to 0.005 ppm. Monthly average background levels of O_3 vary from approximately 50 ppb in the summer to 20 ppb in the winter, but NO_2 background levels appear to be below the detection limit of 5 ppb. Results of total sulfation measurements do not indicate any change in levels of total sulfur over the past three years.

A review of atmospheric chemistry, with particular emphasis on SO_2 as the major pollutant, was carried out to assess the importance of chemical transformation under Alberta conditions [7]. It was concluded that high ozone concentrations and the formation of other oxidants could occur during the summer, provided enough reactive hydrocarbons from the extraction plants, and possibly from natural emission, were present. It has been estimated by computer simulation that an average lifetime of SO_2 of 3.5 days during typical summer conditions and 522 days during winter conditions is to be expected from gas-phase reactions [8].

Field studies were first undertaken in 1977 when only one emission source existed [9-11]. The findings from these studies can be summarized as follows:

1. no distinguishable oxidation of SO_2 during the winter, or summer nights;
2. oxidation rate of SO_2 of approximately 1-3%/hr during summer days;
3. during summer days excess ozone formation was sometimes observed at downwind locations; and
4. general high particulate loading (about 0.2 per gram SO_2) and high relative humidity within the plume.

While the last observation would suggest favorable conditions for oxidation of SO_2 in the liquid phase or possibly via heterogeneous mechanisms, the first three observations strongly suggest homogeneous gas-phase oxidation of SO_2.

These results stimulated the development of a more concise theoretical model based on homogeneous gas-phase oxidation of SO_2. From the onset it was clear, however, that proper model evaluation required additional chemical data specifically related to concentrations of NO_x, O_3, SO_2 and hydrocarbons. Additional field studies are being undertaken to more fully understand the chemistry of plumes.

The plume chemistry model is being designed along the lines of a model originally proposed by Freiberg [12], and expanded on by Lusis [13]. In this model, dispersion is described by an ideal, discrete Gaussian formulation; diffusion within the plume and entrainment of ambient air is included, but chemistry is partly decoupled from dispersion. The chemical model to be used here consists of a compact mechanism which includes essential features of smog chemistry and homogeneous oxidation of SO_2 and which is geared to the chemistry expected to be important in a plume. It was extensively validated against recent smog chamber data.

A first test of the model is presented in Figure 3. Experimental data for June 19, 1977 were used in an attempt to match the O_3 and SO_2 plume profiles [9]. In this case, estimates had to be made for some parameters that were not experimentally determined (notably the NO_2/NO_x ratio and NO_x profile, hydrocarbon levels within and outside the plume, and initial HNO_2 levels). The need to make these estimates, albeit unbiased toward the results, unfortunately severely limits the value at this time of the excellent fit shown in this figure.

The fate of gaseous oxides of sulfur and nitrogen is of great concern in the area because these compounds are characteristically associated with acidity. Removal from the atmosphere can occur by dry and wet deposi-

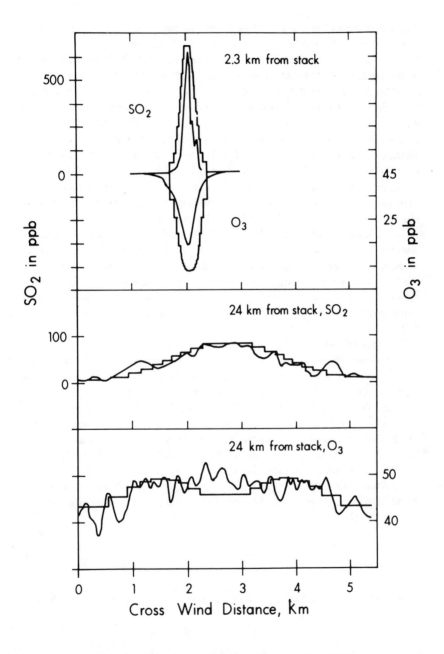

Figure 3.   Comparison between experimental and computed plume profiles for SO_2 and O_3 on June 19, 1977 in the Suncor plume. Stepwise curves are computed.

tion mechanisms, and a literature review of pollution deposition processes was undertaken to provide direction for research studies [14]. An early study to determine sulfur deposition patterns and to estimate how much of the sulfur emitted by the operating Suncor plant was deposited in snow found surprisingly high pH values for the snow pack [15]. Much of the snow in the vicinity of the plant had a pH above 6.0; in remote areas, values between 4.9 and 5.5 were found. It was postulated that the alkaline buffering constituents were from wind-blown dust, products of slash burning, or emissions of alkaline metal oxides by the plant. Furthermore, an overall budget calculation was performed indicating that less than 0.2% of the sulfur emitted comes to the ground within a 25-km radius of the source. The deposition velocity of gaseous sulfur dioxide in a winter situation also was estimated, and the results indicate a value for Vd of 0.3 to 0.4 cm/sec.

Some initial studies into dry and wet deposition in the summer were also undertaken. A lower limit for dry deposition was determined with a network of Harwell collectors [10]. Since such collectors preferentially sample particulates in the 2 to 10 μm diameter range, these data refer mainly to wind-blown dust sources and some particulates from the power plant plumes, while particulate sulfur, which is in the submicron range, will be underestimated.

Wet deposition of SO_2 and particulate sulfate results from two types of processes: those associated with cloud phenomena (rainout) and those concerned with scavenging by falling precipitation (washout). Rain event samples were collected over two seasons at 10 sites in the area, most of them remote from the oil sands development area [16]. Results of analyses are summarized in Table 2, and compared with similar measurements in Saskatchewan and southern Ontario. The pH of rain in the area averages 5.3, which is slightly acidic but well above the southern Ontario value of 4.2, which is attributed to substantial industrial emissions in eastern Canada and the United States. An intriguing and as yet unexplained feature in these data is the high concentration of chloride (Cl), in particular in view of the normal sodium (Na) levels.

Water

In addition to the baseline studies, aquatic ecosystem research has focused on two other main areas:

1. identification and toxicity of potential contaminants; and
2. development of specific applied research areas directed to the selection of monitoring techniques for general ecosystem health.

Table 2. Constituents in Rainwater in the AOSERP Study Area and Northern Saskatchewan Compared to Those in Eastern Canada

Rainwater Constituent	May-Sept. 77 AOSERP Network All Sites		July-Aug. 77 Cansap Network Cree Lake, Sask. Precip. Weighted Mean	June-Aug. 77 Mount Forest Southern Ontario Precip. Weighted Mean
	Arithmetic Mean	Precip. Weighted Mean		
pH	5.2 (77)[a]	5.3 (64)	5.4	4.2
$SO_4^=$-S	0.18 (129)	0.14-0.18 (113)	0.3	2.9
Cl	0.54 (130)	0.48 (114)	0.05	0.19
NO_3^--N	0.06 (109)	0.05 (97)	0.06	0.74
NH_4^+-N	0.12 (89)	0.12-0.13 (79)	0.24	0.51
K^+	0.32 (101)	0.28 (87)	0.10	0.092
Na^+	0.10 (101)	0.06-0.08 (87)	0.06	0.18
Mg^{++}	0.10 (101)	0.06-0.09 (87)	<0.01	0.27
Ca^{++}	0.11 (101)	0.09 (87)	<0.05	0.82
Cond. (μmho/cm)	9.6 (74)	8.9 (61)	6.6	53.1

[a]Numbers in parentheses are the number of data used to calculate the mean.

There are three major sources of potential contaminants which can be related to oil sands developmental stages:

1. initial development: site clearing, drainage of muskeg, lake drainage and overburden removal and storage;
2. mine depressurization waters; and
3. process effluents.

These sources of contaminants are dealt with in more detail in the following sections.

Initial Development

During the initial developmental stages of site clearing and muskeg drainage, the primary effect from these discharge waters is attributed to an increased organic content and/or increased suspended solid load in receiving streams [17]. Impacts on faunal components can be attributed to mechanical, abrasive damage to organisms, clogging of respiratory apparatus or destruction of habitat [18].

The Athabasca River at Fort McMurray has a naturally high suspended solids load, 13.6 x 10^6 metric ton/yr [19]. Less than 0.9 x 10^6 metric ton/yr of sediment are contributed by the tributaries entering the Athabasca River between Fort McMurray and Embarras Channel. This contribution is about 5% of the total sediment entering the Athabasca Delta [18]. Suspended solids resulting from construction activities are not expected to impact the Athabasca River; instead the major effect will occur in the individual receiving tributaries.

Mine Depressurization Water

During surface mining of oil sands it is necessary, for mine access and pit wall stability, to dewater the oil sand formation prior to and during pit excavations. The chemical content of these mine depressurization waters is highly variable; however, two characteristics are consistent: high salinity ($>$100,000 mg/L Cl^-) and potent toxicity [20-24]. A unique feature regarding mine depressurization water is the observed increase in the toxicity of the "whole effluent" with storage time. This presents difficulty when the results of acute and chronic tests are evaluated. The majority of the constituents in mine depressurization water are not toxic individually at the concentrations present in the water. This suggests that a synergistic response is occurring in whole effluent testing.

Process Effluents

Effluents from the operating plants result from two major processes:

1. extraction of the bitumen from the oil sands and
2. upgrading of the bitumen to synthetic crude oil.

Under current provincial regulations, these effluents are held in tailings ponds, as the concept of zero discharge is applied to prevent discharge of process effluents to natural watercourses. This concept creates fairly large water bodies; for example, the tailings ponds for Syncrude covers an area of approximately 28 km^2 and presents a whole series of problems for wildlife biologists, especially those responsible for management of waterfowl. This concept of zero discharge presents concern in three areas:

1. contamination of groundwater,
2. major dike failure, and
3. future reclamation.

These concerns form the basis for ongoing research planning and design.

Trace elements, especially metallic elements, identified as components of bitumen and therefore potential constituents of effluent or emission discharge, have received considerable aquatic bioassay testing. Nickel and vanadium have been identified as the most abundant metallic elements in the raw bitumen [25-27]. Pure solutions of vanadium and nickel were tested under acute and chronic exposures and both metals were assessed to be moderately toxic and possessed minimal bioaccumulation potential [23, 28, 29]. Bioassay tests using binary mixtures of nickel and vanadium indicate that they are antagonistic, whereas tertiary mixtures of V, Ni and phenols were synergistic, with a potency greater than any pure solution or binary mixture of these compounds [29].

Any program which attempts to assess the impact of industrial development on the aquatic ecosystem must develop an understanding of the processes which are occurring and the rate of natural assimilation. Studies to assess heterotrophic assimilation (biological uptake of constituents largely mediated by bacterial cells) have been conducted or are currently ongoing. These investigations have focused on the chemical and microbial characteristics of the Athabasca River and its tributaries, the ability of lotic microorganisms and macrobenthos to degrade and assimilate bitumen and the identification of lower trophic organisms [30-34].

In consideration of the results of all aquatic research conducted to date, no significant impact of current oil sands mining operations on the aquatic environment has been identified.

Land

As noted in the preceding section, most work in the Land System was directed toward describing baseline states for vegetation and animal species. The research relevant to this chapter relates to the effects of industrial emissions on vegetation and soils in the region.

The major research of importance has examined the previsual biochemical and physiological responses, and symptomological effects of vegetation exposure to SO_2 [35-38]. Some investigation of mixtures of SO_2 and NO_2, Ni and V also have been reported in these studies. Although laboratory experiments have revealed various effects at all levels of exposure, field investigations find that, even though there is an apparent decreasing gradient of sulfur in soils and vegetation outward from the source, no appreciable difference in the biochemical or physiological functions of field-collected material could be detected.

It is important to note that research conducted elsewhere has indicated that pollution effects on vegetation, once they are visible, may result in irreversible damage. Consequently, in order to understand more fully the processes occurring at the previsual level, where effects may be transient and the vegetation can recover, the research sponsored by AOSERP concentrated on biochemical and physiological responses. Furthermore, indigenous species were used, as no previous data on these species existed for the oil sands region.

Laboratory experiments under controlled environmental conditions in fumigation chambers permitted several plant species to be exposed to SO_2 at various concentrations [35-38]. The following findings relate to exposure at 0.34 ppm, which is the Canadian permissible one-hour ground level concentration standard.

Based on a comparison of the net CO_2 assimilation rate (NAR) for eight species, the relative sensitivity to fumigation in decreasing order is willow, paper birch, aspen > green alder > Labrador tea > jack pine > white spruce, black spruce [35]. An indication of the assimilation rate change for one species is shown in Figure 4. By the use of photographs, the time required to detect visual symptoms could be recorded. Again, the same relative visual response was found as for the physiological response: deciduous species were more sensitive than Labrador tea, which was more sensitive than conifers; the time for visual effects to be evident varied from approximately 2 days for willow to approximately 20 days for white spruce.

In order to compare the response of species grown on "uncontaminated" native soils to those grown on a tailings pond dike, samples of conifers were recovered from these two sources and exposed to 0.34 ppm SO_2. It was found that the vegetation from the dike was more sensitive, having a much faster decrease in the NAR with time than undisturbed plants. Furthermore,

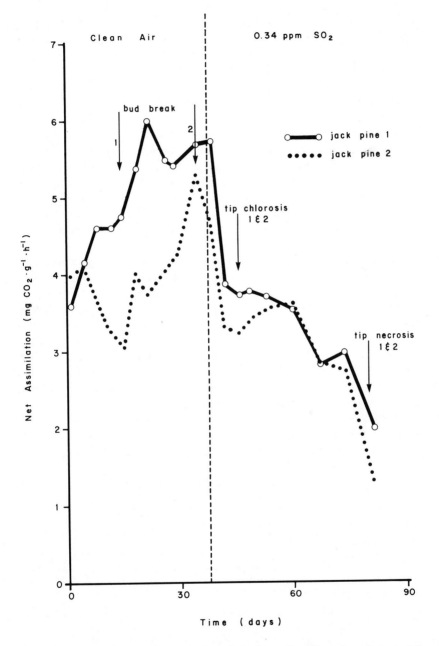

Figure 4. Net CO_2 assimilation rate of two jack pine saplings from the Athabasca Oil Sands area during exposure to SO_2. Plants are in early spring condition and measurements were taken at 600 μE^{-1} at 18°C.

visual symptoms appeared in dike species after an almost complete loss of NAR, but occurred earlier in the undisturbed plants. Although analysis for other constituents in the vegetation was not undertaken, it is believed the dike vegetation may contain other toxic materials and have a decreased resistance to SO_2.

Intensive experiments on jack pine, the most sensitive of the conifers, examined effects on the activities of important enzymes such as peroxidase (implicated in growth and aging), glycollate oxidase (associated with photo-respiration) and ribulose diphosphate carboxylase (primary enzyme in photo-synthesis) [37-38]. All of these activities were altered considerably, and were observed well in advance of the appearance of visual symptoms. Although these biochemical and visual responses can be detected when fumigation is terminated after seven days, recovery of the plant is evident. Other details on effects of V, Ni and NO_2/SO_2 mixtures are contained in the above reports [35-38].

In order to examine actual effects on species exposed in the field, samples were taken from trees in the region, both in an "impingement" area close to the emission source and at a "control" site, and subjected to the biochemical analyses studied above. No significant differences in any of the metabolic responses were observed between the two sites. This was attributed to either the low levels of SO_2 experienced in the region, which are always significantly lower than the 0.34 ppm standard, or the ability of the vegetation to recover its metabolic functions between rare incidents of heavy fumigations. With these field samples, the levels of sulfur were determined and were at the lower end of the normal sulfur range for such tissues.

In a related study [39-41], a set of 11 sites was established to provide baseline information and to be examined for purposes of biomonitoring. These sites were described with respect to their vascular and cryptogamic species, stand density and age, and soil characteristics.

The following results indicate the nature of this project:

1. Al, Fe, S in soils showed some variation, but the variability of the data was so great much more sampling would be necessary to draw any conclusions.
2. Cryptogamic communities quantified in 1976 showed no significant difference in cover in 1979.
3. In almost every plant species there was a significant decrease in Al, Fe, S with distance from the emission source; whether this is due to natural variability or pollutant deposition is not known (the concentrations of Al, Fe, S did not change appreciably over three years, although deposition of S is known to occur).
4. Sulfation plates, which measure total SO_2 in the atmosphere over a period of time, showed S deposition was negatively but significantly correlated with distance from the source; also, sulfation data correlated well with S in jack pine, white spruce, and feathermoss tissues (Table 3).
5. A survey of the two dominant lichen species at 60 locations indicated that the pattern of both sulfur and metal deposition could be determined

Table 3. Sulfur Content (ppm) of Plant Material from Biomonitoring Plots

Distance from Source	Jack Pine	White Spruce	Bearberry	Labrador Tea	Ground Lichen	Feather-moss
<10 km	880±38	833±61	780±66	1101±87	457±40	1229±77
10-25 km	772±36	697±45	473±29	1015±68	445±48	1076±71
>25 km	781±45	675±45	538±41	1116±42	304±15	839±37

using biological indicators. In addition, a significant proportion of the variability in lichen condition could be accounted for by the content of S, V, Ti and Al.

Although this study does indicate that there is a pollutant deposition pattern, there is no evidence of any detectable effects on vegetation.

CURRENT RESEARCH

During the first five years of AOSERP, the program moved from a baseline information-gathering phase to concentrating almost exclusively on research. In the current year, research underway consists mainly of extensions of previous work, with initiation of some new projects; approximately $1,300,000 is being spent on projects. This is a year of evaluation of present knowledge and assessment of what future directions should be taken.

In the Air System, research is continuing in the characterization of plume chemistry—in particular, that existing under midwinter conditions. Air quality data collection and analysis of emissions from a pilot in situ operation continue. Because of the meteorological conditions existing in the region, pollutant transport eastward out of Alberta and into neighboring Saskatchewan is known to occur. In order to examine this situation, the governments of Alberta, Saskatchewan and Canada are sponsoring a study which will review all relevant information, prepare an inventory of industrial emissions with estimates to the year 2000, identify sensitive areas, and recommend research and monitoring projects.

In the Land System, effects research continues, with emphasis on gaseous mixtures. As it will be important to ensure that suitable sites are readily available for future research or the establishment of biomonitoring systems, a network of permanent sample plots is being defined and will be set aside as restricted areas.

The suitability of several aquatic biomonitoring techniques is presently being evaluated in one small project in the Water System. Negotiations continue to progress on a major project, to be jointly funded with AOSTRA, which will examine contaminants in in situ operations. Again this year regional water quality data are being gathered.

At the present time, a major report reviewing the results of the past five years of research is being prepared. As this report will be the only major synthesis of research information that is available, it is expected to be useful in providing guidance on future research needs.

ACKNOWLEDGMENTS

The authors thank their colleagues for providing research information and for useful discussions.

REFERENCES

1. Smith, S.B., Ed. "Alberta Oil Sands Environmental Research Program Interim Report Covering the Period April 1975 to November 1978," prepared by A.S. Mann, R.A. Hursey, R.T. Seidner, and B. Kasinska-Banas, Edmonton, Alberta, AOSERP Report 22 (1979).
2. "Canada-Alberta Agreement for the Alberta Oil Sands Environmental Research Program, Amended September 1977."
3. Sandhu, H.S. "Industrial Sulphur Emissions for Alberta, 1974 to 1978," Alberta Department of the Environment, Research Secretariat, Edmonton, Alberta (1979).
4. Peters, R., and H.S. Sandhu. "Nitrogen Oxide Emissions for Alberta," Alberta Department of the Environment, Research Management Division, Edmonton, Alberta (1980).
5. "The LRTAP Problem in North America: a Preliminary Overview," prepared by the United States-Canada Research Consultation Group on the Long-Range Transport of Air Pollutants, Atmospheric Environment Service, Environment Canada (1979).
6. Strosher, M.M., "Ambient Air Quality in the AOSERP Study Area, 1977." prepared for the Alberta Oil Sands Environmental Research Program by Alberta Environment, AOSERP Report 30 (1978), 74 pp.
7. Bottenheim, J.W., and O.P. Strausz. "Review of Pollutant Transformation Processes Relevant to the Alberta Oil Sands Area," prepared for the Alberta Oil Sands Environmental Research Program by the University of Alberta, Hydrocarbon Research Centre, AOSERP Report 25 (1977), 166 pp.
8. Bottenheim, J.W., and O.P. Strausz. "Computer Modelling on Polluted Atmospheres and the Conversion of Atmospheric Sulfur Dioxide to Sulfuric Acid," prepared for Alberta Environment, Research Secretariat Report 1978/5 (1978).
9. Fanaki, F., Compiler. "Air System Field Study in the AOSERP Study Area," prepared for the Alberta Oil Sands Environmental Research

Program by Fisheries and Environment Canada, Atmospheric Environment Service, AOSERP Report 24 (1979), 182 pp.

10. Fanaki, F., Compiler. "Air System Summer Field Study in the AOSERP Study Area," prepared for the Alberta Oil Sands Environmental Research Program by Atmospheric Environment Service, AOSERP Report 68 (1979), 248 pp.

11. Lusis, M.A., K.G. Anlauf, L.A. Barrie and H.A. Weibe. *Atmos. Environ.* 12: 2429 (1978).

12. Freiberg, J. *Atmos. Environ.* 10: 121 (1976).

13. Lusis, M.A., Proceedings of the 7th International NATO/CCMS Technical Meeting on Air Pollution Mod. and Its Appl., Airlie, VA, (September 7-10, 1976), pp. 831-855.

14. Denison, P.J., R.A. McMahon and J.R. Kramer. "Literature Review on Pollution Deposition Processes," prepared for the Alberta Oil Sands Environmental Research Program and Syncrude Canada Ltd. by Acres Consulting Services Limited and Earth Science Consultants, Inc., AOSERP Report 50 (1979), 264 pp.

15. Barrie, L.A., and D.M. Whelpdale. "Background Air and Precipitation Chemistry," in "Meteorology and Air Quality Winter Field Study in the AOSERP Study Area, March 1976," F. Fanaki, Compiler, AOSERP Report 27 (1978), 249 pp.

16. Barrie, L.A., V. Nespiak and J. Arnold. "Chemistry of Rain in the Athabasca Oil Sands Region: Summer 1977," prepared for the Alberta Oil Sands Environmental Research Program by Atmospheric Environment Service, Internal Report ARQT 3-78, AOSERP Project ME 1.4 (1978).

17. Mayhood, D.W. "Monthly Report, May 1980. Aquatic Monitoring Study, Alsands Project Area," Aquatic Environments Ltd. Progress Report to Alsands Project Group (June 10, 1980), 10 pp.

18. Thurber, Crippen and Northwest Hydraulics Consulting Ltd. "Athabasca River Power Development: Crooked Rapids Dam Feasibility Study," prepared for Alberta Environment (1975).

19. Griffiths, W.H., and B.D. Walton. "The Effects of Sedimentation on the Aquatic Biota," prepared for the Alberta Oil Sands Environmental Research Program by Renewable Resources Consulting Services Ltd., AOSERP Report 35 (1978), 86 pp.

20. Machniak, K. "The Impact of Saline Waters Upon Freshwater Biota (A Literature Review and Bibliography)," prepared for the Alberta Oil Sands Environmental Research Program by Aquatic Environments Ltd., AOSERP Report 8 (1977), 258 pp.

21. Tsui, P.T.P. "The Effects in Mine Depressurization Groundwater on Aquatic Organisms: A Literature Review," being prepared for the Alberta Oil Sands Environmental Research Program by Aquatic Environments Ltd., AOSERP Project WS 2.6.1.

22. Lake, W., and W. Rogers. "Acute Lethality of Mine Depressurization Water on Trout Perch and Rainbow Trout," prepared for the Alberta Oil Sands Environmental Research Program by Alberta Environment, AOSERP Report 23 (1979), 44 pp.

23. Giles, M.A., J.F. Klaverkamp and S.G. Lawrence. "The Acute Toxicity of Saline Groundwater and of Vanadium to Fish and Aquatic Inverte-

brates," prepared for the Alberta Oil Sands Environmental Research Program by Department of Fisheries and the Environment, Freshwater Institute, AOSERP Report 56 (1979), 216 pp.

24. Tsui, P.T.P., B.R. McMahon and P.J. McCart. "A Laboratory Study of Long-Term Effects of Mine Depressurization Groundwater on Fish and Invertebrates," prepared for the Alberta Oil Sands Environmental Research Program by Aquatic Environments Ltd., AOSERP Project WS 2.6.1.

25. Lutz, A., and M. Hendzel. "A Survey of Baseline Levels of Contaminants in Aquatic Biota of the AOSERP Study Area," prepared for the Alberta Oil Sands Environmental Research Program by Fisheries and Environment Canada, Freshwater Institute, AOSERP Report 17 (1977), 51 pp.

26. Allan, R., and R. Jackson. "Heavy Metals in Bottom Sediments of the Mainstem Athabasca River System in the AOSERP Study Area," prepared for the Alberta Oil Sands Environmental Research Program by Fisheries and Environment Canada, Freshwater Institute, AOSERP Report 34 (1978), 74 pp.

27. Korchinski, M. "Interaction of Humic Elements with Metallic Elements," prepared for the Alberta Oil Sands Environmental Research Program by Fisheries and Environment Canada, Water Quality Branch, AOSERP Project HY 2.3.

28. Sprague, J.B., D.A. Holdway and D. Stendahl. "Acute and Chronic Toxicity of Vanadium to Fish," prepared for the Alberta Oil Sands Environmental Research Program by the University of Guelph, AOSERP Report 41 (1978), 92 pp.

29. Anderson, P.D., et al. "The Multiple Toxicity of Vanadium, Nickel and Phenol to Fish," prepared for the Alberta Oil Sands Environmental Research Program by Department of Biology, Concordia University, AOSERP Report 79 (1979), 109 pp.

30. Nix, P.G., et al. "A Preliminary Study of Chemical and Microbial Characteristics in the Athabasca River in the Athabasca Oil Sands Area of Northeastern Alberta." prepared for the Alberta Oil Sands Environmental Research Program by Chemical and Geological Labs Ltd., Microbios Ltd., and Xentox Services Ltd., AOSERP Report 54 (1979), 135 pp.

31. Lock, M.A., and R.R. Wallace. "Interim Report on Ecological Studies on the Lower Trophic Levels of Muskeg Rivers Within the Alberta Oil Sands Environmental Research Program Study Area," prepared for the Alberta Oil Sands Environmental Research Program by Fisheries and Environment Canada, AOSERP Report 58 (1979), 104 pp.

32. Hickman, M., S.E.D. Charlton and C. G. Jenkerson. "Interim Report on a Comparative Study of Benthic Algal Primary Productivity in the AOSERP Study Area," prepared for the Alberta Oil Sands Environmental Research Program by Department of Botany, University of Alberta, AOSERP Report 75 (1979), 107 pp.

33. Costerton, J.W., and G. G. Geesey. "Microbial Populations in the Athabasca River," prepared for the Alberta Oil Sands Environmental Research Program by Department of Biology, University of Calgary, AOSERP Report 55 (1979), 66 pp.

34. Nix, P. G., et al. "Assimilative Capacity of the Athabasca River Due to Microbial Degradation of Organic Compounds," prepared for the Alberta Oil Sands Environmental Research Program by Chemical and Geological Laboratories Ltd., Microbios Ltd., and Xentox Services Ltd., AOSERP Project WS 2.3.

35. Malhotra, S.S., and P.A. Addison. "Symptomology and Threshold Levels of Air Pollutant Injury to Vegetation, 1975 to 1978," prepared for the Alberta Oil Sands Environmental Research Program by Canadian Forestry Service, Northern Forest Research Centre, AOSERP Report 44 (1979), 13 pp.

36. Malhotra, S.S., and A.A. Khan. "Interim Report on Physiology and Mechanisms of Airborne Injury to Vegetation, 1975 to 1978," prepared for the Alberta Oil Sands Environmental Research Program by Canadian Forestry Service, Northern Forest Research Centre, AOSERP Report 45 (1979), 38 pp.

37. Malhotra, S.S., and A.A. Khan. "Interim Report on Physiology and Mechanisms of Airborne Injury to Vegetation, 1978 to 1979," prepared for the Alberta Oil Sands Environmental Research Program by Canadian Forestry Service, Northern Forest Research Centre, AOSERP O.F. 2 (1979), 23 pp.

38. Malhotra, S.S., P.A. Addison and A.A. Khan. "Phisology and Mechanisms of Airborne Injury to Vegetation, 1979 to 1980," prepared for the Alberta Oil Sands Environmental Research Program by Canadian Forestry Service, Northern Forest Research Centre, AOSERP Project LS 3.3.

39. Addison, P.A., and J. Baker. "Interim Report on Ecological Benchmarking and Biomonitoring for Detection of Airborne Pollutant Effects on Vegetation and Soils 1975 to 1978," prepared for the Alberta Oil Sands Environmental Research Program by Canadian Forestry Service, Northern Forest Research Centre, AOSERP Report 46 (1979), 40 pp.

40. Addison, P.A., and J. Baker "Interim Report on Ecological Benchmarking and Biomonitoring for Detection of Airborne Pollutant Effects on Vegetation and Soils 1978 to 1979," prepared for the Alberta Oil Sands Environmental Research Program by Canadian Forestry Service, Northern Forest Research Centre, AOSERP O.F. 3 (1979), 17 pp.

41. Addison, P.A. "Ecological Benchmarking and Biomonitoring for Detection of Airborne Pollutant Effects on Vegetation and Soils," being prepared for the Alberta Oil Sands Environmental Research Program by Canadian Forestry Service, Northern Forest Research Centre, AOSERP Project LS 3.4.

CHAPTER 2

MANAGEMENT OF AQUATIC ENVIRONMENTS AT THE SYNCRUDE OIL SANDS PROJECT

M. Aleksiuk, J.T. Retallack and K.S. Yonge

Aquatic Environment Section
Department of Environmental Affairs
Syncrude Canada Ltd.
Edmonton, Alberta, Canada T5J 3E5

INTRODUCTION

Large deposits of oil in the form of bitumen exist in northern Alberta's oil sands region. Although those deposits were known two centuries ago, development of the oil sands is still very much an industry of the future. The first commercial development began in the 1960s (Suncor Inc.), a second commercial project was initiated in the 1970s (Syncrude Canada Ltd.), and additional developments are expected to take place in the 1980s and 1990s. Full development will not occur until the early part of the next century or perhaps later. Because the industry has several potential environmental problems, and because society (through government) is likely to demand that environmental controls on oil sands projects be increasingly stringent, it is essential that environmental considerations be incorporated into project design at this early stage. Environmental planning is clearly a prerequisite for full-scale oil sand development. The decisions of environmental managers today will set standards which will help safeguard the environment in coming decades.

The objective of this chapter is to review the approach to management of aquatic environments within Syncrude Canada Ltd., an existing oil sand development which began production in 1978. We outline Syncrude's approach to meeting its zero discharge objective, and secondly we describe some potential problems associated with a zero discharge policy.

ENVIRONMENTAL BACKGROUND

The Syncrude project is located in the Mixedwood Section of the Boreal Forest Region, a nonagricultural area of Alberta. Terrestrial habitats are dominated by black spruce bogs and aspen/white spruce forests. An overview of surface water patterns in the area prior to development (Figure 1) shows the Athabasca River on the eastern boundary of Syncrude's lease and the MacKay River near the western boundary. Beaver Creek meanders through the central part of the lease, with four small tributary streams entering it from the west. Poplar Creek flows into the Athabasca just south of the lease. Three small lakes are present. The dominant feature of the watershed is the Athabasca River, which originates near Jasper, Alberta and reaches the Arctic Ocean via the Mackenzie River. Mean flow rates for the major streams are given in Table 1.

SYNCRUDE'S ENVIRONMENTAL PHILOSOPHY

In planning the Syncrude project, a commitment was made to sound environmental management. Syncrude recognized that, in order to succeed, the project must meet environmental standards as well as engineering and economic standards. A clear mandate was given to environmental managers in the company to ensure that environmental standards were met and that the project be environmentally sound. Environmental personnel were required to undertake detailed advanced planning, along with engineers and economists, to ensure the success of the venture.

The field of environmental science is in its infancy. There is frequent disagreement among environmental managers as to the objectives of environmental work, and, unfortunately, environmental standards vary. In the absence of clear guidelines, environmental goal-setting fell largely to the discretion of the company. The goal we selected was to prevent environmental damage beyond the immediate confines of the project site, with an emphasis on prevention rather than on repair of damages.

ZERO DISCHARGE

In relation to aquatic environments specifically, it was decided the most

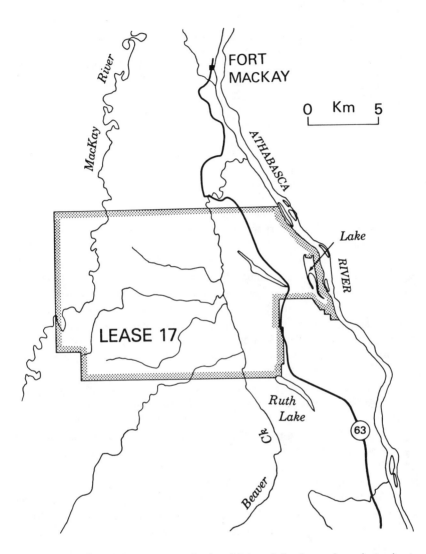

Figure 1. Surface drainage patterns in the vicinity of the Syncrude project prior to development.

effective approach to prevention of damage was to prevent all process-affected water from leaving the project site. In other words, a "zero discharge" policy was adopted.

We strived for zero discharge in two major ways: through the diversion of surface waters away from the site and the retention of process-affected waters on the site.

Table 1. Mean Annual Flow Rates of Streams in the Vicinity
of the Syncrude Oil Sands Development

Stream	Mean Annual Flow Rate (m^3/sec)
Athabasca River	677
MacKay River	24.4
Beaver Creek	1.9
Poplar Creek	0.6

1. Diversion of Surface Drainage Patterns. First, the surface drainage pattern was altered to divert water away from the plant area. Drainage patterns on the area prior to development are depicted in Figure 1. A stream, called Beaver Creek, originally crossed the areas proposed for mining oil sand and for extracting and upgrading bitumen (maximum and minimum annual flow rates of ~40 and 0.02 m^3/sec, respectively). An earthen dam, 3 km long, was constructed at the south end of the development area to divert Beaver Creek water away from the site (Figure 2). The dam was designed so that water backed up to form Beaver Creek Reservoir. A canal carries the impounded Beaver Creek water into Ruth Lake. Ruth Lake water passes, via a second canal, into the Poplar Creek Reservoir, over a spillway into Poplar Creek and ultimately into the Athabasca River. Four small tributary streams that originally entered Beaver Creek in the development area were intercepted by constructing a new channel we call the West Interception Ditch (Figure 2). Water from three of those streams now flows northward into the original Beaver Creek channel downstream of the development area, and that from the fourth stream flows southward into Beaver Creek Reservoir. This alteration of the surface drainage pattern effectively diverted water away from the development area. Although the primary purpose of the diversion was to permit mining and creation of a tailings pond, it secondarily reduced the potential of industrial effluents entering natural waterways by way of surface drainage.

2. Creation of Tailings Pond. The second major procedure aimed at achieving the zero discharge objective is the retention of all process-affected waters in a tailings pond (Figure 2). Currently, two years after start-up, the tailings pond is approximately 12 km^2 in surface area. The pond will reach a maximum size of 28 km^2 in 1983, with a maximum dike height of 79 m and a maximum water depth of 62 m in the Beaver Creek valley. The geographical relationship of the major features of the Syncrude project to the surface water drainage pattern is depicted in Figure 2.

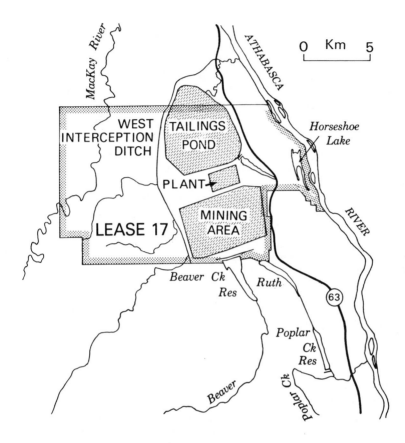

Figure 2. Surface drainage patterns in the vicinity of the Syncrude project after construction and start-up of operation.

Water Quality Monitoring

A surface water quality monitoring program is in place, with sampling stations at key locations. A range of parameters, defined by Alberta Environment, is measured on samples collected at weekly intervals. Analyzed contaminants entering surface water from the project would show up in this monitoring program. To date there has been no evidence of contamination, suggesting effluent retention has been effective.

Impact of Diversion

The diversion per se, which has been instrumental in minimizing damage to aquatic ecosystems, has had no unacceptable effects on the surrounding environment. Intensive baseline studies [1-4] and impact studies [5-9] of the diversion system have demonstrated that the animal and plant communities are healthy. Indeed, the diversion has had beneficial effects on the natural environment. Wetland areas have increased considerably as a direct result of the diversion, with a resultant increase in habitat for muskrat, beaver, otter, moose and waterbirds. Probably most important, there has been an increase of lake habitat in an area of few lakes. The small lakes that existed in the area prior to development were in fact marshes, generally less than 2 m deep. Beaver Creek Reservoir and Poplar Creek Reservoir, both created by the diversion, have depths in excess of 8 m and 16 m, respectively.

Disposal of Natural Groundwater

Because groundwater posed a threat to mine wall stability, a number of wells were drilled in the mine area to depressurize the aquifers and permit the safe excavation of the mine. Discharge of groundwater into Beaver Creek Reservoir began in April 1977 under Alberta Government permit, which stipulates that the chloride ion concentration in the water discharged over the Poplar Creek Spillway must not exceed the prevailing natural background concentration of chloride in Poplar Creek upstream of the confluence by more than 400 mg/L.

Pumping rates of mine depressurization water to Beaver Creek Reservoir in 1978 and 1979 are presented in Table 2. The quality of the groundwater which is pumped to the surface is highly variable. A breakdown of the chemical constituents of the groundwater is given in Tables 3 and 4. The water is quite saline. Chloride ion concentrations range from 50 mg/L in the surface

Table 2. Seasonal Changes in Mean Daily Discharge of Diluted (10%) Groundwater to Beaver Creek Reservoir

Month	Mean Daily Discharge (m^3/day)	
	1978	1979
April	0	1814
May	259	2419
June	682	1296
July	432	1209
August	1425	518
September	1944	691
October	1892	1036
November	0	510

Table 3. Ionic Composition of Groundwater in the Syncrude Mine Area

Parameter	Mean Value (mg/L) (n = 72)
Na	3517
K	33
Ca	56
Mg	69
Cl	4560
HCO$_3$	2016
SO$_4$	66
CO$_3$	494
pH	7.0

Table 4. Concentration of Selected Metals in Groundwater from the Syncrude Mine Area

Metal	Mean Value (mg/L) (n = 72)
Al	1.60
Ba	1.00
B	5.80
Fe	11.40
Pb	0.05
Mn	0.90
Zn	0.65

of the overburden aquifer to more than 20,000 mg/L in the basal aquifer. The salinity of the depressurization water may decrease with time as the highly saline water is removed and more dilute water flows in by recharge from other groundwater layers.

Concerns were raised initially as to whether the chloride content of the water leaving the diversion system would exceed the limit set by government or whether the concentration of groundwater remaining in the diversion system would affect aquatic life. An evaluation of the problem prior to the discharge [10] projected chloride concentrations within the system as a function of time. The projected scenarios ranged from rapid and complete mixing, with no buildup of salinity in the reservoir, to gradual "ponding" of heavily saline water at the bottom of the reservoir. The most probable scenario was considered to be something between those two extremes.

Because of the possibility of "ponding", toxicity studies of the effluent were conducted [11]. The most sensitive organism tested was the walleye *(Stizostedion vitreum)*, a local game fish. Walleye exhibited a 96-hr LC50 of about 10% (Figure 3). LC50 is the concentration of effluent that is lethal to

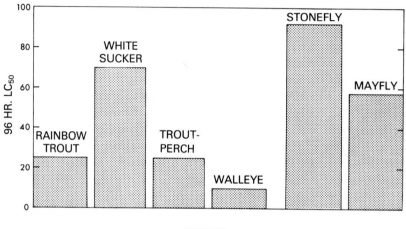

Figure 3. Toxicity of saline groundwater from the Syncrude mine area.

50% of the test organisms within a specified period of time, indicating the water is highly toxic.

The observation that the groundwater is toxic prompted a detailed study of the dispersal of the water in the Beaver Creek Diversion System [12]. Field surveillance of a network of 26 monitoring stations in the Beaver Creek Diversion System commenced in December 1976 and continued until October 1978 (Figure 4). Vertical profiles of temperature and conductivity were obtained at one-meter intervals. Additionally, water samples were obtained at one-meter intervals at six locations in the diversion system to determine the concentration of the following ions: calcium, magnesium, sodium, potassium, bicarbonate, sulfate and chloride.

The two main sources of water entering the diversion system are Beaver Creek and the diluted mine depressurization water. Although the ionic concentration of each parent water type varies with time, significant differences can be noted between the two water types. The upstream Beaver Creek water is typical of inland surface waters of the region (i.e., rich in bicarbonate) while the water overboarded to the reservoir is more typical of seawater (i.e., rich in chloride). The anionic composition of the water flowing out of the diversion system in both years was dominated by bicarbonate ion.

The effectiveness of the diversion system in dispersing groundwater is dependent primarily on regional rainfall and resultant stream flow rates. The water that is discharged into the Beaver Creek Diversion is already greatly dilute by surface runoff (by a factor of 10). This effluent is pumped into

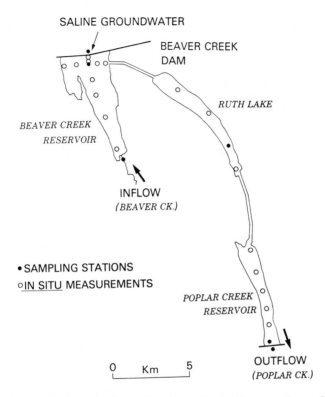

SALINE GROUNDWATER

BEAVER CREEK
DAM

RUTH LAKE

BEAVER CREEK
RESERVOIR

INFLOW
(BEAVER CK.)

• SAMPLING STATIONS
○ IN SITU MEASUREMENTS

POPLAR CREEK
RESERVOIR

0   Km   5

OUTFLOW
(POPLAR CK.)

Figure 4. Network of monitoring stations for study of saline groundwater dispersal in the Beaver Creek diversion system.

the reservoir from a shallow holding pond north of the dam during the ice-free season.

In both years of detailed study [12] salinity in the diversion system increased during the open water season. The increase occurred throughout the summer and early fall, probably as a combined result of groundwater pumping and low levels of surface runoff. The highest buildup of salinity occurred in the Beaver Creek Reservoir, nearest the pumping site. Although similar seasonal trends of saline buildup occurred in the rest of the system, the concentrations were much reduced (dependent on the bulk residence time of each water body). Fall overturn and high autumn rainfall effectively flushed the system of the entrained saline water (Figures 5 and 6). Chloride and salinity values of the water leaving the diversion system were not significantly different from those of the receiving waters (Poplar Creek).

Although laboratory tests showed the groundwater to be toxic, Carmack

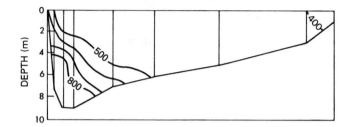

Figure 5. Stratification of conducitivity (μmho) in Beaver Creek Reservoir, July, 1978, during stable summer period.

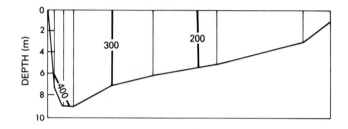

Figure 6. Stratification of conductivity (μmho) in Beaver Creek Reservoir, October 1978, after fall overturn.

Table 5. Chloride Ion Concentration (mg/L) in Beaver Creek Reservoir Near the Inflow of Diluted Groundwater

Depth (m)	May	July	October
2	31	56	22
4	39	93	23
6	120	168	26
8	109	102	28

and Killworth [12] found that it quickly dispersed in the diversion system, and chloride did not reach toxic levels. Concentrations of chloride ion within the top four meters of the reservoir only rarely exceeded 100 mg/L (about 100X dilution of the undiluted mine depressurization water) (Table 5). To be certain that those concentrations are biologically safe, controlled experiments were conducted in situ with local test organisms [13]. No toxic effects could be attributed to the presence of diluted groundwater. As a final check that this disposal method is ecologically safe, Poplar Creek benthic macro-

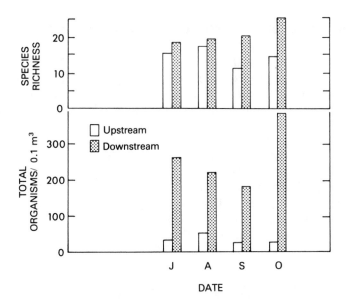

Figure 7. Density and species richness of benthic macroinvertebrates in Poplar Creek in relation to saline groundwater discharge, 1979.

invertebrates were studied on natural and artificial substrates both upstream and downstream from the point of inflow of water from the diversion. No adverse effects of the groundwater on aquatic organisms were detected (Figure 7). In fact, probably due to the presence of suitable rocky substrate (artificially placed to reduce erosion) and the availability of additional food (large volumes of nutrient-rich water from the reservoirs of the diversion system), the density and variety of organisms were always higher downstream of the inflow.

Saline groundwater in the diversion system has behaved much as predicted in preliminary studies. The dense saline water tends to sink to the bottom of Beaver Creek Reservoir, and the development of haloclines enhances the buildup of salinity over the summer months. But convective currents during fall overturn eliminate this stratification. As a result, the diversion system functions as an effective mixing machine for the dispersal of groundwater. Although undiluted groundwater is toxic to aquatic organisms, our data show that the ultimate concentrations of effluent which result from dilution by surface waters pose no threat to the short- or long-term well being of the aquatic environment. Due to interannual variations, periodic buildup of salinity within the diversion system could conceivably occur in the future. However, the apparent decrease in salt concentration of the aquifer, annual spring flushing and annual fall overturn will preclude any serious long-term buildup of salinity.

POTENTIAL PROBLEMS ASSOCIATED WITH ZERO DISCHARGE

The diversion of natural waters away from the development area and the retention of contaminated water in a tailings pond clearly facilitate the achievement of zero discharge. In our assessment, discharge of aqueous contaminants with resultant damage to surrounding aquatic environments is essentially nonexistent. Thus, in our case pursuit of the zero discharge objective as an approach to management of aquatic environments has been successful. However, the approach is not without shortcomings. The concentration of effluents in a tailings pond creates three potential environmental problems: (1) bitumen on the surface of the tailings pond is hazardous to migratory birds; (2) seepage from the tailings pond, if it occurs, could contaminate groundwater; and (3) the tailings pond must be reclaimed at the termination of the project. These three potential problems are discussed in detail below.

Management of the Tailings Pond as a Wildlife Concern

The waterbodies of the Fort McMurray region are few and relatively unproductive for waterfowl [14, 15]. Lakes near the Syncrude project are not important as waterfowl breeding areas [1, 16], or as migratory staging areas [17]. However, the region lies within a major migration corridor, and large numbers of waterfowl and shorebirds pass through, to and from the Peace-Athabasca Delta (Figure 8). The Delta, roughly 200 km north of Fort McMurray, is the convergence of several major flyways. Therefore, although the number of birds which normally stop in the region is low, the potential number is high each spring and fall as millions of waterbirds pass through during migration [18].

The formation of a large tailings pond created a major potential impact on wildlife. The tailings pond will be the largest waterbody within a radius of 75 kilometers by 1983, and is viewed as a serious environmental concern by provincial authorities [19]. Bitumen, on the surface of the tailings pond, is hazardous to birds (brief exposure of birds to other constituents of the tailings have proven relatively harmless [20]). Although most of the bitumen in the oil sand is removed during extraction, some residual bitumen concentrates on the surface of the tailings pond. Most oils on water are a hazard to birds; bitumen is particularly hazardous due to its high viscosity [21]. Bitumen contaminates and disrupts the plumage, and if feathers accumulate more than trace amounts of oil, the bird generally becomes trapped on the pond and dies. The effects of various oils on birds are well documented [22].

The tailings pond is unattractive to most types of birds, and most instinctively avoid the pond. However, many thousands of birds could become

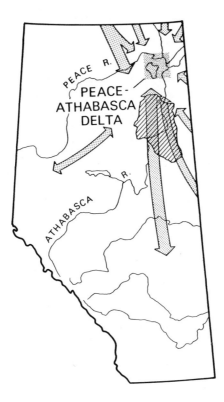

Figure 8. Migratory flyways in the Athabasca Oil Sands region (after Bellrose 1976 [18]).

oiled if even a small fraction of the birds that pass annually near the pond were to land.

The potential for birds to enter the pond is greatest during spring breakup of lakes when only the lack of open water limits the northward migration of waterfowl. The discharge of warm effluent into the tailings pond causes sections of the pond to open in advance of surrounding lakes and ponds. Thus, for a brief period in the spring, the tailings pond may be the only area of open water on which early migrants can land. During this critical period, certain weather systems could force many birds to land on the tailings pond.

Conceptual Development of a Bird Protection Strategy

In 1976, Syncrude initiated a research program aimed at mitigating the problem of migratory birds landing on the tailings pond. This led to the development of a Bird Protection Program which consists of four measures, each dealing with the problem at a different level.

The first measure is to maintain the tailings pond habitat in a condition which is unattractive to aquatic birds. This is done by removing debris and preventing the invasion of aquatic vegetation. By reducing the attractiveness of the habitat, the number of birds which enter the tailings pond area can be minimized. The second measure is to reduce the hazardous nature of the pond by restricting the extent and movement of bitumen on the surface of the pond through the use of booms and skimmers. By confining surface bitumen to less than 10% of the pond area, birds which land briefly are less likely to be contaminated or to become fatally oiled. The third measure is to prevent birds from landing on the pond by maintaining mechanical deterrents on the surface of the tailings pond. Syncrude developed a unique deterrent for this purpose (described below). The principle of a bird deterrent is to further reduce the attractiveness of the tailings pond, and therefore to discourage landing by birds which do not avoid it instinctively. The fourth measure was developed in response to birds which land and become oiled despite the above measures. Recognizing that no program will exclude all birds from the tailings pond, a procedure for systematic search and humane disposal of oiled birds was developed to minimize suffering. As the cleaning and rehabilitation of oiled birds is seldom successful [23, 24], only endangered species are to be turned over to wildlife authorities for care.

Development of Bird Deterrent Measures

As it is unlikely that the tailings pond will ever be clean enough for prolonged habitation by birds, it was necessary to develop an active means to prevent waterfowl from landing. Because no conventional deterrent was suitable for tailings pond use [25], Syncrude developed a deterrent device which incorporated the most effective components of several existing devices.

The deterrent device, which mimics a hunter, consists of a floating raft which supports a propane birdscare cannon (trademarked by Zon Mark II, D.M. Lawrence and Co., California) and a conspicuous mobile scarecrow. The scarecrow is 2 m high, is bright orange in color, and moves erratically in wind like a kite. The synergism of the scarecrow and cannon heightens the deterrent effect.

Field tests of the prototype deterrent device on natural lakes in 1977 showed it was highly effective in displacing birds [26]. Hunted species were deterred most strongly. The devices were more effective in displacing birds from one part of a lake to another, than in excluding birds from a lake entirely (Table 6). The deterrent devices reduced activity of game (hunted) species by more than 90% and activity of nongame species by more than 50%.

Table 6. Change in Bird Numbers[a] on Natural Lakes Following Activation of
Mechanical Deterrent Devices

| | Alternate Landing Area | |
	Available	Not Available
Dabbling Ducks	−86	−58
Scaup	−95	−84
Bufflehead	−85	−25
Coots	+5	−19

[a] Percent change from control.

A network of deterrent devices was deployed on the tailings pond in
1978, when the pond was still relatively small. During the fall migration of
1978, waterbird activity on the tailings pond was compared to that observed simultaneously at two adjacent natural lakes. There were consistently
fewer birds flying over the tailings pond than over a control lake [27], although migration patterns were similar over all waterbodies studied (Figure 9).
Lower numbers over the tailings pond were probably due mainly to the
unattractiveness of the tailings pond habitat. Avoidance of the tailings pond
by birds was most dramatic when numbers which landed on the tailings pond
were compared to numbers on adjacent lakes (Figure 10). Although many

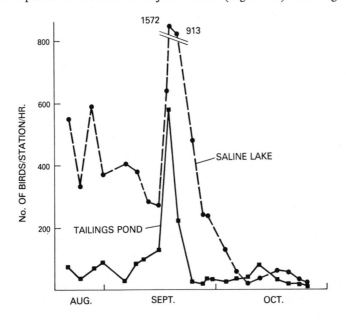

Figure 9.   Observations of bird flight over the Syncrude tailings pond and Saline Lake
during fall migration, 1978.

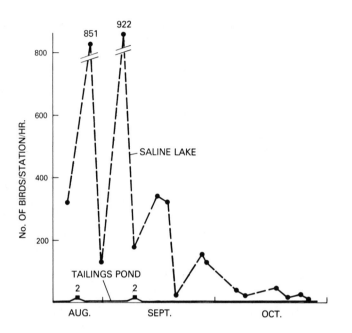

Figure 10. Observations of birds on surface of Syncrude tailings pond and Saline Lake during fall migration, 1978.

thousands of birds were seen on natural lakes, only four were observed on the tailings pond during two months of daily study. This result, attributed in part to the unattractive nature of the tailings pond, was due largely to the effect of the deterrent devices.

Although minor improvements to the deterrent system will continue, a full-scale deterrent program is now operational. The deterrent device density is highest near boomed areas where the hazard is high and near shoreline areas where birds are most likely to land.

Effectiveness of the Program

Figure 11 shows the planned (1983) arrangement of booms, skimmers and deterrents on the tailings pond when it has reached full size. At present the placement of booms is complicated by irregularities in depth of the tailings pond. Although the skimmer has proven to be effective in collecting bitumen, a system for transporting recovered bitumen back to the plant still needs to be developed.

The eventual number of deterrent devices required on the tailings pond (85 units are currently used) will be determined largely by the effectiveness of the oil control program. In our projected system, 4 tailings outfalls will be enclosed by booms, confining bitumen to less than one tenth of the total pond surface area (Figure 11). Each boom will feed one skimmer, which will be serviced by barges to return skimmed bitumen to the plant. Depending on the effectiveness of the oil containment program, a secondary cleaning of the pond surface may be undertaken with rope mops. We believe that the pond cleaning program is unique in that it not only reduces an environmental problem, but has the potential to provide a financial return as well.

To ensure that the Bird Protection Program continues to function effectively, an annual evaluation is conducted. Pond conditions are monitored daily and the pond is searched for birds several times each week during migration. Overall we regard the Bird Protection Program as a practicable solution to the problem of oiled birds on the tailings pond.

From the outset, our policy has been to encourage the involvement of government and university scientists at all developmental stages. This has led to a comprehensive solution, and one which is strongly supported by those involved. This cooperation was formalized in 1979 in the form of an Oil Sands Bird Protection Committee, with representation from the federal and provincial governments, the university community and three resource development companies.

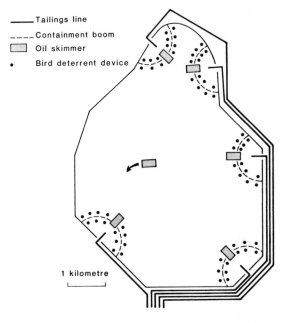

Figure 11. Conceptual deployment of bird deterrent devices and surface booms on Syncrude tailings ponds, 1983.

Groundwater Contamination

Concentration of effluents in the tailings pond effectively prevents contamination of surface water by way of surface runoff. However, that same concentration of effluents theoretically can result in contamination of groundwater by seepage through the substrates at the bottom of the tailings pond, and there is concern that contaminants in the tailings pond may find their way into the Athabasca River and other streams via the groundwater flows.

Tailings pond water contains a high concentration of fine suspended matter (clay fines) and variable quantities of residual bitumen (Table 7). The fines are expected to penetrate and seal the pore spaces of the underlying strata. Thus, the tailings pond is expected to seal itself.

Table 7. Physical Composition of Tailings Pond Water at 5 m Depth, May 1980

Total Solids	0.98%
Total Suspended Solids	0.79%
Total Dissolved Solids	0.19%
Total Bitumen	0.049%

To determine whether this is indeed the case, a groundwater monitoring program has been implemented. A system of 32 wells was drilled to allow groundwater to be sampled on a quarterly basis. Groundwater samples are analyzed for specified parameters on a regular basis, under government permit. If changes in groundwater pressure or chemical composition suggest tailings pond seepage is occurring, an extensive system of wells will be drilled between the tailings pond and the Athabasca River, and groundwater will be pumped back to the tailings pond. Syncrude works closely with regulatory agencies to ensure adequate controls are in place.

Tailings Pond Reclamation

Aqueous contaminants can indeed by prevented from entering aquatic ecosystems by placing the contaminants in a retention pond. However, a concentration of contaminants in this manner will result in a waterbody of poor quality. The quality of water in the tailings pond is already low, after only two years of operation (the 96-hr LC50 is now 4%), and we believe it will be even lower in 25 or 30 years. The tailings pond, ultimately covering 28 km^2 and containing 10^9 m^3 of water, must be reclaimed as either a viable lake or dry land before the site is abandoned. Ultimately there will be an enormous quantity of toxic material that must be detoxified or disposed of safely.

Although actual reclamation of the tailings pond is more than 30 years in the future, development of reclamation options and methods has already begun in Syncrude. The problem is a very difficult one, perhaps the most difficult water management problem facing the oil sands industry.

We expect two major problems in reclamation of the tailings pond. First, the high concentration of suspended solids ("fines") must be reduced to a level that is acceptable in a lake or in water entering a river system. Secondly, we must identify the toxic components of tailing pond water, and develop methods of neutralizing them in large quantities. The large volume of the tailings pond will also be an obstacle in detoxification. Methods that are effective on a bench scale may be impractical on the massive scale of the tailings pond. As the first step in researching tailings pond reclamation techniques, we are studying in detail the chemical and physical properties of the pond (oil sands tailings ponds are essentially unknown as chemical and physical entities). Studies of water balance, sediment structure, reclamation methods, etc., will follow.

CONCLUSION

The zero discharge policy has been effective in minimizing the impact of the Syncrude project on surrounding aquatic environments. It is true that, while the zero discharge approach eliminates certain potential problems, it creates others, such as bird protection, groundwater contamination and tailings pond reclamation. However, the problems created are fewer in number and probably more manageable than those prevented. The problem of bird protection is under control. Groundwater contamination remains a concern, but the tailings pond is expected to self-seal. Although the problem of tailings pond reclamation is a very difficult one, 30 years are available for its solution. Thus, from the point of view of environmental management, it has been advantageous to use the zero discharge approach in Syncrude.

There are other advantages. A policy of zero discharge places the emphasis on prevention rather than cure. Potential problems often can be readily averted by incorporating appropriate features into the design and operating plans of the development. Emphasis on prevention rather than on repair results in improved rapport with regulatory agencies, which leads to better cooperation in responding to unforeseen problems or requests for future approvals. Finally, it is usually less expensive to prevent environmental damage from occurring than to repair the damage after it has occurred. Thus prevention is more cost-effective than repair.

REFERENCES

1. Sharp, P. L., D. A. Birdsall and W. J. Richardson. "Inventory Studies of Birds on and near Crown Lease Number 17, Athabasca Tar Sands, 1974," Syncrude Environmental Research Monograph 1975-4 (1975).
2. Syncrude Canada Ltd. "Baseline Environmental Studies of Ruth Lake and Poplar Creek," Syncrude Environmental Monograph 1975-3 (1975), 120 pp. + App.
3. Penner, D. F. "Preliminary Baseline Investigations of Furbearing and Ungulate Mammals Using Lease No. 17," Syncrude Environmental Research Monograph 1976-3 (1976), 181 pp.
4. Noton, L. R. and N. R. Chymko. "Aquatic Studies of Upper Beaver Creek, Ruth Lake and Poplar Creek, 1975," report prepared for Syncrude Canada Ltd. by Renewable Resources Consulting Services Ltd., Edmonton, Alberta (1977), 203 pp.
5. Noton, L. R. and N. R. Chymko. "Water Quality and Aquatic Resources of the Beaver Creek Diversion System, 1977," Syncrude Environmental Research Monograph 1978-3 (1978), 340 pp.
6. Tsui, P., D. Tripp and W. Grant. "A Study of Biological Colonization of the West Interceptor Ditch and Lower Beaver Creek," Syncrude Environmental Research Monograph 1978-6 (1978), 144 pp.
7. Westworth, D. A. "Beaver and Muskrat Aerial Survey, October 1979," report prepared for Syncrude Canada Ltd. by D. A. Westworth & Associates, Edmonton, Alberta (1978), 8 pp.
8. Westworth, D. A. "Surveys of Moose Populations in the Vicinity of the Syncrude Development, Winter 1979-80," report prepared for Syncrude Canada Ltd. by D. A. Westworth & Associates, Edmonton, Alberta (1978), 13 pp.
9. O'Neil, J. P. "Fisheries Survey of the Beaver Creek Diversion System, 1978," Syncrude Environmental Research Monograph 1979-3 (1979), 63 pp + App.
10. Carmack, E. C. "Physical Limnology of Saline Water Disposal in the Beaver Creek Diversion," unpublished report, Syncrude Canada Ltd., 1976, 26 pp+ App.
11. McMahon, B., P. McCart, A. Peltzner and G. Walder. "Toxicity of Saline Groundwater from Syncrude's Lease 17 to Fish and Benthic Macroinvertebrates," Syncrude Environmental Research Monograph 1977-3 (1977), 99 pp.
12. Carmack, E. C. and P. D. Killworth. Observations on the Dispersal of Saline Groundwater in the Beaver Creek Diversion System. 1976-1978," Syncrude Environmental Research Monograph 1979-2 (1979), 83 pp.
13. Jantzie, T., L. R. Noton and N. R. Chymko. "An *in situ* Study of the Potential Toxicity of Saline Mine Depressurization Water," Syncrude Environmental Research Report 1980-2 (1980), 78 pp.
14. Schick, C. D. and K. R. Ambrock. "Waterfowl Investigations in the Athabasca Tar Sands Area," unpublished manuscript, Canadian Wildlife Service, Ottawa (1974), 44 pp.
15. Syncrude Canada Ltd. "Migratory Waterfowl and the Syncrude Tar Sands Lease: A Report," Syncrude Environmental Research Monograph 1973-3 (1973).

16. Sharp, P. L. and W. J. Richardson. "Inventory Studies of Birds on and near Crown Lease Number 17, Athabasca Oil Sands, 1975," unpublished L. G. L. Ltd. report to Syncrude Canada Ltd. (1976).

17. Ward, J. G., G. R. Dyke and P. L. Sharp. "Bird Migration Watches on Crown Lease Number 17, Athabasca Oil Sands, Alberta, 1976," unpublished LGL Ltd. report to Syncrude Canada Ltd. (1976).

18. Bellrose, F. C. "Ducks, Geese and Swans of North America," (Harrisburg, PA: Stackpole Books, 1976).

19. Alberta E. C. A. "Review of Interaction Between Migratory Birds and Athabasca Oil Sands Tailings Ponds," Alberta Government Publication (1975).

20. Howell, D. H. "Environmental Studies, Waterfowl and Effluent Ponds," Syncrude Canada Ltd., Internal Report No. 28 (1972).

21. Boag, D. A., and V. Lewin. "Effectiveness of Three Waterfowl Deterrents on Natural and Polluted Ponds," J. Wild. Manag. 44: 145-154 (1980).

22. L. G. L. Ltd. "Review of Current Knowledge on Reducing Bird Mortality Associated with Oil Spills," P. A. C. E. Report No. 75-4 (1974).

23. Clark, R. B. "Oiled Seabird Rescue and Conservation," J. Fish. Res. Bd. Can. 35: 675-678 (1978).

24. Stanton, P. B. "The Hard Truth about Oil Pollution," Mass. Wild. 26: 16-19 (1975).

25. Koski, W. R., and W. J. Richardson. "Waterbird Deterrent Systems for Oil Spills: Review of Methods," report to P. A. C. E. by LGL Ltd. (1976), 122 pp.

26. Ward, J. G. "Tests of the Syncrude Bird Deterrent Device for Use on a Tailings Pond," unpublished report to Syncrude Canada Ltd. (1978), 115 pp.

27. Yonge, K. S. "Development of a Bird Protection Strategy for Tar Sands Tailings Ponds," Proceedings of the 8th Bird Control Seminar, Bowling Green, OH (1979).

CHAPTER 3

IMPACT OF TAR SANDS DEVELOPMENT
ON ARCHAEOLOGICAL RESOURCES

H. L. Diemer

Archaeological Survey of Alberta
Department of Alberta Culture
Edmonton, Alberta, Canada

INTRODUCTION

The inherent value in understanding the past and the related need for the conservation of information leading to this understanding in the face of a rapidly changing physical world has been the subject of much concern for many years. This can be exemplified by the conclusion of the Ninth Session of the General Conference of UNESCO in New Delhi, 1956, concerning the value of the past and its preservation.

> Being of the opinion that the surest guarantee for the preservation of monuments and works of the past rests in the respect and affection felt for them by the peoples themselves . . . Convinced that the feelings aroused by the contemplation and study of works of the past do much to foster mutual understanding between nations . . . Considering that the history of man implies the knowledge of all different civilizations; and that it is therefore necessary, in the general interest, that all archaeological (and historical) remains be studied, and where possible, preserved and taken into safe keeping . . . The General Conference recommends that Members States should . . . take whatever legislative or other steps may be required to give effect . . . to the principles and norms formulated in the present recommendations.[1]

This preservation ethic has been developing for many years on the North

American continent. Preserved and respected monuments of and to history and prehistory may be found in many areas of Mexico, the United States and Canada. These monuments are a physical manifestation of the value placed on the heritage of the involved nations.

The concern expressed by UNESCO indicates by implication that there are lingering threats to the physical integrity of evidence of the past and that insofar as some of these threats can be controlled by man, UNESCO has resolved that efforts be directed toward the conservation of this evidence.

In any nation the physical evidence of the past has become part of an inherited environment and is a naturally affiliated cultural resource base which in the case of archaeology is:

> the totality of information sources that can be used to understand past human activities. This base includes not only cultural remains such as artifacts, structures, features, activity areas, and so forth, but any parts of the natural and cultural environment that were either used or modified by people in the past or which can aid in understanding the basic relationship between people and the environment in the past. [2]

A key to implementing the study and preservation of archaeological resource sites is the recognition, as implied by McGimsey, that prehistorical information contained in those sites is related therefore to the physical, natural and cultural environment reflecting the relationship between man and the environment in prehistoric times. This intimate past relationship, in itself, has become a legacy and is an integral part of the contemporary natural and cultural environment, just as any other modification perpetrated by man becomes a part thereof [3].

A logical conclusion to be drawn, therefore, is that if archaeological resources are a part of the contemporary natural and cultural environment, any impact on that environment could possibly impact those valuable resources, resulting in a potential loss of archaeological information, hence a loss in the better understanding of the prehistoric past.

Archaeological resources are particularly sensitive to physical impact due to the fact that their value in providing prehistoric data does not rest solely with the cultural materials left behind, but also with the relative location—vertically and horizontally—of artifacts and activity areas to one another in and on the land surface. In other words, the land surface and subsurface is the medium in which prehistoric information is being held and through which this information may be passed onward. Therefore concern is generated when any major surface-disturbing activities are contemplated because the resulting impact could destroy this cultural documentation.

Contemporary major impacts on the environment have much to do with the exploration for, the extraction of and processing of the world's natural resources. Much deliberation in that regard over the years has been directed toward the intensive and extensive nature of the impact of surface mining and the related difficulties of post-mining reclamation.

Contemporary proven modes of producing crude bitumen from tar sands are by the surface mining method because only bitumen from tar sands buried by 46 m or less of overburden can economically be extracted [4]. Accordingly, the production of crude bitumen from tar sands areas must be considered as posing a significant threat to archaeological resources and the information contained therein.

The Province of Alberta, Canada, contains vast areas of tar sands deposits and a rich archaeological resource base. This situation produces a significant potential for conflict between the development of tar sands and the conservation of archaeological resources. Since tar sands development has only recently commenced in Alberta, it will be valuable for all those involved in such development to consider its impact on the culturally related aspects of the environment and to appreciate the approaches taken by the Province of Alberta to ameliorate this conflict. Furthermore, since archaeological research in tar sands areas of Alberta is relatively embryonic, it will be valuable to look briefly at the preliminary results of research done to date on a representative selection of tar sands–related development projects.*

Clearly, due to a paucity of statistically valid data, this will be a descriptive discussion intended to establish one of a number of environment concerns relating to tar sands development.

LEGISLATION AND HISTORICAL RESOURCES CONSERVATION

In Canada, the responsibility for archaeological resource management is vested in the province in which the resources are located, except where they are located on federally owned lands. Accordingly, in 1973 the Alberta Historical Resources Act was promulgated in response to a demand by the public, which had expressed a strong conservation ethic at extensive public hearings held during 1972 by the Environment Conservation Authority of Alberta [5]. Perhaps the most powerful and comprehensive provincial historical resources legislation in Canada, the Act is specifically aimed at:

---

*It should be noted that although this discussion pertains for the most part to prehistoric (archaeological) resources, some mention will be made of the impacts on historic resources as well, since these too, are a significant cultural and historical feature of the environment.

(a)   the coordination of the orderly development,
(b)   the preservation,
(c)   the study and interpretation, and
(d)   the promotion of appreciation of Alberta's historic resources.
       ([6], Section 2)

Based on these specific mandates, the Department of Alberta Culture has established a policy wherein all major forms of land development must be preceded by historical resources investigations,

> ... Where the Minister is of the opinion that any operation or activity which may be undertaken by any person will, or is likely to, result in the alteration, damage, or destruction of historic resources ... ([6], Section 22(2))

Specifically, a development proponent initiating a project in Alberta may be ordered by the Minister:

(a)   to carry out an assessment to determine the effect of the proposed operation or activity on historic resources in the area where the operation or activity is carried on.
(b)   to prepare and submit to the Minister in accordance with the order, a report containing the assessment of the effect of the proposed operation or activity referred to in clause (a), and
(c)   to undertake all salvage, preservative or protective measures or take any other action which the Minister considers necessary. [7]

In recognizing that there is a potential conflict between historical resources and tar sand development, any proponent of a tar sands project will therefore become subject to those requirements of the Alberta Historical Resources Act.

Typically, it is required that the project area be intensively surveyed by a professional archaeolgist in possession of an Archaeological Research Permit. This person is hired as a consultant by the development proponent to determine the location of archaeological sites in the development area, to assess and evaluate them and to make recommendations as to what forms of mitigative action, if any, will be required in the face of the destruction of the sites to conserve all of the information contained therein. This investigation, called an Historical Resources Impact Assessment, will provide sufficient data on any existing archaeological sites in the development area so that the province may classify those sites as:

1.   having been sufficiently studied by the impact assessment to warrant no further investigation,

2. having further investigative value as all of the archaeological information available in the site has not been extracted, or
3. having such significance to Alberta archaeology, or the field of archaeology that the site should be preserved in perpetuity under the Alberta Historical Resources Act as a Provincial or Registered Historic Resource. [6, Sections 17, 18]

Although sites in the last classification are not frequently encountered, it is not uncommon to discover sites that require further investigation before they may be impacted by development.* The Assessment and any follow-up investigations on valuable sites must, by the Act, be conducted prior to the commencement of any land surface disturbance activities related to the project [6, Section 22(2)].

In this manner, most archaeological data can be retrieved from the development area before it is altered by land surface disturbances related to exploration, extraction and production of crude bitumen.

THE PHYSICAL IMPACT OF TAR SANDS DEVELOPMENT ON THE LAND SURFACE

Since any land surface disturbance in an area of historical resources sensitivity potentially threatens those resources, a broad range of physical impact types could be considered when assessing the effects of complex land development projects. To be sure, the mine site development is but one of a myriad of identifiable surface impacts that may be directly or indirectly related to tar sands development.

The historical resources conservation program in Alberta therefore recognizes that the geographical extent of tar sands development impacts is not solely limited to the resource development area (lease) and to the exploration, extraction and production activities occurring in that area, but that it also includes offsite impacts resulting from the need for the provision of utilities systems, transportation facilities and other systems.**

It is also recognized that major natural resource development, especially in relatively remote areas having little infrastructural development, will have complex and physically expressed economic and social impacts on the region where the resource is located. Table 1 summarizes these issues where three distinct classifications as to the location of potential impacts are presented in relation to their type. They are: (1) impacts occurring on the lease area,

---

*To date only 0.3% of all historical resources sites in Alberta have been classified as requiring outright preservation.

**In Alberta, rights to tar sands deposits are allocated to potential developers through the sale of specific geographical areas referred to as Bituminous Sands Leases, and Oil Sands Leases.

Table 1. Development Impact Type and Impact Location Matrix
Tar Sands Development

| Impact Location | Land Surface Modification | | |
	Exploration, Extraction and Processing	Support and Transport Systems	Spin-Off Effects
Lease	seismic activities drilling construction camp mine site mine site roadways tailing ponds plant site storage areas ancillary drainage channels	oil/gas pipelines power lines water lines rail lines roadways airport construction camp ancillary	N/A
Off-Lease Vicinity	N/A	products pipelines roadways o/g pipelines power line rail line airport water line exploration granular/other resource areas drainage channels	N/A
Off-Lease Regional/Provincial	N/A	rail lines pipelines transmission-lines highways airports energy corridors	new towns urban expansion

(2) those found off the lease but still project-specific, and (3) those occurring off the lease area and having regional overtones being the result of more than one tar sands development project.

Those activities on the lease area relate directly to the exploration, extraction and processing of crude bitumen and the provision of support systems for those operations. Off-lease vicinity impacts are those created by the outward extension of the on-lease support systems relating to one project. Off-lease regional impacts are those relating to the support systems that supply more than one project and to those indirect spin-off impacts affecting the environmental, economic and social fabric of the region in which the development occurs. The construction of new towns or major residential expansion of existing urban areas are two examples of such effects.

This classification specifies the functional and geographical areas of concern being subject to the attention of Alberta's historical resources conservation program and offers statement as to the wide range of impacts that can occur. It is clear that the conservation issue becomes increasingly complex as more participants are brought into the development scenario. Although on-site impacts are clearly the responsibility of the proponents of individual projects, once offsite impacts are initiated, responsibilities regarding the related effects on historical resources may change from the lease developer to the pipeline, the transmission line, the rail line builders and other private agencies, or to government departments that are responsible for the establishment of the needed infrastructure.

THE ALBERTA TAR SANDS

The geographical context of the province of Alberta in western North America is shown in Figure 1. Figure 2 establishes the distribution of the tar sands resources in the province [8, pp. 1-5]. As can be seen, most of the resource is located in the northern half of the province, with a slightly eastward bias, and with the southernmost portion of the area being approximately 500 km north of the 49th parallel and the state of Montana. In total area, the resource is found in approximately 69,120 km^2, in four main deposit areas. Depths of the deposits range from surface outcrops to beneath 763 m of overburden. In total, there are about 1 trillion barrels of bitumen in place, 78% of which is under more than 152 m of overburden (in situ), 15% of which is beneath 46-153 m of overburden and 7% of which has less than 46 m of overburden and is surface-mineable [8, pp 1-5].

The largest of the four resource areas—and that with which this discussion is concerned—is the Athabasca-Wabiskaw-McMurray deposit containing some 720 billion barrels of crude bitumen, 74 billion of which may be surface mined [8, p.11].

Although this deposit was discovered as early as 1778, it was not commercially mined until 1967 when Great Canadian Oil Sands Limited (now Suncor Limited) began the first commercial production through surface mining, producing 65,000 barrels of synthetic crude per day. Syncrude Canada Limited recently constructed a second plant in the deposit area, based on surface mining, and produced 103,000 bbl/day in 1978. Subsequently, the Alsands Project Group has initiated the development of the third mining project with a projected capacity of 140,000 bbl/day [8, p. 1].

From an historical resources perspective, the Athabasca-Wabiskaw-

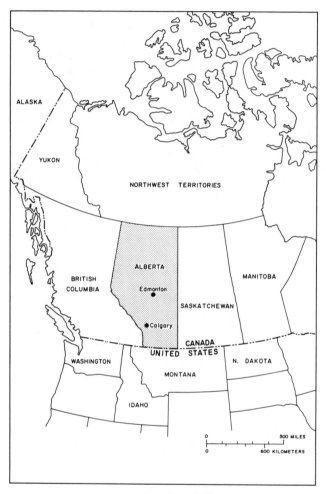

Figure 1. Location of Alberta.

McMurray deposit has remained relatively untouched, in today's standards, by man's impact on the environment, although it is not devoid of urban, agricultural and natural resource exploration and extraction related impacts.

In association with this relatively undisturbed nature, the area has a high expected sensitivity from an historical resources perspective. As far back as the late 18th century, it was visited and settled by fur seekers whose existence is imprinted on the area in the form of fort remains, cabins, trading posts, trails and the like. More recently, historic remains reflect the agricultural, resource exploration and resource extraction activities of the area.

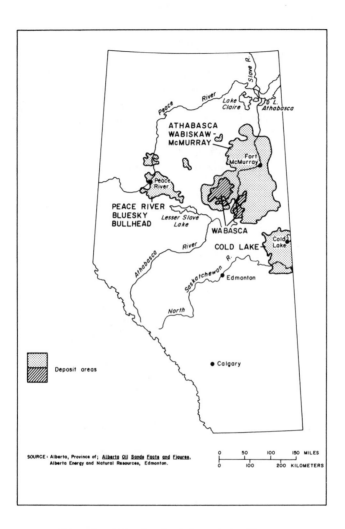

Figure 2. Major Tar Sands Deposits in Alberta.

Prehistorically, the natural resources of the area attracted groups of nomadic peoples who left behind evidence of their presence in the form of campsites, workshops, quarry sites and other activity impacts reflecting their lifestyles.

This expected high historical resources sensitivity was a sufficient basis for the province to direct its attention to tar sands development from the perspective of minimizing expected impacts on historical resources. It is difficult to estimate specifically the total area of land surface that is expected to be disturbed as a result of complete development of the mineable resources in the Athabasca-Wabiskaw-McMurray deposit as only the three mentioned specific proposals have been initiated. To appreciate, at least, the impact of the complete development of the mineable resource on-lease, it is quite possible that in this deposit as much as 2920 km^2 could be completely disturbed by mining alone [8, p. 27]. The mineable portion of the deposit is shown in Figure 3. Such vast physical impacts to the land surface must be considered as an equally significant impact on historical resources.

The Province of Alberta therefore entered discussions with the proponents

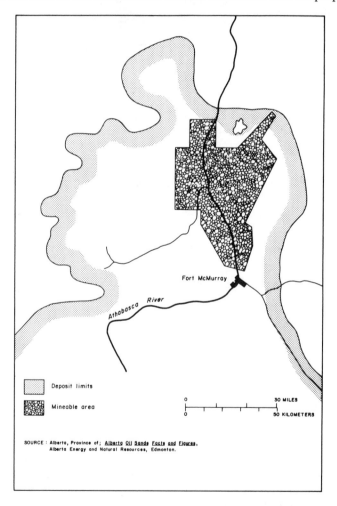

Figure 3. Mineable Area in the Athabasca-Wabiskaw-McMurray Deposit.

of development in the deposit area, with a view to the requirements regarding historical resources legislation.

Implementation of the legislation has resulted in a considerable amount of in-field research, which has led to the conclusion that concern over historical resources in response to the potential impact of tar sands development is well founded.

PROJECT-SPECIFIC INVESTIGATION

In-field investigations have been conducted on a number of development projects associated with tar sands development in the Athabasca-Wabiskaw-McMurray deposit. Those which have been sufficiently completed to warrant discussion involve the following projects:

1. Syncrude Canada Ltd. Project
2. Alsands Project
3. Highway 963 Construction
4. Residential expansion in the town of Fort McMurray.*

The locations of these project areas are shown on Figure 4. In consideration of the impact location and type matrix (Table 1), the investigations on the Syncrude Canada Ltd. and Alsands Project activities must be classified as being lease-specific and involve the exploration, extraction and processing phases. The Highway 963 project, a government program, is classified as an off-lease regional project, as it is aimed at supplying effective road transportation access to more than one tar sands project. The Fort McMurray Townsite residential expansion is classified as off-lease regional impact; the town being a recipient of considerable pressure to expand the residential capacity of the region due to an influx of population. In that manner it represents a spin-off that can be tied to the industrial boom of the Fort McMurray area due to tar sands development. It should be pointed out that there are a number of other projects, as shown in Figure 4, that are being investigated archaeologically at the present time to which Table 1 is applicable. A new townsite (regional spin-off) to handle increasing residential needs in a more acceptable geographic location is being proposed north of Fort McMurray. An airport (regional support and transport systems) is being planned just south of the new town site, and an energy corridor

---

*It should be noted that the Syncrude Canada Ltd., the Alsands, and the Fort McMurray projects were investigated by consultants hired by the companies and the town respectively, and the highway project was investigated by a consultant hired by the Province of Alberta. These investigations are listed as References 9–21.

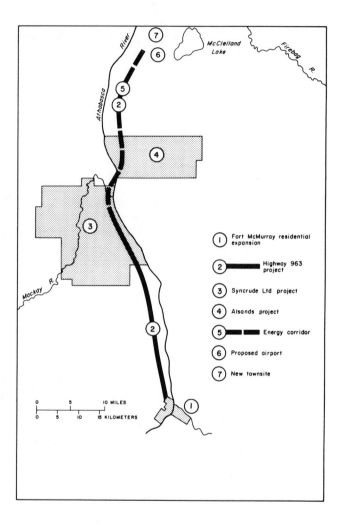

Figure 4. Major tar sands-related projects in the Athabasca-Wabiskaw-Fort McMurray area.

(regional support and transportation systems) is being planned in order to concentrate all offsite impacts resulting from tar sands development.

Basically, each project area was investigated in the field by professional research teams that, as mentioned previously, were directed to locate any archaeological sites that may be impacted by land surface disturbance associated with the projects. They were then to assess and evaluate the sites discovered and make recommendations as to their ultimate disposition in the face of destruction, and to conduct any salvage archaeology if necessary.*

Table 2 summarizes the results of those investigations as conducted on the four projects. Clearly, the investigations were not conducted in vain, as research in each project area encountered a number of historical resources. Of 134 historical resources sites discovered, 16 were historic in nature and the remaining 118 are classified as prehistoric. While it is clear that the most prevalent historical resource site type was prehistoric, historic sites also must be considered significant.

Table 2. Frequency of Sites Encountered by Project
Source: Archaeolgical Survey of Alberta

	Historic	Prehistoric	Total
Syncrude	1	34	35
Alsands	6	55	61
Highway 963	3	24	27
Fort McMurray	6	5	11
Total	16	118	134

General Site Value

As previously mentioned, sites encountered in such investigations have varying degrees of historical value depending on the quantity, quality and type of information they contain. The value placed on them determines their ultimate disposition and the type of conservation attention that must be given to them.

Table 2 summarizes the general conservation approaches taken to the 134 sites encountered in the subject project areas and serves to give at least an indication of what could be expected in future development projects in the area. Of particular note is the fact that of the 134 sites discovered in the project areas 82 (61.2%) were sufficiently investigated through the initial assessment phase so that no further study (salvage) was required before they could be impacted by development. Only 52 (38.8%) of the 134 sites required investigation in greater detail than provided by the survey prior to their being impacted by construction of the project. Finally, only 1 of the 134 sites was of such unusually high value that it was viewed as requiring preservation in perpetuity through direct protection under the Alberta Historical Resources Act by having it designated as a Provincial Historical Resource. This site, the

---

*Salvage archaeology is an approach which rapidly removes all of the available data in an archaeological site before it is destroyed, in a manner so that the site can be statistically reconstructed and analyzed.

Beaver Creek Quarry Site, is located on Syncrude Lease #22 and has been subjected to considerable investigation [22].

Although the sample population of projects studied is small, the results shown in Table 3 are generally representative of investigations conducted on development projects in Alberta, i.e., the majority of historical resources sites encountered need only to be discovered, recorded and assessed and they will have been sufficiently investigated to warrant no further concerns as to their disposition. The balance will require further investigations to suf-ficiently record the available data. A very small proportion of sites will require preservation. Indeed, although in the case studied above 0.8% of the sites required preservation, in Alberta as a whole only 0.083% of all histor-ical resources sites have required such legislative attention.

Table 3. Conservation Requirements on Sites Encountered
Source: Archaeological Survey of Alberta

	No Further Investigation Required	% Total	Further Investigation Required	% Total	Preservation Required	% Total	Total
Syncrude	21	60.0	13	37.0	1	3.0	35
Alsands	41	67.3	20	32.7			61
Highway 963	13	48.1	14	51.9			27
Fort McMurray	7	63.6	4	36.4			11
Total	82	61.2	51	38.0	1	0.8	134

Site Density

When coupled with the dimensions of the project areas, the site frequency data can be employed to produce site density estimates, or the number of sites/unit area that might be expected to be encountered in a given project area. Table 4 tabulates the known site density on a per project basis. To be sure, this is an underestimation of the actual site density due to the fact that most investigations were based on predictive models rather than probabil-istic approaches, and therefore the research for the most part cannot be said to be spatially representative. Indeed, in the Alsands Lease #13, Shell Canada estimated that there were 185 sites in the area, approximately three times the number actually encountered by the predictive survey [17, p. 165].

Table 4 indicates that the four projects represent a disturbance area of over 465 km^2, in which 134 sites were found. The mean site density of these

Table 4. Site Density Estimates: Encountered Sites

	Project Area (km^2)	(mi^2)	Sites per km^2	Sites per mi^2	Total Sites
Syncrude	190.0	75.4	0.18	0.45	35
Alsands	196.4	77.9	0.31	0.77	61
Highway 963	39.1	15.5	0.69	1.73	27
Fort McMurray	40.3	16.0	0.27	0.70	11
Total	465.8	184.9	0.29	0.70	134

project areas was therefore 0.29 sites/km^2 and the range was from 0.18 sites/ km^2 (Syncrude) to 0.69 sites/km^2 (Highway 963).

Although the sample is not large, a general appreciation can be gleaned from the data that will establish the approximate expectation of the number of historical resources sites present in a given development project area. Given that there may be as much as 2920 km^2 of mining area in the Athabasca deposit alone, the potential minimum number of historical resources sites that could be encountered based on the results of the research on the four projects is approximately 850 sites.

Specific Site Impact

As mentioned previously, historical resources sites may be impacted by various types of surface disturbance associated with the development types mentioned in Table 1. Table 5 indicates the type of impact that threatened the sites encountered during investigation of the four subject projects.

Of the 10 types of identified impacts it would appear that the linear type of impact represents the greatest threat to historical resources because 65 of the 134 sites (48.5%) were threatened or already impacted by either a roadway or cutline. All other disturbance types accounted for impact to 17 sites (12.7%). Of the remaining sites, 34 sites (25.4%) were to be avoided by development, and the impact on 18 sites (13.4%) could not be determined. At present it can only be surmised as to why such a large proportion of sites are threatened by linear disturbances. It may involve the effect on site discovery of going into the field after clearing activities have been conducted where surface site visibility is enhanced, or the fact that linear developments represent a more spatially random type of disturbance as it represents an environmental transect. Large disturbance areas such as mine sites represent major spatial samples that can only be practically investigated through a sampling model which in these cases was usually predictive, and which therefore may not represent the true situation.

Table 5. Impact Type Threatening Sites

Project	Road	Cutline	Drill Site	Campsite	Mine Area	Granular Resource Area	Airport	Plant Site	Sludge Pond	Barge Landing	None	Unknown
Alsands	28	10	2	3	2	4	2	1	0	1	2	6
Syncrude	1	12	1	0	0	0	0	0	1	0	18	2
Fort McMurray	0	1	0	0	0	0	0	0	0	0	0	10
Highway 963	13	0	0	0	0	0	0	0	0	0	14	0
Total	42	23	3	3	2	4	2	1	1	1	34	18

If these are significant factors in the high number of sites encountered by linear developments, the impact of tar sands development may in fact be much greater in degree than is now anticipated and the prediction model of Shell Canada may in fact be quite applicable. The discovery of historical resources in these areas, then, may reflect more the methodology employed in field research than the actual site distribution or density. This is an unknown which must be determined to increase the effectiveness of historical resources conservation in the area.

CONCLUSION

The development of tar sands deposits in Northern Alberta represents the first major and concentrated thrust of man into a relatively pristine environmental setting. Since archaeological resources are a part of this setting, and since there is a recognized value in preserving and understanding these manifestations of the past, the potential impact of tar sands development on historical resources had become an area of concern to the Province of Alberta. The result was the implementation of an historical resources conservation program aimed at ensuring that information concerning the past would not be lost or destroyed. Accordingly, tar sands and associated development have initiated the first main thrust of concentrated archaeological investigation in this area of Alberta.

Research was conducted in all major tar sands and associated development areas and the results have shown that the historical resource concerns of the province are well founded in that there is a definite high potential for development projects encountering historical resources. Past conservation efforts, however, in response to this potential have only been partially successful in that the general research methodologies employed may have resulted in low site discovery rates. This is likely attributable to:

1.   a lack of knowledge of Boreal Archaeology in Alberta,
2.   the characteristics of legislative requirements and
3.   the spatial extent of tar sands developments.

The combination of these three factors has generally led to the employment of predictive rather than probabilistic research models where a predictive model will employ existing archaeological knowledge of other areas and apply the same in extrapolated form to project areas of concern. These areas, which must legally be subject to investigation, can only be investigated to a degree that is practicable given the extensive nature of tar sands development projects. Accordingly, in large projects, only those areas of expected high sensitivity and within direct impact zones can generally be investigated. There are difficulties inherent in this:

1.   A predictive model will bias any research results, where only site-specific conclusions can scientifically be drawn and can only confirm or refute the predictive model. It cannot establish or prove the real situation.
2.   Sites in areas not expected to have sensitivity in the predictive model and that are avoided due to project scale may exist and will be lost to development, thereby destroying any possibility of determining the existing settlement pattern in the area.
3.   The fact that legislation requires that only impact areas be investigated compounds the spatial bias in the investigation as that impact area may not represent a statistically valid representative sample of the area.

It is therefore clear that existing conservation measures could be improved to ensure that no historical data are lost, and to ensure that retrieved information is collected in methodological format that allows for a broader geographical base for analysis. The result would be a more effective conservation program, as statistically valid predictive models may be established. Furthermore, these models could be employed to reduce the amount of time, effort and hence cost required by industry to meet the needs of the Alberta Historical Resources Act.

It may in fact be appropriate for government and industry to cooperate even further in the short-term to establish research programs that go beyond minimum legislative requirements as an investment to lessening the cost of future endeavors on given projects. These programs would be directed toward the development of a settlement pattern model that can be employed in proposed project areas.

Regardless, the Alberta experience has shown that tar sands development and historical resources conservation can be accomplished concomitantly and that continuing cooperation between developers—whether private firms or government agencies—and those vested with the responsibility for historical resources conservation is a major part in ensuring that knowledge of our past is not lost.

REFERENCES

1.   UNESCO. *General Conference, 9th Session.* New Delhi, December, 1956; Paris, France, 1957.
2.   McGimsey III, C. R. and H. A. Davis, Eds. *The Management of Archaeological Resources* (The Airlie House Report). The Society for American Archaeology, Virginia (1977), p. 27.

3. Moos, H. R. *The Human Context—Environmental Determinants of Behavior,* John Wiley and Sons (1976), p. 63.
4. Province of Alberta. "Alberta Oil Sands Facts and Figures," Report #110, Department of Energy and Natural Resources, Edmonton, p. 1.
5. Province of Alberta. "The Conservation of Historical and Archaeological Resources in Alberta," Environment Conservation Authority, Edmonton (1973), 120 pp. (Note: This agency is now the Environment Council of Alberta.)
6. Province of Alberta. "The Alberta Historical Resources Act," C.5 (1973).
7. Province of Alberta. "Historical Resources Conservation Work Occasioned by Development Programs," Department of Alberta Culture, Edmonton (1977).
8. Province of Alberta, "Alberta Oil Sands Facts and Figures," Energy and Natural Resources, Edmonton (1978).
9. Conaty, G. T. "Alsands Lease—Archaeological Survey," Alsands Project Group, Calgary (1980).
10. Fromhold, J. and Associates. "New Town of Fort McMurray, Timberlea and Area Four Subdivisions—Historical Resources Impact Assessment," Town of Fort McMurray (1979).
11. Gryba, E. "Highway Archaeological Survey (North) 1978," Archaeological Survey of Alberta, Edmonton (1978), pp. 104–119.
12. Lifeways of Canada Ltd. "Archaeological Reconnaissances—Alberta Transportation Highway Construction Program," Archaeological Survey of Alberta, Edmonton (1977).
13. Lifeways of Canada Ltd. "Historical Resources Impact Assessment Syncrude, Canada Ltd., Western Portion of Lease No. 17," Syncrude Canada Ltd., Edmonton (1977).
14. Lifeways of Canada Ltd. "Conservation Archaeology—Alberta Transportation Highway Construction Program Project 963," Alberta Transportation, Edmonton (1978).
15. Losey, T. C. "Alberta Highways North," Archaeological Survey of Alberta, Edmonton (1974).
16. Losey, T. C. "Archaeological Reconnaissance—Alberta Highways North," Archaeological Survey of Alberta, Edmonton (1974), p. 27–38.
17. Losey, T. C. "Archaeology, Environmental Impact Assessment—Lease 13 Mining Project, Alberta Oil Sands," Shell Canada Ltd., Calgary, Alberta (June 1975), p. 153–166.
18. Losey, T. C. and C. Sims. "Syncrude Lease No. 17: An Archaeological Survey," Syncrude Canada Ltd., Edmonton (1973).
19. Paleo-Sciences Integrated Ltd. "1979 Archaeological Investigation on Highway Project 963," Alberta Transportation, Edmonton (1980).
20. Sims, C. "An Archaeological Survey of Certain Boreal Forest Highway Projects in N.E. Alberta," Archaeological Survey of Alberta, Edmonton (1975), p. 1–45.
21. Sims, C. "The Boreal Highway Archaeological Survey," Archaeological Survey of Alberta, Edmonton (1976), p. 15–19.
22. Syncrude Canada Ltd. "The Beaver Creek Site: A Prehistoric Stone Quarry," Environmental Research Monograph, 1974-2, Edmonton (1974), 113 pp.

CHAPTER 4

TRACE ELEMENTS IN OIL SHALE MATERIALS

J. P. Fox, A. T. Hodgson and D. C. Girvin

Lawrence Berkeley National Laboratory
University of California
Berkeley, California

INTRODUCTION

Concern about toxic element mobilization and potential public health and environmental consequences has led to investigations of elemental distributions in Green River Formation oil shales and in the products of oil shale processing [1-16]. The results of those studies and other previously unreported data are presented and reviewed here as they relate to the characterization of the environmentally important elements As, B, Cd, Co, Cr, Cu, F, Hg, Mn, Mo, Ni, Pb, Sb, Se, V and Zn in spent shale, shale oil, process waters and offgases for representative surface and in situ retorting processes.

PROCESS DESCRIPTIONS

This chapter reviews characterization data for the Fischer Assay, Paraho direct, Lawrence Livermore National Laboratory (LLNL) simulated in situ, Occidental modified in situ, and Geokinetics horizontal true in situ retorting processes. These processes were selected because they represent those which are presently under commercial development and/or because they have been the subject of elemental partitioning studies. Characteristics of these processes are discussed below and are summarized in Table 1.

Table 1. Processes and Retort Operating Conditions Considered in This Study

Process	Retort Capacity (kg)	Shale Source	Operational Mode	Shale Charge Size (mm)	Retorting Atmosphere	Maximum Retorting Temperature
Surface						
TOSCO Modified Fischer Assay	0.1	Dow Mine, CO	Batch	1.3	N_2	500°C
Paraho Direct Mode	3.18×10^5	Anvil Points Mine, CO	Continuous	6–76	Air + Recycle Gas	750°C
In Situ						
LLNL Small Retort (Run S-11)	125	Anvil Points Mine, CO	Batch	13–25	Air	1000°C
LBNL Retort	4	Anvil Points Mine, CO	Batch	0.6–6.4	N_2	500°C
Occidental Modified In Situ						
Retort 3	$\simeq 3.6 \times 10^6$	Logan Wash, CO	Batch	a	Air + Recycle Gas	b
Retort 6	3.28×10^8	Logan Wash, CO	Batch	a	Air + Steam	b
Geokinetics Horizontal In Situ		Book Cliff, UT	Batch	a	Air	b

[a] A satisfactory method has not been developed for characterizing the particle size range in a field in situ retort.
[b] Field retorting temperatures are not accurately known due to corrosion problems with thermocouples. However, mineral analyses of spent shales from the Geokinetics and Occidental processes suggest temperatures may reach 1000°C locally.

Surface Retorts

Surface retorts are aboveground, vertical kilns in which crushed oil shale is thermally decomposed. There are two broad classes of surface retorting processes: direct and indirect. The indirect processes are those in which heat is transferred to the crushed shale by an externally or indirectly heated gas or solid. In these processes, the atmosphere within the retort is reducing, and carbon or char remains on the spent shale. Indirect-heated processes include Fischer Assay, TOSCO II, Lurgi, Union B, Paraho indirect, Superior, Galoter and Petrosix. The direct processes are those in which heat is supplied by the combustion of char and recycle gases within the retort. Thus, both reducing and oxidizing atmospheres exist within direct-heated retorts. Examples of direct-heated retorts include Paraho direct and Union A.

Selection of the Fischer Assay and the Paraho-direct processes as representative of surface technologies was based on the availability and quality of elemental partitioning data.

Fischer Assay

A Fischer Assay is a standardized oil yield test for oil shale. Several variations of the assay have been proposed and used [17–19]. The studies by Shendrikar and Faudel [4], Wildeman and Meglen [7] and Fox et al. [20], which are reviewed here, used the TOSCO Modified Fischer Assay [18].

In the TOSCO Modified Fischer Assay, about 100 g of finely ground raw shale are heated in a steel retort with an external electric heater. A programmed temperature-time profile with a final temperature of 500°C is followed. The vapors distilled from the sample are cooled and condensed at 0°C. The products are then collected and weighed.

Paraho-direct

The Paraho retort consists of a vertical kiln that may be operated in either the direct or indirect mode (Figure 1) as described by Jones [21]. The studies reviewed here used products from the 200-bbl/day Paraho direct-mode 23-m high semiworks retort [10, 11] operated by Development Engineering, Inc. at Anvil Points, Colorado. Process shale is crushed, screened and top loaded into the retort. The retort operates as a countercurrent, plug-flow reactor; recycle gas moves upward through the retort while the solids move downward. Shale entering the top of the retort is initially heated by rising hot gases. It next passes into the pyrolysis zone where the organics are thermally distilled by heat produced in the combustion zone from burning of char and recycle gas. The distilled vapors are swept upward, out of the retort, and the

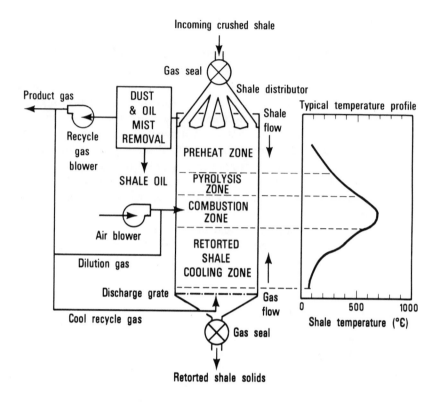

Figure 1. Schematic of the Paraho retort.

oil is separated from the gas by a coalescer and electrostatic precipitator. The oil-water emulsion is separated by settling. The hot spent shale passes through the cooling zone and is discharged at the bottom of the retort. In this retort, the maximum temperature is 750°C and both oxidizing and reducing conditions occur.

In Situ Retorts

In situ processes are those in which oil shale is thermally decomposed in the geological formation. Permeability must first be developed to permit the flow of gases and liquids. This is accomplished either by mining out part of the formation and rubblizing the balance (modified in situ processes) or by creating void space without shale removal by surface uplift or other means (true in situ processes). Modified in situ processes are presently under development by Occidental Oil Shale at Logan Wash and tract C-b in Colorado and by the Rio Blanco Oil Shale Project at tract C-a. True in situ processes are under development by Geokinetics, Inc. in Utah and by Equity Oil in Colorado. In addition to these commercial projects, several organizations have built and operate simulated in situ retorts which are surface retorts designed and operated to simulate in situ conditions. These experimental reactors are used to develop an understanding of the retorting process under controlled conditions. A side benefit has been the production of valuable characterization and partitioning data that could not otherwise have been obtained.

The LLNL simulated in situ retorts, the Occidental modified in situ process, and the Geokinetics true in situ process were selected for review because of the relative abundance and quality of available data compared to other processes.

Simulated In Situ Retorts

Much of the trace element characterization and partitioning data for in situ processes have been developed using simulated in situ retorts [1, 9]. This chapter focuses on the results obtained from the LLNL small retort due to the abundance of available data and to the fact that this retort simulates in situ conditions more nearly than others. Gas data from the Lawrence Berkeley National Laboratory (LBNL) retort and the LLNL large retort are also presented.

The LLNL retorts have been previously described [22]. The small retort is 0.30 m in diameter by 1.5 m high and has a capacity of 125 kg of shale; the large retort is 0.91 m in diameter by 6.1 m high and has a capacity of 6000 kg. The retorts are surrounded by a contiguous series of 15-cm vertical electrical-resistance heaters to prevent heat loss through the walls. These retorts can be operated for a wide range of conditions by varying the input gas composition and flow rate, particle size, packing density, retorting rate and other variables. To date, LLNL has completed 24 runs of the small retort for a range of conditions, and characterization and partitioning studies have been conducted on seven of these. We focus on results from run S-11 of the small retort, an air run in which temperatures reached 1000°C. We also present some gas data from run L-3 of the large retort, an air-steam run with graded lean and rich Anvil Points shale at a maximum temperature of 1000°C.

The LBNL retort is a 4-kg Fischer Assay-type retort designed to evaluate the effect of various gas environments, flow rates and heating conditions on trace element volatilization and partitioning. The retort vessel is a stainless steel tube, 8.9 cm i.d. by 90.5 cm long. The retort is heated by a single zone furnace designed for a 1200°C maximum operating temperature and a programmer capable of providing a heating ramp of 0.4-9.9°C/min.

Occidental Modified In Situ Process

The in situ recovery of shale oil by the Occidental modified in situ (MIS) process involves the underground pyrolysis of large chambers of rubblized shale. These chambers are constructed by mining out $\sim 20\%$ of the volume of the retort and blasting the balance so that the entire chamber is filled with fractured rock. A commercial-sized retort will measure ~ 100 x 50 m in plan and 120 m high. Oil is recovered from such a retort (shown in Figure 2) by initiating combustion at the top of the retort with an external fuel supply and propagating the reaction zone, which consists of a pyrolysis zone and a trailing combustion zone, down the packed bed of shale with input gas. The volatile hydrocarbons condense in the cool region at the bottom of the retort and are pumped to the surface.

Occidental has tested six experimental retorts at Logan Wash, Colorado. Environmental studies have been conducted on retorts 3 and 6, which are briefly described here. Retort 3 was operated between February and July of 1975 in an air-recycle gas mode. Yield was around 60% Fischer Assay. This retort was cored in December 1978 and the spent shale characterized. An intensive, onsite sampling program was conducted during operation of Retort 6. This retort was operated in an air-steam mode. Failure of the sill pillar during operation resulted in channeling of the flame front. Yield was around 40% Fischer Assay. The run took place between August 1978 and the summer of 1979 and environmental sampling occurred from March through July of 1979.

Geokinetics Horizontal In Situ Process

The Geokinetics process, shown in Figure 3, is designed for areas where oil shale beds are relatively thin and close to the surface. Permeability is created by explosive fracturing, which uplifts the surface by about 3 m. A commercial-sized retort may measure ~ 70 m long by 70 m wide by 10 m thick. Oil is recovered from these retorts by igniting the shale bed near the air injection well and driving the reaction zone horizontally through the fractured bed with input gas. The volatile hydrocarbons condense ahead of the reaction zone and collect in the recovery wells where they are pumped to the surface.

Geokinetics has burned 24 experimental retorts to date. A field sampling program was carried out during the operation of several retorts, and raw and spent shale cores have been taken from Retort 16. Some of these data are reviewed here.

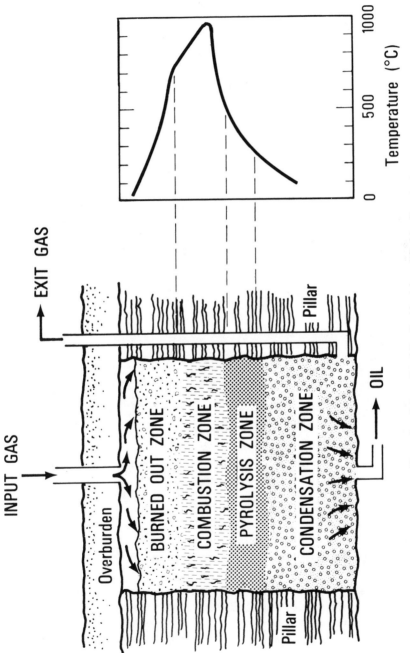

Figure 2. Schematic of the Occidental modified in situ retort.

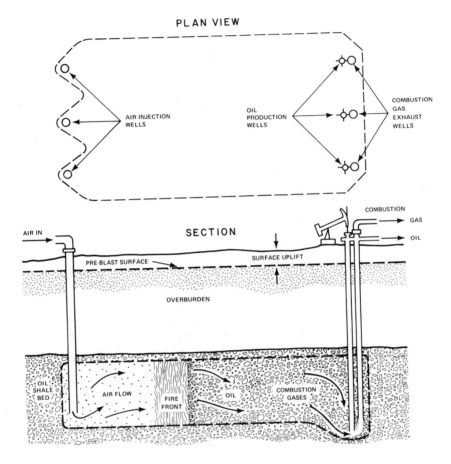

Figure 3. Schematic of the Geokinetics process.

CHARACTERIZATION

The elemental abundances of toxic elements in raw and spent oil shales, shale oils, process waters and offgases from the above processes are reviewed. These abundances are compared with the elemental abundances of selected reference materials and with environmental quality standards in order to assess their significance.

Raw Oil Shale

All of the selected processes use oil shale from the Green River Formation of the Piceance Basin of Colorado, Utah and Wyoming. Green River oil shale is a fine-grained sedimentary rock that contains ~20 wt % organic material known as kerogen. The inorganic portion is primarily dolomite and calcite with lesser amounts of quartz, analcime, Na-feldspar and K-feldspar. In addition, Mg-siderite, illite, pyrite and aragonite are frequently detected minerals [2, 3]. The toxic elements of interest may occur as substitution products in the major mineral phases; they may be organically bound or they may occur as minor mineral phases.

The elemental abundances of some toxic elements in feedstocks for five retorting processes are summarized in Table 2 and compared with Green River concentration ranges and average crustal rock abundances. The first column presents the range in elemental concentrations measured in 280 samples (555 m of core) of two core holes from the Naval Oil Shale Reserve No. 1 [2]. These are the most extensive data presently available on Green River oil shale and are presented here for comparison with specific feedstocks. These investigators found that the elemental compositions of two core holes from the Naval Oil Shale Reserve No. 1 were very similar even though the core holes were from the center and edge of the depositional basin and 10 km apart. This lateral uniformity is particularly important for environmental studies of MIS retorting because it means that several core holes can be well characterized and the results extrapolated to MIS retorts located throughout the formation. Giauque et al. [2] also found that the concentrations of most of the 57 elements they studied did not vary with depth by more than a factor of three or four in either of the core holes. However, variations of a factor of 10 or greater were observed for As, Cd, Hg, Mo, Se, B and F. Vertical variations in As, Se, Pb, Sb, Hg, Cu and Co for core hole 25, near the Logan Wash site, are shown in Figure 4. This figure demonstrates the vertical uniformity of each element throughout much of the core as well as a strong correlation among the seven elements.

The second and third columns of Table 2 present average elemental concentrations for the two core holes for the retorting horizon or the portion of the formation that is economically recoverable by MIS retorting. The composition of core hole 25 (118- 215-m depth) from the depositional edge of the basin approximates that of shale from the retorting horizon at Logan Wash (Occidental MIS retorts 3 and 6); the composition of core hole 15/16 (261- to 447-m depth) from the basin center represents shale from the retorting horizon of tracts C-a and C-b (MIS processes presently under development here). Column 4 summarizes the results of preliminary analyses by Wilkerson [23] for five samples of weathered raw shale from the Geokinetics site

Table 2. Elemental Abundances of Some Toxic Elements in Raw Oil Shale Charges to Retorting Processes and in Average Crustal Rocks (ppm)

	Range in Green River Oil Shale[a] (1)	Retorting Horizon (118–215 m depth) of Logan Wash, CO (Core Hole 25 from Edge of Depositional Basin)[a] (2)	Retorting Horizon (261–447 m depth) at Tracts C-a and C-b, CO (Core Hole 15/16 from Center of Depositional Basin)[a] (3)	Geokinetics True In Situ[b] (4)	LLNL and LBNL Simulated In Situ Retorts[c] (5)	Paraho Direct-Mode Retort[d] (6)	TOSCO Modified Fischer Assay[c] (7)	Crustal Abundance[e] (8)	Concentration Ratio C_{RS}/C_{CR} (9)
As	5–134	39±19	42±23	43±23	42±1	44±1	75±2	1.8	26
B	25–250	82±32	74±38	—	108±11	94±2	80±8	10	8.8
Cd	0.2–1.4	0.61±0.25	—	—	0.72±0.07	0.64±0.03	1.1±0.1	0.2	3.8
Co	4–14	8.3±1.9	8.6±1.9	—	8.8±0.2	9.0±0.1	9.7±0.2	25	0.36
Cr	18–60	33±9	33±8	36±9	46±1	34±1	29±1	100	0.35
Cu	11–63	31±9	30±10	40±12	39±3	40±2	46±2	55	0.69
F	100–3200	1260±510	1340±630	—	990±20	—	1020±100	625	1.8
Hg	0.01–0.50	0.065±0.032	0.077±0.048	—	0.08±0.004	0.09±0.01	0.14±0.01	0.08	1.1
Mn	190–460	335±54	317±63	244±60	377±6	315±12	272±9	950	0.32
Mo	4–54	18±9	20±9	—	20±1	22±2	28±2	1.5	14
Ni	11–32	21±4	20±4	28±5	24±2	28±1	31±3	75	0.34
Pb	8–41	20±7	19±8	27±10	24±1	26±2	29±1	12.5	1.9
Sb	0.5–5.2	1.7±0.6	1.6±0.7	—	2.0±0.1	2.1±0.1	3.2±0.1	0.2	11
Se	0.5–12.3	1.6±0.6	2.0±1.0	1.5±0.3	2.5±0.3	2.0±0.1	4.3±0.3	0.05	46
V	29–203	95±36	94±30	<50–110	93±5	94±2	127±30	135	0.75
Zn	38–153	68±13	71±17	52±12	65±2	64±2	74±4	70	0.94

[a]Giauque et al. [2].
[b]Wilkerson [23].
[c]Fox et al. [20].
[d]Fruchter et al. [10, 11].
[e]Taylor [24].

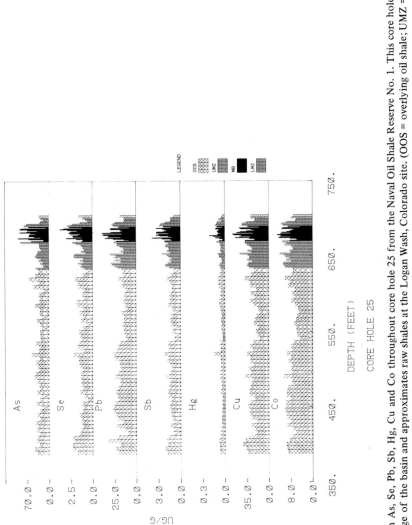

Figure 4. Vertical variation in As, Se, Pb, Sb, Hg, Cu and Co throughout core hole 25 from the Naval Oil Shale Reserve No. 1. This core hole is near the depositional edge of the basin and approximates raw shales at the Logan Wash, Colorado site. (OOS = overlying oil shale; UMZ = Upper Mahogany Zone; MB = Mahogany Bed; LMZ = Lower Mahogany Zone) [2].

at Book Cliff, Utah. Columns 5 and 6 present the elemental concentrations in feedstock for the LLNL and LBNL simulated in situ and Paraho direct-mode retorts, respectively. Both of these feedstocks are from the Anvil Points Mine, Colorado. The two sets of values are within two standard deviations of each other for all elements. This is remarkable considering the different preretorting shale preparation, differences in sampling and analytical methodologies, and the possibility that the samples came from different horizons within the mine. Column 7 presents the elemental abundances in feedstock for the Fischer Assay retort runs discussed in this chapter. The samples are from the Dow Mine of Colony Development Co. Note that this shale differs from the others reported here. Column 8 summarizes the elemental concentration of average crustal rock as computed by Taylor [24]. These average abundances were calculated on the basis of a 1:1 mixture of granite and basalt abundances. Average crustal rock is used here as a reference material to identify elements which are enriched or depleted in oil shale relative to other rocks. This is achieved by using the concentration ratio (CR) which is summarized in column 9 and which is defined as:

$$CR = C_s/C_r$$

where C_s = the elemental concentration in the sample of interest, i.e. raw shale (RS), spent shale (SS), shale oil (SO)

C_r = the concentration of the same element in a reference material, CR = crustal rock (CO = conventional oil)

The concentration ratios in column 9 represent the ratio of the average elemental abundances in the six raw shales presented in Table 2 (columns 2-7) to the crustal rock concentrations.

The data in Table 2 illustrate several important differences between Green River shales and average crustal rocks. The concentration ratios in column 9 indicate that the conventionally sulfide-forming elements As, Cd, Mo, Sb and Se as well as B are significantly enriched in Green River oil shales and that the elements Co, Cr, Mn and Ni are significantly depleted compared with average crustal rocks. The other elements, viz., Cu, F, Hg, Pb, V and Zn, are similar to average crustal rocks. This agrees with results previously reported by Fox [1] except that Cd was not reported as enriched in that study.

The high concentrations of the toxic elements As, B, Cd, Mo, Sb and Se could lead to environmental problems. The elements As, Cd and Se have been demonstrated to be released in significant quantities to oils, waters and gases during oil shale retorting [1, 11]. Boron and Mo are readily leached from spent shales [25] and thus may find their way into surface and ground-waters. All of these elements will occur in raw and spent shale-derived fu-gitive dust. This dust is viewed as a major environmental concern for oil shale development because high atmospheric dust concentrations could lead to violations of the Clean Air Act and contribute to worker and public health problems and ecological disturbances. Fugitive dust is discussed in a separate section of this chapter.

The chemical form of elements in oil shale is of critical importance be-cause it controls the mobility of each element during retorting and leaching. Unfortunately, little work has been done in this area. The data which are available are summarized in Table 3. Desborough et al. [26] hypothesized the mineral residence of several toxic elements based on their occurrence in other crustal rocks. Giauque et al. [2, 3] investigated mineral residence by corre-lating mineral phases and elemental abundance data for 280 samples of oil shale. This work revealed that more than half of the 57 elements studied correlated well with the minerals Na-feldspar and K-feldspar while As, Cd, Hg, Mo and Se correlated with the organic fraction of the oil shale matrix. These investigators suggested that Co, Cu, Ni, Pb, Sb and Zn occurred as sulfides, which is consistent with findings by Saether et al. [27]. Saether et al. [28] demonstrated that F is associated with micaceous clay minerals, especially illite, in the Mahogany Zone.

Table 3. Postulated Residences of Some Toxic Trace Elements[a]

Sulfides	
Co, Cu, Ni, Pb, Sb, Zn, As, Mo, Se, Hg	
Illite	
F	
Feldspars and Mg-siderite	
B, Cr, Mn, V	
Organic Phase[b]	
As, Mo, Se, Hg	
Unknown	
Cd	

[a] Based on data presented by Giauque et al. [2], Saether et. al [27, 28], and Des-borough et al. [26].
[b] Recent analyses show a strong correlation between these elements and total sulfur, which suggests a dual residence in sulfide minerals and/or in the organic phase.

Spent Oil Shale

Spent oil shale is the solid residue that remains after the organic material has been removed by pyrolysis and combustion. The elemental compositions of raw and spent oil shale are similar for most elements, although the mineral compositions may vary considerably. Differences in the elemental and mineral composition of raw and spent shales are due to high-temperature mineral reactions and to the loss of organics and carbon dioxide from the shale matrix during retorting. The major mineral phases of spent shales vary, but they generally include survivors of the retorting process such as quartz and Na- and K-feldspars and the decomposition products periclase (MgO) and calcite ($CaCO_3$). Synthesis products, which are the results of high-temperature mineral reactions, are also present in spent shales from in situ processes where maximum retorting temperatures may exceed $1000°C$. Synthesis products identified to date include augite, mellilite, kalsilite and monticellite.

The elemental concentrations of nonvolatile elements are higher in spent shale than in raw shale because 20% or more of the oil shale is removed, primarily as volatilized organics and carbon dioxide, leaving behind ~ 100% of the nonvolatile elemental masses. On the other hand, the volatile elements, including Hg, Cd and, under some circumstances, Se, may be depleted in the spent shale relative to the raw shale due to their loss from the matrix [1, 29]. The elemental abundances of some toxic elements in the spent shale from the five retorting processes under investigation are summarized in Table 4 and are compared with the average crustal rock abundances presented previously. Variations in elemental abundances in spent shale from a core through Occidental Retort 3 are shown in Figure 5. These concentrations are not directly comparable to concentrations shown in Figure 4 because the vertical positions of the spent shale samples were not carefully controlled during coring.

Comparison of Table 4 and Table 2 shows that there is greater variability in elemental concentrations among the spent shales than among the raw shales. The increased variability is the result of retorting as well as the difficulties inherent in sampling spent shales, especially those from field in situ retorts. These difficulties include contamination, poor recovery during coring and nonuniform spent shale. These sampling difficulties preclude accurate assessments of the effects of in situ retorting processes as employed by Geokinetics and Occidental on trace element mobilization. For example, coring of Occidental Retort 3 at Logan Wash recovered <20% of the in-place material, and of that a majority came from highly localized zones within the retort. Additionally, the samples were contaminated by drilling fluids, and chemical changes occurred in the spent shale due to water leakage into the retort [14]. Nonuniform retorting, due to different porosity and permea-

Table 4. Comparison of Elemental Abundances of Some Toxic Elements in Spent Shale with Average Crustal Values (ppm)

	Occidental Retort 3 at Logan Wash (1)	Geokinetics True In Situ[a] (2)	LLNL Simulated In Situ Retort[b] (3)	Paraho Direct-Mode[c] (4)	TOSCO Modified Fischer Assay[b] (5)	Crustal Abundance[d] (6)	Concentration Ratio[e] C_{SS}/C_{CA} (7)
As	34±24	80±22	58±1	59±1	82±3	1.8	37
B	252±138[f]	—	140±15	107±2	110±9	10	12
Cd	—	—	0.77±0.08	0.91±0.04	1.28±0.13	0.2	4.9
Co	—	—	11.9±0.2	11.1±0.2	11.5±0.2	25	0.46
Cr	—	78±41	50±1	44±1	36±1	100	0.43
Cu	36±11	86±15	49±2	56±1	67±5	55	1.0
F	—	—	980±60	—	1420±200	625	1.9
Hg	0.041[f]	—	<0.01	0.035±0.003	0.040±0.001	0.08	0.47
Mn	—	316±62	480±10	396±14	342±10	950	0.43
Mo	10±7[f]	—	27±1	34±1	37±2	1.5	22
Ni	25±6	57±19	38±3	32±2	38±1	75	0.51
Pb	26±14	40±30	39±2	36±2	38±3	12.5	3.0
Sb	—	—	2.9±0.1	2.6±1.5	3.6±0.1	0.2	15
Se	—	—	1.6±0.3	2.3±0.1	5.1±0.3	0.05	60
V	—	182±42	128±6	129±6	161±36	135	1.0
Zn	148±87	91±30	123±3	82±2	99±4	70	1.5

[a]Wilkerson [23].
[b]Fox et al. [20].
[c]Fruchter et al. [11].
[d]Taylor [24].
[e]Excludes Occidental Retort 3 and Geokinetics samples because of sampling problems noted in the text.
[f]Fruchter [12].

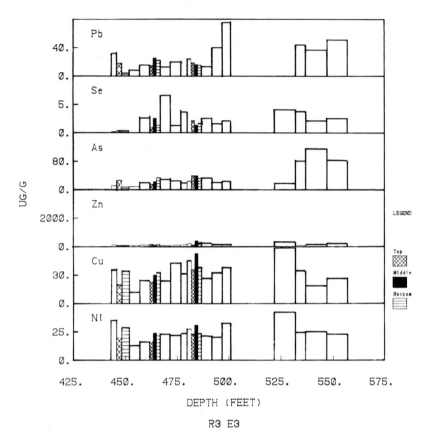

Figure 5. Variation in Pb, Se, As, Zn, Cu and Ni throughout core hole 3E3 taken through Occidental's Retort 3 at Logan Wash, Colorado. Large samples of core material were segregated into "top", "middle" and "bottom" samples as shown in the legend. The open bars were unsegregated.

bility distributions within in situ retorts, results in considerable heterogeneity in spent shales. Thus, large numbers of cores are required to obtain a representative sample. The required number is rarely available. Thus, elemental abundances in column 1 of Table 4 are not representative of the entire retorted horizon of retort 3 and cannot be reliably compared with elemental abundances for the raw shale reported in column 2 of Table 2. Similar difficulties exist with the Geokinetics data, and the values reported in column 2 of Table 4 cannot be reliably compared with the raw shale values reported in column 4 of Table 2. An extensive coring program at the Geokinetics site was recently completed by LETC using methods designed to mitigate some of these sampling problems. Detailed elemental characterization of the new cores is in progress at LBNL.

Other spent shales in Table 4, however, are comparable to the raw shales shown in Table 2. These other spent shales are from smaller-scale laboratory reactors in which retorting was closely controlled and sampling was carefully conducted. Because they are laboratory reactors, rather than large underground chambers of shale, sampling is considerably simplified; contamination from drilling fluids does not occur, 100% recovery of the spent shale is possible, and the spent shale is of a uniform composition. Additionally, the other samples in Table 4 (Paraho, LLNL, Fischer Assay) were the subject of a careful scientific investigation in which critical sampling parameters were controlled [11, 20].

The elemental concentration ratios presented in Table 4 are slightly higher than the corresponding ratios reported in Table 2 for raw shales, with the exception of the ratio for Hg. The Hg ratio is lower because Hg is volatilized during retorting, depleting it in spent shales relative to raw shales.

The elevated concentrations of As, B, Cd, Mo, Sb and Se in spent shales may pose some of the same hazards previously discussed for raw shales, namely release of these elements to the environment as a consequence of leaching and fugitive dust exposure and fallout. Preliminary investigations on the leachability of spent shale from Occidental Retort 3 indicate that the concentrations of F, B, Mo, As and Se in the initial leachates are high enough to warrant further study [14]. Static leaching experiments on 16 spent shale samples from Retort 3 revealed As concentrations ranging from 0.005 to 0.15 mg/L; B concentrations of <0.3-6.7 mg/L; Mo concentrations of 0.04-0.05 mg/L; and Se concentrations of <0.001-0.040 mg/L. The leachability of some elements may be significantly modified in spent shales, relative to raw shales, due to mineralogical changes that occur during retorting. Stollenwerk and Runnels [25] found that the amount of Mo and B leached from TOSCO II spent shale was significantly greater than the amount leached from the raw shale, but the amount of F was lower in the spent shale leachate than the raw shale leachate. There was no significant difference in the amounts of Se and As leached from raw and spent shales.

Shale Oil

Shale oil is the hydrocarbon product produced by the condensation of vapors from the pyrolysis of oil shale. Typically, it is a black waxy liquid that has a high pour point. Shale oils produced by surface retorts such as Paraho or TOSCO retorts are different from in situ-produced oils. The nitrogen content of surface-produced oils usually ranges from 1.4 to 2.3% and the pour point ranges from 18 to 32°C. The nitrogen content of in situ-produced oils is 1.4-1.8% and the pour point may range from -1.1 to 21°C [1]. Other differences have also been noted in yields of distillation products, viscosity and elemental abundances [16, 30, 31]. These differences are primarily due to differences in retort operations. Surface retorts typically use short rapid shale heatup, small shale particles, retorting temperatures of 500-700°C and continuous shale feed. In situ retorts, on the other hand, are batch reactors that typically employ slow shale heatup, low retorting rates, temperatures which can reach 1000°C, and larger particles. In in situ retorts, the relatively longer residence time of the oil in high temperature zones and the holdup of heavy components on the cooler shale ahead of the reaction front result in in situ oil cracking. Thus, in situ oils have a lower nitrogen and residuum content, a lower pour point and viscosity, and a higher light distillate content than surface oils.

Shale oils, in general, differ in some significant ways from crude oils. Relative to conventional crudes [32], shale oils have a high nitrogen content, a moderate sulfur content and a lower hydrogen content. Hydrocarbon composition is markedly different, and distillation yields vary accordingly. The refining of shale oil will produce less gasoline and more kerosene and diesel fuel than refining of an equivalent volume of crude oil. Since shale oil has a larger percentage of distillation products in the higher boiling fractions, its viscosity and pour point are higher. Another difference is that shale oil may contain process-derived inorganic solids and pyrolytic products [11, 16].

The elemental compositions of shale oils from the five processes studied here are compared with the elemental composition of average conventional crude in Table 5. This tabulation shows that shale oils from different processes vary considerably with respect to some elements. Fox [1, 30] investigated the effect of shale source and retort operating conditions on the elemental composition of 24 oils from three simulated in situ retorts and concluded that shale source and the retort type are the major determiners of elemental composition. On the other hand, retorting temperature and atmosphere did not significantly affect elemental composition of shale oils within the limits of experimental error. Fox also found elevated concentrations of Cu, Pb and Zn in oils produced in the LLNL simulated in situ retorts. The conclusion was drawn that these elevated concentrations were due to contamination from copper alloys used in the product collection system.

The concentration ratios in column 7 of Table 5 indicate that As, Co, Hg, Se and Zn are enriched in shale oils compared to conventional crude oils and that Cr, Cu, Mo, Ni and V are depleted. The elevated concentrations of As, Co, Hg, Se and Zn could result in both process and environmental problems, depending on the chemical form of the element and the use of the oil.

Shale oil may be directly combusted in oil-fired power plants, refined into gasoline, diesel fuel and other distillation products, or used as a feedstock for chemical production. Combustion of shale oil may volatilize greater quantities of Hg, As and Se than combustion of conventional crudes. Tests of

Table 5. Comparison of Elemental Abundances of Some Toxic Elements in Shale Oils with Conventional Crude Oils (ppm)

	Occidental Retort 6[a] (1)	Geokinetics True In Situ[a] (2)	LLNL Simulated In Situ Retort[a] (3)	Paraho Direct-Mode[a] (4)	TOSCO Modified Fischer Assay (5)	Conventional Crude Oil[b] (6)	Concentration Ratio C_{SO}/C_{CO} (7)
As	15.1±0.1	8.75±0.09	14±3	28.2±0.1	20±4	0.01	1720
B	–	–	–	–	–	–	–
Cd	<0.14	<0.06	0.025±0.003	<0.17	<0.01	<0.01	–
Co	3.59±0.02	1.84±0.01	0.35±0.08	1.32±0.01	0.22±0.03	0.2	7.3
Cr	0.033±0.001	<0.01	0.14±0.09	0.023±0.005	0.41±0.08	0.3	0.41
Cu	0.10±0.01	<0.29	2.5±0.2	0.05±0.05	0.04±0.01	0.14	0.45[d]
F	–	–	–	–	–	–	–
Hg	0.06[c]	0.12[c]	0.22±0.04	0.11[c]	0.23±0.02	0.07	2.1
Mn	0.0123±0.0002	<2	0.039±0.001	0.085±0.001	0.08±0.02	0.1	0.54
Mo	2.5±0.1	3.0±0.1	0.86±0.26	0.85±0.86	–	10	0.18
Ni	8.8±0.4	3.8±0.6	2.9±0.4	3.0±0.3	1.2±0.6	10	0.39
Pb	<2	<2	1.5±0.6	<2	0.03±0.01	0.3	–
Sb	0.026±0.001	0.016±0.001	0.03±0.01	0.046±0.001	0.035±0.003	–	–
Se	0.81±0.01	0.80±0.01	2.9±0.3	1.20±0.03	0.8±0.1	0.17	7.7
V	0.45±0.01	<6	0.29±0.01	0.224±0.003	1.2±0.5	50	0.11
Zn	1.5±0.1	3.4±0.1	5.4±1.5	2.0±0.1	0.98±0.40	0.25	8[d]

[a]Fox [30].
[b]Bertine and Goldberg [33].
[c]Wilkerson [16].
[d]Excludes LLNL oil due to contamination.

a Paraho shale oil in a utility boiler equipped with a scrubber indicated that 21, 31 and 100% of the As, Pb and Hg, respectively, were emitted with the stack gases [34]. If these increased emissions are not controlled by stack gas cleaning processes such as desulfurization or electrostatic precipitation, elevated atmospheric and soil-column concentrations of these elements, particularly in the vicinity of the power plant, may result.

The high concentrations of As in shale oils are known to interfere with catalytic reactions during prerefining (hydrotreating) and with subsequent refinery operations [35, 36]. As a result, As must be removed from shale oil before it is refined. Proposed processes for As removal employ an absorbent [37] or an alumina guard bed [36]. The used absorbents will contain high concentrations of As and other toxic elements and probably will be classified as hazardous wastes necessitating special handling and disposal.

The fate of these trace elements in shale oils depends on their chemical form. There has been very little work done in this area. Shaw [38], in analyses of distillation fractions of a Rock Springs true in situ shale oil, found that the concentrations of Cr, Co, Mo, Mn, As and Se increased with the temperature range of the cut, but V was concentrated in the naphtha and residuum cuts. The elements concentrated in the residuum are probably associated with nonvolatile asphalt ends. This suggests that they would not be volatile during combustion. Some volatile compounds, however, are certain to exist in the lighter fractions.

In other work, Wilkerson [16] fractionated Paraho, Geokinetics, and Occidental shale oils using a solvent precipitation method. This work showed that most of the Ni, Co, Mo, Zn, V, Hg and Sb were present in the asphaltic fraction. The majority of the As was associated with the resin and n-pentane/methanol-soluble fractions which contain the nonpolar, lower-molecular-weight components of shale oil. Wilkerson postulated that inorganic forms of As in shale oils may include arsenic oxide (As_4O_6) and/or metallic arsenic (As_2^0 or As_4^0). Alkaline extracts of shale oils in our laboratory suggest that methylarsonic acid may be present. Wilkerson [16] also postulated that Ni and Co may be present as porphyrins or multidentate mixed ligands.

Process Waters

Oil shale retorting will generate effluents from conventional water uses, such as steam and power production, mining, and domestic and service consumption of water. Additionally, two process waters are produced during retorting. These process waters, known as retort water and gas condensate, are unique to the oil shale industry and will require special study and consideration by the industry because they are chemically complex and pose difficult waste treatment problems [39]. We will consider only the process waters, and, of them, we will focus on retort waters. The other waters have been discussed previously [31].

The quantities, types and nature of waters produced during oil shale retorting depend on the process and the retort operating conditions. The sources of these waters are combustion, mineral dehydration, input steam and groundwater intrusion (in situ processes). Combined production of water ranges between 0.1 and 10 bbl H_2O/bbl oil [31]. Water production for surface processes is at the lower end of this range (0.1–0.5 bbl H_2O/bbl oil) and at the upper end for in situ processes (0.4–10 bbl H_2O/bbl oil) due to groundwater intrusion. Several types of waters are generated during retorting, as illustrated by the Occidental MIS process shown in Figure 6. In this process, water and oil are collected together in an underground sump at the bottom of the retort. These are pumped into a separator tank where the majority of the water, termed "retort water," is separated from the oil by decantation. The oil from this tank is then introduced into the heater-treater to remove any residual moisture. The oil is heated and the water, termed "heater-treater water," is collected by decantation. Both retort water and heater-treater water may then be introduced into a low-pressure boiler and a portion blown down to control solids within the boiler. This is termed "boiler blowdown water." Water condensed from the offgas stream is termed "gas condensate."

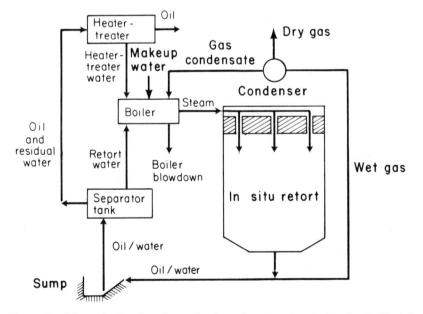

Figure 6. Schematic showing the production of wastewaters during the Occidental modified in situ process.

Table 6. Comparison of Elemental Abundances of Some Toxic Elements in Retort Waters with Water Quality Criteria (ppm)

	Occidental Retort 6[a] (1)	Geokinetics True In Situ[b] (2)	LLNL Simulated In Situ Retort[c] (3)	Paraho Direct Mode[d] (4)	TOSCO Modified Fischer Assay (5)	Constraining Water Quality[e] Criteria (6)
As	1.1±0.3	2.6±3.9	2.6±0.1	5.8±1.0	8.0±2.4	0.1 (D)
B	12.9±0.4	61±68	—	43±6	—	0.75–2.0 (I)
Cd	—	0.084±0.088	0.005±0.0005	—	—	0.03–0.0004 (A)
Co	<0.02	0.56±0.30	<0.03	0.032±0.001	<0.12	0.05–5.0 (L, I)
Cr	<0.07	0.078±0.039	0.39±0.22	<0.060	<7.4	0.05 (D, A)
Cu	0.03±0.01	0.21±0.17	133±6	0.66	<0.10	0.5 (L)
F	35±1	35±6	—	25	—	1.4–2.4 (D)
Hg	<0.02	0.0038±0.011	0.085±0.011	0.0023±0.0001	—	0.00005 (A)
Mn	<0.04	0.94±1.9	0.19±0.03	0.17±0.05	0.3±0.1	0.05 (D)
Mo	0.3±0.1	11.9±8.7	—	0.15	0.13±0.04	0.010–0.050 (I)
Ni	<0.02	1.6±2.1	1.57±0.10	0.54±0.10	0.31±0.08	0.20–2.0 (I)
Pb	<0.04	0.64±0.92	1.15±0.16	<0.40	<0.36	0.03 (A)
Sb	0.018±0.008	0.011±0.003	<0.07	0.020±0.004	—	0.2 (A)
Se	0.04±0.01	0.22±0.47	1.1±0.1	9.8±1.1	3.1±0.4	0.01 (D)
V	<0.04	0.43±0.25	0.08±0.01	0.044±0.008	—	0.1 (L)
Zn	<0.1	0.095±0.067	13.3±0.6	0.41±0.04	0.20±0.07	2.0–10 (D, I)

[a]U.S. DOE [14].
[b]Geokinetics [41].
[c]Fox [1].
[d]Fruchter et al. [10].
[e]NAS [42]; D = domestic, I = irrigation, A = aquatic life, L = livestock watering.

The only process water for which there are abundant characterization data is retort water. The concentrations of some toxic elements in retort waters from five processes are presented in Table 6. This table shows significant differences among waters from the five processes. These differences have been hypothesized to be due to sampling and analytical problems [1]. In addition, the problems attendant with the analysis of retort waters are legendary [40]. Fox [1] concluded that water composition depends, among other things, on contact time between the oil and water and on operation of the product collection system. Variability also may be caused by changes in product water composition during retorting. Significant temporal variations in the concentrations of many constituents have been noted in unpublished data from several field in situ experiments. Because of these problems, the concentrations shown in Table 6 should not be construed as accurate representations of elemental abundances in waters from commercial processes. Considerable additional work is required in order to resolve the variability issue.

The data in Table 6 demonstrate that the concentrations of As, Se and Zn are elevated in surface process waters relative to in situ process waters. This difference was previously noted in work by the U.S. DOE [14] and was hypothesized to be due to operating conditions and to the fact that surface processes produce much less water, and thus soluble species partitioned from the oil would be more highly concentrated. Another noticeable feature of the data presented in Table 6 is the unusually high Mo value for Geokinetics retort water. A similar concentration was also observed in the U.S. DOE study; the source of this Mo is unknown. The high concentrations of Cr, Cu and Zn in the LLNL water are presumably due to contamination from copper alloys used in the product collection system [1].

The last column of Table 6 presents constraining water quality criteria. These criteria are compared with retort water concentrations to evaluate the significance of the release of this water to the environment. These water quality criteria were developed under the aegis of the National Research Council at the request of the U.S. Environmental Protection Agency. Water quality criteria are recommended concentrations that should be maintained in the main water mass to protect a specified water use. The "constraining" criteria in Table 6 are the most stringent among domestic supply, irrigation water, aquatic life or livestock watering.

Comparison of the five retort waters with constraining water quality criteria indicates that the concentration of As, B, F, Hg, Mn, Mo, Pb and Se exceed these criteria for most waters. The environmental significance of these concentrations depends on the projected use of the water. Some developers of surface processes presently plan to codispose of retort water and spent shale. Thus, some of these elements may be released to the environment through leaching or volatilization from the disposal pile. Other developers may evaporate the retort water in surface ponds, creating an opportunity for loss of volatile species.

Only limited data exist for other types of process waters. Analyses of boiler blowdown water for the Occidental process have been reported by the U.S. DOE [14], and analyses of other Paraho effluents have been reported by Fruchter et al. [10].

The environmental significance of toxic elements in process waters depends on the chemical form of the element, since some chemical forms are highly toxic while others are benign. The classical case, of course, is methyl mercury, which is considerably more toxic than elemental Hg. Very few chemical speciation investigations have been performed due to the difficulty of identifying species in complex matrices. So far, the focus of speciation investigations has been on arsenicals because As is elevated in all shale products. The results of these investigations are summarized in Table 7. Fox et al. [15] found that the major arsenical in several in situ process waters was arsenate and that lesser but significant amounts of methylarsonic acid also were present. Other data [15] reveal that the major arsenicals in Paraho retort water are arsenite and arsenate. The presence of arsenite has been corroborated by researchers at Battelle Pacific Northwest Laboratory [10].

The potential presence of these arsenic species is significant from an industrial and public health standpoint. Arsenite is a known carcinogen [44]; the methylated form is low boiling and may be volatile under ambient conditions. Microorganisms are known to produce toxic gaseous arsenical products, such as arsine [45], from arsenate, arsenite, and mono- and dimethyl arsenate. Thus, these forms may escape into the atmosphere if the waters are evaporated or codisposed with spent shales.

Offgases

The pyrolysis of oil shale produces from 70 to 14,000 scf of low-Btu gas per ton of shale retorted [1]. Surface processes produce less gas per ton of retorted shale than do the in situ processes, due to the lower temperatures required for surface retorting. The major constituents in these gases are CO_2, CO, CH_4, H_2 and C_2 through C_5 hydrocarbons. Principal gaseous pollutants include H_2S, SO_2, NO_x, COS, thiophene and certain volatile trace elements [1, 11].

There are few direct offgas data available on the concentration of trace elements. This is primarily due to the difficulties of sampling and performing trace element analyses on a matrix of such complex and variable composition. For example, the standard EPA reference methods for sampling gaseous

Table 7. Tentatively Identified Arsenicals in Oil Shale Process Waters[a] [15]

	LLNL Simulated In Situ Retort Water	Occidental Retort 6			Heater Treater	Geokinetics True In Situ Water	Paraho Direct-Mode Water
		Retort Water	Gas Condensate	Boiler Blowdown			
Arsenite (AsO_2^-)	---	---	---	---	---	---	(+)
Arsenate ($AsO_4^=$)	(+)	(+)	---	(+)	(+)	(+)	(+)
Methylarsonic Acid ($CH_3AsO(OH)_2$)	(+)	(+)	---	(+)	(+)	(+)	(+)
Dimethylarsinic Acid (($CH_3)_2AsO(OH)$)	---	---	---	---	---	---	---
Phenylarsonic Acid ($\phi - AsO(OH)_2$)	---	---	---	---	---	---	---

[a] --- signifies that the species was looked for but not detected; (+) signifies that the species was tentatively identified by high-performance liquid chromatography coupled with atomic absorption spectroscopy using known retention times of authentic arsenicals.

Hg in stack gases [46] have been found to be unsuitable for sampling Hg in oil shale offgases during retorting experiments at LETC and LLNL [29, 47]. Attempts to use acid gas scrubber solutions to sample Hg in Paraho-direct recycle gas also have proven to be unsatisfactory [10].

Available data for direct measurements of total Hg and As and speciation measurements of these two elements in gases from the Paraho-direct and Occidental retort experiments are summarized in Table 8. Time-weighted average threshold limit values (TLV) for the workplace are included in Table 8 for comparison. The Au-amalgamation sampling technique was used for total Hg and has proven to be a reliable sampling method. The chemical forms of Hg and As were determined using the techniques of Braman and Johnson [48] and Johnson and Braman [49], respectively.

Arsenic speciation measurements indicate that the major forms of As in thermal oxidizer gas are gaseous arsenic trioxide (As_2O_3) and particulate As. In the recycle gas, arsenic trioxide predominates. The estimated error

Table 8. Mercury and Arsenic Species in Paraho-direct and Occidental Retort 6 Offgas Compared to Time-Weighted Average Threshold Limit Values (TLV) ($\mu g/m^3$)[a]

	Paraho Direct-Mode		Occidental Retort 6[c]	TLV for 40-hr Work Week[d]
	Thermal Oxidizer[b]	Recycle[b]		
Mercury				
Hg^0	4.5–9.4	–	–	50
$HgCl_2$	<0.1	–	–	50
CH_3HgCl	<0.2	–	–	10
$(CH_3)_2HgCl$	<0.1	–	–	10
Total Hg	4.5–12	46–75	10–26	–
	(n=5)[d]		(n=13)	
Arsenic				
$As_2O_3(g)$	12–36	112–149	<0.8–1.9	50
AsH_3	<0.2–0.3	2	0.9–3.5	200
CH_3AsH_2	<0.2	3	–	–
$(CH_3)_2AsH$	<0.2	1.0–1.7	<0.4	–
Other Organic As	<0.2	0.5	–	–
Particulate As	19	1.2	–	–
	(n=1)	(n=1)		
Total As	12-55	120–155	–	500

[a]Unless otherwise indicated in parentheses, two discrete samples were analyzed to determine the concentrations.
[b]Fruchter et al. [10].
[c]U.S. DOE [14].
[d]ACGIH [44].

in all of the Paraho-direct As data (Table 8) is ± 50%. The Occidental arsine (AsH_3) values are near blank and reported as not significant by the authors. Considering the errors, only tentative identification of arsine and methyl- and dimethylarsine can be made in the recycle gas. Comparison of these values with workroom TLVs indicate that all reported concentrations except that of As_2O_3 in the Paraho recycle gas are less than the TLVs. Arsenic trioxide is a human carcinogen [44] and could pose a worker health problem if recycle gas were vented in the workplace. Considering the limited data available, any general statements regarding the levels and chemical form of As and Hg in surface and MIS retort offgas are premature.

Mercury speciation measurements in the flared offgas (thermal oxidizer) show that only elemental Hg is emitted to the atmosphere. This is as expected since the vast majority of Hg compounds decompose to Hg° at the high flare temperatures. Although Hg levels in the flared offgas are less than TLVs, additional characterization is necessary to determine the concentration, identity and stability of mercury compounds in the recycle and unflared offgases.

Recent work at LBNL suggests that organically bound Hg is the pre- dominant chemical form of Hg vapor in the offgas from the LBNL retort. Mercury measurements were made with carbosive (activated carbon) columns upstream of and in series with Au-amalgam tubes. These carbosive columns selectively adsorb organic Hg to the exclusion of inorganic forms. The results suggest that organically bound Hg is the predominant chemical form of Hg. Additional experimentation is necessary to verify these observations and the specificity of carbosive columns for organic/inorganic Hg in oil shale offgas.

Measurements of total Hg in the offgases during simulated in situ retort ex- periments at LETC, LLNL, and LBNL have been made. At the LETC 10-ton retort, 10 discrete samples using Au-amalgam tubes were collected and analyzed. For the other retort experiments, continuous on-line analyses were made using Zeeman atomic absorption spectroscopy (ZAA). This method is described elsewhere [50]. Except for the 10-ton experiment, the reported Hg concentrations in Table 9 are orders of magnitude higher than those reported for Hg in Table 8. The discrepancy between offgas Hg con- centrations measured at field retorts [9, 10, 14] and those measured by Fox et al.[29] and Girvin et al. [5, 50] are real and not thought to be due to differences in analytical techniques. The Au-amalgamation and the ZAA techniques generally agree to within ± 20% or better for the measurement of total Hg in LBNL retort offgas.

Two factors contribute to the difference between offgas Hg values re- ported in Tables 8 and 9: condensation and sampling time. First, during LBNL retort experiments, it has been found that ∼ 50% of the total Hg vapor originally present in the offgas stream is removed when offgases

Table 9. Measurements of Total Hg in the Offgas of Some Simulated
In Situ Retorts ($\mu g/m^3$)

	Hg Concentration	Retort
Girvin et al. [5]	42–540	LLNL 6000-kg Run L–3
Fox et al. [29]	80–8000	LETC controlled state Run CS–69
Girvin et al. [6]	6–2200	LBNL retort Run LBNL–02
Fruchter et al. [9]	2–5	LETC 10-ton

contact cool ($< 40°$ C) surfaces. This is presumably due to condensation, and, when the surface is stainless steel, subsequent amalgamation. It was found that stainless steel surfaces must be maintained at temperatures $> 150°$C to prevent losses. The elimination of cool surfaces in contact with the offgas resulted in the recovery of 100% of the Hg originally present in the shale [47]. It has also been observed during LBNL, LLNL and LETC retort experiments that Hg vapor is almost totally removed from the offgas stream when passing through raw oil shale whose temperature is $140°$ C or less [29, 47].

These condensation opportunities may partially explain why Hg concentrations in Paraho and Occidental offgases are much lower than in offgases from simulated in situ retorts. At Paraho, the offgas passes sequentially through the cool raw oil shale entering the retort, an oil coalescer, an electrostatic precipitator, and finally through extensive plumbing before reaching the gas sampling point. Offgases from Occidental retort 6 contacted cool raw shale in the lower portion of the retort and a section of cool plumbing prior to reaching the point of sampling. In both cases, loss of Hg vapor to cold or cool surfaces may have resulted in the loss of Hg and thus the observation of lower concentrations in both the recycle gas and gas discharged to the atmosphere. This offers a de facto control technology but complicates comparison of data among investigators.

The second factor affecting the difference in observed Hg concentrations is the time during the oil shale retort experiment when sampling for Hg vapor is conducted. This factor will primarily affect batch-type processes, i.e., all of the in situ processes. The continuous ZAA Hg measurements by Fox et al. [29] and Girvin et al. [6] have shown that Hg emissions during simulated in situ retorting are nonuniform. During the final stage of retorting, when the last segment of raw shale exceeds $200°$C, a pulse of Hg is observed in the offgas which is two to three orders of magnitude higher than concentrations observed during the initial and intermediate stages of retorting. The Occidental measurements were not made during the final stage of the retorting process.

The mechanism responsible for this Hg pulse has been suggested by Fox et al. [29] ; Hg originally present in raw shale is volatilized by the approaching pyrolysis front, swept ahead of the front by the offgas, and then condenses on a layer of cool raw shale. If this process continues as the front propagates down or through the retort bed, Hg becomes progressively more concentrated in the successive layers of raw shale. When the final layer is heated, a large pulse of Hg appears in the offgas. No measurements have been made during the final stages of field retorting experiments to verify the existence of a final Hg pulse. This has been due to limited sampling during field experiments that last many months.

Except for the two measurements of As presented above, direct quantitative measurements in the offgas of the other volatile toxic elements under study here have not been made. However, elemental mass balance measurements of product streams (excluding offgas) from simulated in situ retorting experiments at LETC and LLNL attributed mass deficiencies to the offgas [1]. On this basis, ~30% of the Cd in the raw shale may be volatilized into the offgas, and at retorting temperatures in excess of 900°C, a maximum of 1 and 13% of the original As and Se may be mobilized to the gas phase. It should be noted that the maximum retorting temperatures reached for the Paraho retort run discussed above was 750°C. For this temperature, As volatilization into gas phase would be expected to be exceedingly low.

Cadmium was recently detected for the first time in oil shale offgases. The successful qualitative on-line measurement of Cd vapor in offgas was made by directly aspirating offgas from the LBNL retort into a conventional flame atomic absorption spectrometer. Significant quantities of Cd were volatilized at retort temperatures above 500°C. An effort is currently under way to make quantitative measurements using this technique.

As can be seen, a paucity of direct trace element offgas data exists, even for Hg. Mechanisms analogous to those affecting Hg may result in vapor phase losses to surfaces contacting offgases and nonuniform emission rates (pulse formation) during retorting for the other volatile elements, namely Cd, Se, Sb and As. The possibility that such mechanisms exist should be examined.

Fugitive Dust

Fugitive dust and particulates, largely raw and spent shale fines and suspended soil material, will be produced by oil shale production. Activities that produce fugitive particulates include blasting to release the ore; raw and spent shale loading, haulage, crushing, screening and stockpiling; loading and unloading of surface retorts; and disposal pile development. Particulates from these activities will have a chemical composition similar to that of raw and spent shales shown in Tables 2 and 4. Considerable dust is also produced by vehicular traffic and construction activities, because the soils in the oil shale region are light and disintegrate into a fine powder when disturbed during dry seasons. Additional particulate material may be produced by wind erosion of spent shale disposal piles and other waste dumps. The quantity of these emissions is roughly proportional to the amount of material handled and the areal extent of the plant and disposal pile. Emission quantity additionally depends on the particle size of the raw and spent shale and on mining methods. Surface mining and retorting produce higher particulate emissions than underground processes. Retorting processes that use a fine shale charge or produce a fine spent shale, such as Lurgi or TOSCO, would be expected to generate more particulates. Open pit mining operations, in which activities are exposed to the atmosphere, generate more fugitive particulates than underground mining methods.

The quantities and composition of fugitive dust at the Paraho semiworks plant have been investigated by Cotter et al. [51] and Fruchter et al. [10]. Some of these data are summarized in Table 10. These studies indicate that atmospheric dust concentrations in the mining, crushing and loading areas are significantly elevated above background levels (~ 5 mg/m^3), ranging from 1 to 35 mg/m^3. Both studies observed a bimodal particle size distribution, consisting largely of particles > 7 μm and < 0.6 μm with relatively few particles in the size fractions between these limits. Cotter et al. [51] found 20–37% of the total loading was in the size range > 8 μm and even greater percentages, 33–63%, in the < 0.6 μm fraction. Therefore, respirable dust, defined as particles < 3 μm, is a significant fraction of fugitive dust at the Paraho plant. Elemental analyses suggest that most of this material is raw or spent shale fines. However, Fruchter et al. [10] showed that As, Cr, Co, V and Sb were elevated in particles in the 1.1- to 2.0-μm size fraction compared to raw and spent shales, suggesting another source such as thermal oxidizer emissions or diesel particulates. This fugitive dust could be inhaled by workers or transported away from the site and deposited in the surrounding countryside. The resultant health and environmental impacts have not been investigated.

REFERENCES

1. Fox, J. P., "The Partitioning of Major, Minor, and Trace Elements during Simulated In-Situ Oil Shale Retorting," Ph.D. Dissertation, University of California, Berkeley (1980).

Table 10. Elemental Composition of Particulates Collected from the Paraho Semiworks Retort in November 1977 by Fruchter et al. [10]

Particle Size (μm)	Elemental Concentration in Particulates (ng/m^3)				
	As	Co	Cr	Mn	Sb
Retort Site					
>7.0	158±9	24.2±0.6	76±4	610±20	4.4±0.5
3.3–7.0	60±4	9.8±0.3	32±2	193±9	2.5±0.3
2.0–3.3	31±2	5.8±0.3	24±2	98±5	1.3±0.3
1.1–2.0	30±2	5.2±0.3	23±2	76±5	1.8±0.3
<1.1	159±11	52.6±1.4	280±15	700±30	13±1.0
Crusher Site					
>7.0	77±11	11.0±0.2	44±1	326±11	3.3±0.7
3.3–7.0	23±4	4.7±0.2	19±1	101±5	1.4±0.3
2.0–3.3	13±2	2.3±0.1	13.2±0.8	46±3	0.71±0.25
1.1–2.0	12±2	2.0±0.1	21±1	37±3	0.56±0.21
<1.1	254±34	20.9±0.6	146±4	1070±30	8.22±1.8
Tailings Pile					
>7.0	125±20	20.7±0.6	107±4	605±35	5.9±1.3
3.3–7.0	71±12	12.1±0.4	69±3	260±12	3.4±0.8
2.0–3.3	46±8	7.9±0.3	50±3	169±9	3.4±0.8
1.1–2.0	46±8	8.2±0.4	48±3	150±9	2.2±0.6
<1.1	301±48	61.3±1.5	378±14	1360±50	18.3±3.9
Background Site					
>7.0	1.56±0.33	0.33±0.02	1.89±0.09	9.2±0.4	0.066±0.016
3.3–7.0	0.64±0.14	0.18±0.01	1.81±0.08	4.3±0.2	0.046±0.012
2.0–3.3	0.52±0.13	0.13±0.01	1.46±0.07	2.0±0.2	0.030±0.014
1.1–2.0	0.43±0.10	0.12±0.01	1.42±0.08	2.1±0.2	0.057±0.013
<1.1	3.77±0.83	0.81±0.04	13.7±0.3	16.2±0.7	0.29±0.04

2. Giauque, R. D., J. P. Fox and J. W. Smith. "Characterization of Two Core Holes from the Naval Oil Shale Reserve Number 1," Lawrence Berkeley Laboratory Report No. LBL–10809 (1980).

3. Giauque, R. D., J. P. Fox and J. W. Smith. "Characterization of Two Core Holes from the Naval Oil Shale Reserve Number 1," Lawrence Berkeley Laboratory Report No. LBL–11450, submitted to *Geochimica et Cosmochimica Acta.*

4. Shendrikar, A. D., and G. B. Faudel. "Distribution of Trace Metals during Oil Shale Retorting," *Environ. Sci. Technol.* 12(3):332 (1978).

5. Girvin, D. C., A. T. Hodgson and S. Doyle. "On-Line Measurement of Trace Elements in Oil Shale Offgases by Zeeman Atomic Absorption Spectroscopy," Energy and Environment Division 1979 Annual Report, Lawrence Berkeley Laboratory Report No. LBL–11486 (1980).

6. Girvin, D. C., et al. "Partitioning of As, Cd, Hg, Pb, Sb, and Se During Simulated In-Situ Oil Shale Retorting," Energy and Environment Division 1980 Annual Report, Lawrence Berkeley Laboratory Report No. LBL–11989 (1981).

7. Wildeman, T. R., and R. R. Meglen. "Analysis of Oil Shale Materials for Element Balance Studies," in *Analytical Chemistry of Oil Shales and Tar Sands,* Advances in Chemistry Series No. 170 (Washington, DC: American Chemical Society, 1978), p. 195.

8. Wildeman, T. R., and R. N. Heistand. "Trace Element Variations in an Oil-Shale Retorting Operation," *ACS Division of Fuel Chemistry Preprints,* Vol. 24 (Washington, DC: American Chemical Society, 1979).

9. Fruchter, J. S., et al. "High Precision Trace Element and Organic Constituent Analysis of Oil Shale and Solvent-Refined Coal Materials," in *Analytical Chemistry of Oil Shales and Tar Sands,* Advances in Chemistry Series No. 170 (Washington, DC: American Chemical Society, 1978).

10. Fruchter, J. S., et al. "Source Characterization Studies at the Paraho Semiworks Oil Shale Retort," Pacific Northwest Laboratory Report No. PNL–2945 (1979).

11. Fruchter, J. S., et al. "Elemental Partitioning in an Aboveground Oil Shale Retort Pilot Plant," *Environ. Sci. Technol.* 14(11):1374 (1980).

12. Fruchter, J. S., et al. "Analysis of a Spent Shale Core from a Modified In-Situ Oil Shale Retort," Pacific Northwest Laboratory, Annual Report for 1979 to the DOE Assistant Secretary for Environment, Part 4, Physical Sciences, PNL–3300 (1980), p. 107.

13. Fruchter, J. S., and C. L. Wilkerson. "Characterization of Oil Shale Retort Effluents," in *Proceedings of Oil Shale, the Environmental Challenges* (Golden, CO: Colorado School of Mines Press, 1981).

14. "Environmental Research on a Modified In-Situ Oil Shale Process: A Progress Report from the Oil Shale Task Force," U.S. DOE/EV–0078 (1980).

15. Fox, J. P. et al. "Inorganic Arsenic and Organoarsenic Compounds in Some Oil Shale Process Waters" (in preparation).

16. Wilkerson, C. L., "Trace Metal Composition of Green River Retorted Shale Oil," Accepted for publication in *Fuel.*

17. Heistand, R. N., "The Fischer Assay, A Standard Method?," preprints of papers presented at the 172nd National Meeting of the American Chemical Society, Division of Fuel Chemistry in San Francisco, August 29–September 3, 1976.

18. Goodfellow, L., and M. T. Atwood. "Fischer Assay of Oil Shale—Procedures of the Oil Shale Corporation," *Qtr. Colo. Sch. Mines,* 69(2): 205 (1974).

19. Smith, J. W. draft of Standard Method of Test for Oil from Oil Shale, Resource Evaluation by the USBM Fischer Assay Procedure (June 1979).

20. Fox, J. P., et al. "Intercomparison Study of Elemental Abundances in Raw and Spent Oil Shales," *Proceedings of Sampling, Analysis, and Quality Assurance Symposium,* U.S. EPA–600/9–80–022 (1979).

21. Jones, J. B., Jr., "The Paraho Oil-Shale Retort," *Qtr. Colo. Sch. Mines,* 71(4):39 (1976).

22. Sandholtz, W. A. and F. J. Ackerman. "Operating Laboratory Oil Shale Retorts in an In-Situ Mode," paper presented at Society of Petroleum Engineers 52nd Annual Meeting, Denver, CO, October 9–12, Lawrence Livermore Laboratory Report UCRL–79035 (1977).

23. Wilkerson, C. L. Letter to Dr. Willard R. Chappell transmitting Battelle Pacific Northwest preliminary analyses of Geokinetics samples (July 18, 1980).

24. Taylor, S. R., "Abundance of Chemical Elements in the Continental Crust: A New Table," *Geochim. Cosmochim. Acta,* 28:1273 (1964).

25. Stollenwerk, K. G., and D. D. Runnells. "Leachability of Arsenic, Selenium, Molybdenum, Boron, and Fluoride from Retorted Oil Shale," *Proceedings of the Second Pacific Chemical Engineering Congress* (New York: American Institute of Chemical Engineers, 1977), p. 1023.

26. Desborough, G. A., J. K. Pitman and C. Huffman, Jr. "Concentration and Mineralogical Residence of Elements in Rich Oil Shales of the Green River Formation, Piceance Creek Basin, Colorado, and the Uinta Basin, Utah—A Preliminary Report," *Chem. Geol.* 17:13 (1976).

27. Saether, O. M., D. D. Runnells and R. H. Meglen. "Trace and Minor Elements in Colorado Oil Shale, Concentrated by Differential Density Centrifugation," submitted to *Environ. Sci. Technol.* (1980).

28. Saether, O. M., D. D. Runnells, R. A. Ristinen and W. R. Smythe. "Fluorine: Its Mineralogical Residence in the Oil Shale of the Mahogany Zone of the Green River Formation, Piceance Creek Basin, Colorado, U.S.A.," *Chem. Geol.* 31:169 (1981).

29. Fox, J. P., et al. "Mercury Emissions from a Simulated In-Situ Oil Shale Retort," in *Proceedings of the Eleventh Oil Shale Symposium,* (Golden, CO: Colorado School of Mines Press, 1978).

30. Fox, J. P., "The Elemental Composition of Shale Oils," Lawrence Berkeley Laboratory Report No. LBL–10745 (1980).

31. Fox, J. P. "Water-related Impacts of In-Situ Oil Shale Processing," Lawrence Berkeley Laboratory Report No. LBL–6300 (1980).

32. Magee, E. M., J. J. Hall and G. M. Varga, Jr. "Potential Pollutants in Fossil Fuels," U.S. EPA–R2–73–249 (1973).

33. Bertine, K. K., and E. D. Goldberg. "Fossil Fuel Combustion and the Major Sedimentary Cycle," *Science* 173:233 (1971).
34. Southern California Edison. "Emission Characteristics of Paraho Shale Oil as Tested in a Utility Boiler" (1976).
35. Burger, E. D., et al. "Prerefining of Shale Oil," ACS Division of Petroleum Chemistry preprints, 20(4): 765 (1975).
36. Sullivan, R. F. et al. *Refining Shale Oil,* 43rd Midyear Meeting American Petroleum Institute, Toronto, Canada, May 10, 1978, API, Washington, DC Preprint No. 25-78 (1978).
37. Shih, C. C. et al. "Technological Overview Reports for Eight Shale Oil Recovery Processes," Supplement to the Fifth Quarterly Report, TRW Redondo Beach, CA (September 1976).
38. Shaw, P., "Analysis and Characterization of Trace Elements in Shale Oil and Shale Oil Products by Instrumental Neutron Activation Analysis," NTIS PB-291 421 (1978).
39. Fox, J. P. and T. E. Phillips. "Wastewater Treatment in the Oil Shale Industry," in *Proceedings of Oil Shale, the Environmental Challenges* (Golden, CO: Colorado School of Mines Press, 1981).
40. Farrier, D. S., J. P. Fox and R. E. Poulson. "Interlaboratory, Multimethod Study of an In-Situ Produced Oil Shale Process Water," *Proceedings of the Oil Shale Symposium: Sampling, Analysis and Quality Assurance,* U.S. EPA-600/9-80-022 (1979).
41. Geokinetics, "Water Quality Studies Progress Report," Concord, CA (1979).
42. National Academy of Sciences. "Water Quality Criteria 1972, A Report of the Committee on Water Quality Criteria," U.S. EPA-R3-73-003 (1973).
43. Cheng, C. N., and D. D. Focht. "Production of Arsine and Methylarsine in Soil and in Culture," *Appl. Environ. Microbiol.* 38:494-498 (1979).
44. "Threshold Limit Values for Chemical Substances and Physical Agents in the Workroom Environment with Intended Changes for 1977," American Conference of Governmental Industrial Hygienists (1977).
45. Sax, N. I. *Dangerous Properties of Industrial Materials,* 5th ed. (New York: Van Nostrand Reinhold Co., 1979).
46. "National Emission Standards for Hazardous Air Pollutants, Appendix B, Test Methods 101 and 102," U.S. Code of Federal Regulations, Title 40—Protection of the Environment, Chapter 1 (Revised July 1, 1977).
47. Hodgson, A. T., et al. "Mercury Mass Distribution During Laboratory Oil Shale Retorting," Lawrence Berkeley Laboratory Report No. LBL-12908 (1981).
48. Braman, R. S., and D. L. Johnson. "Selective Absorption Tubes and Emission Technique for Determination of Ambient Forms of Mercury in Air," *Environ. Sci. Technol.* 8:996 (1974).
49. Johnson, D. L., and R. S. Braman. "Alkyl- and Inorganic Arsenic in Air Samples," *Chemosphere,* 4:333 (1975).
50. Girvin, D. C., and J. P. Fox. "On-Line Zeeman Atomic Absorption Spectroscopy for Mercury Analysis in Oil Shale Gases," Lawrence Berkeley Laboratory Report No. 9702 (1980).
51. Cotter, J. E., D. J. Powell and C. Habenicht. "Fugitive Dust at the Paraho Oil Shale Demonstration Retort and Mine," U.S. EPA-600/7-29-208 (1979).

CHAPTER 5

EVALUATION OF MUTAGENICITY TESTING OF EXTRACTS FROM PROCESSED OIL SHALES

Judith G. Dickson and V. Dean Adams
Department of Civil and Environmental Engineering
and the Utah Water Research Laboratory
Utah State University
Logan, Utah 84322

INTRODUCTION

The water pollution potential of processed (spent) oil shale residues is a major concern regarding the environmental impact of oil shale development [1-3]. Salts, trace elements and organics including carcinogenic polycyclic aromatic hydrocarbons (PAH), may leach out of spent shale as a result of weathering. These materials, when entering the surface water and groundwater, may pose a serious problem to public health and the environment. Despite the relatively low concentration of the benzene-soluble organics in spent shale, 200-2000 ppm [4], the large quantities of retorted shale that will be produced during a full day of production (estimated to be 86,000 metric tons at one facility, the White River Shale Project [5]) will contain 17-170 metric tons of benzene-soluble materials per day. Schmidt-Collerus [4] found that 20-40% of the total benzene-soluble organic matter in shale can be leached by water and that most of the carcinogens in the shale were dissolved and concentrated in the salt residue.

The present study was initiated to assess the carcinogenic potential of spent shale leachates. The Ames test [6], a well-known microbial mutagenicity bioassay, was chosen for this purpose because it had previously been shown to detect carcinogens as mutagens and noncarcinogens as nonmutagens with good accuracy [7]. There were three phases of this research: (1) Ames test assays of fifteen PAH compounds of known carcinogenic activity were performed to confirm the reported accuracy; (2) assays of both organic solvent extracts and aqueous leachates were carried out to estimate the maximum potential of mutagens/carcinogens which could be derived from this source, in the former case, and to consider the carcinogenic hazard of spent shale under simulated field conditions, in the latter case; and (3) assays of various mixtures of PAH compounds of known mutagenicity were conducted to begin to evaluate the capability of the Ames test to give an integrated response to the full composition of mutagens/carcinogens present in environmental samples.

MATERIALS AND METHODS

In applying the Ames test [6], the *Salmonella typhimurium* mutant strains employed for the screening (provided by Professor B. N. Ames, University of California at Berkeley) were TA 98, 1537, 1538 and 100. Strain TA 1535 also was used initially but later excluded from general testing because it responded to so few of the known mutagens. The known PAH compounds and the samples resulting from spent shale extraction were assayed using the standard plate incorporation assay over a wide concentration range (2.5-100 μg/plate for known PAH, and quantities representing from 0.1 to 10 g of spent shale/plate for unknown extracts). These assays were performed with and without the addition of mammalian microsomes (rat liver used in the homogenate (S-9) had been induced with Aroclor 1254 [6]). The general procedures followed for all tests included positive controls to test for S-9 activity, and blanks to test for solvent effect and for sterile technique. Four replicate plates were used initially for each compound or sample tested (coefficient of variance (C.V.) = 13.4%, in 27 experiments), but two replicates were later used with little loss of precision (C.V. = 14.0%, in 22 experiments).

Known PAH compounds assayed in the Ames test were: benzo(a)pyrene, triphenylene, benzo(g,h,i)perylene, 1-aminopyrene, acridine, 13-H-dibenzo-(a,i)carbazole, fluoranthene, phenanthrene, anthracene, carbazole, pyrene, dibenzothiophene, 4-azafluorene, and thianthrene, obtained from Aldrich Chemical Co. (Milwaukee, Wisconsin); and benz(a)anthracene and 7,12-dimethylbenz(a)anthracene from Eastman Kodak Co. (Rochester, New York).

The majority of samples for mutagenicity testing (provided by Maase [8]) were obtained by Soxhlet extraction. A diversity of extracts was attained by varying the sequence of solvents, the composition of solvents in mixtures, and the extraction time. Samples developed in methanol following a period of extraction with benzene gave the greatest success in mutagenicity assays. The use of solvents other than methanol frequently elicited a toxic response during mutagenicity testing (benzene) or were incompatible with the bioassay procedure due to volatility (pentane) or hydrophobicity (cyclohexane). Attempts were made to evaporate the original solvent and redissolve the sample in a compatible solvent, dimethylsulfoxide (DMSO); however, this procedure met with little success, for a majority of samples still showed toxicity or failed to become resolubilized.

Aqueous extracts were obtained by providing extended contact of water with shale in both an upflow column and in a large Teflon®*-lined drum equipped with a motor-driven mixer. The organic compounds suspended in the water were then either concentrated on an exchange resin (XAD-2 and XAD-7, Rohm-Haas Co.) and eluted with solvents of varying polarity, or extracted directly with organic solvent (liquid-liquid extraction). As with the organic solvent extracts obtained from the Soxhlet, these samples were concentrated by either roto-evaporation or Kuderna-Danish concentration. Thin layer chromatography was used to further separate the components of certain samples exhibiting a great amount of mutagenic activity.

The potential complexity of obtaining an integrated mutagenic response to unknown chemical mixtures was investigated by assaying two known PAH compounds at various concentrations (equivalent to 5, 10 and 50 μg each per plate) in a one-to-one solution and in two other weight ratios: 10% compound A with 90% compound B and vice versa. The response was compared to those of each of the compounds assayed alone at the same concentrations.

RESULTS

Mutagenicity Testing of Known Compounds

Prior to presenting the data, a brief explanation of the normalization procedure used to express the results follows. Since the spontaneous reversion rate peculiar to a strain enhances its ultimate response, the average mutational response at each chemical concentration for a given strain was normalized to the average spontaneous response (solvent control) determined for each experiment (number of revertants per plate for each treatment/ number of spontaneous revertants); thus, a comparison of the relative response of each strain to mutagens could be made. This ratio, henceforth

---

*Registered trademark of E. I. du Pont de Nemours & Company, Inc., Wilmington, Delaware.

referred to as the revertant ratio or R.R., proved to be a simple way to determine a positive mutagenic response, which according to codevelopers of the Ames test [7] is defined by R.R. >2.0. The relative magnitude of the R.R. (and therefore mutagenic strength) is indicated by the – or + according to the scheme presented in Table 1.

The results of the Ames test assay of the fifteen known PAH compounds are compiled in Table 2. These data illustrate several features of the Ames test which make it suitable for mutagenicity screening of PAH-containing samples. First, the combinations of five mutant *Salmonella* strains (each differing in types of mutations) allow the detection of a wide variety of chemical mutagens. Of the compounds for which the carcinogenic activity is known (Table 2), the Ames test correctly identified 86% of the known carcinogens as mutagens and 71% of the known noncarcinogens as nonmutagens. These figures are slightly lower than those put forth by McCann and co-workers [7] (100% accuracy in detecting 26 PAH carcinogens as mutagens; 88% of 8 noncarcinogens were found to be nonmutagens). Second, a comparison of the relative carcinogenic activity and mutagenic activity demonstrates that the magnitude of the mutagenic response of the bacterial strains was closely correlated with the intensity of carcinogenic activity. Examples which display this correlation are the inactive carcinogens: acridine, anthracene, carbazole, phenanthrene and pyrene, which were nonmutagens; the weakly-to-moderately active carcinogens: benz(a)anthracene and dibenz-(a,h)anthracene, which were moderately strong mutagens; and the highly active carcinogens: benzo(a)pyrene and 7,12-dimethylbenz(a)anthracene, which were strong mutagens. And third, the results presented in Table 2 suggest the type of mutagen a compound may be. Some of the compounds, for example benzo(a)pyrene and 7,12-dimethylbenz(a)anthracene are general mutagens, i.e., they can cause mutations in both the strains having base pair substitutions (TA 100, 1537) and those having frame-shift mutations (TA 1537, 1538, and 98). Other PAH compounds cause specific mutations of one

Table 1. Description of Symbols Used to Indicate Relative Mutagenicity

Symbol	Average Revertant Ratio Value	Interpretation
–	<2.0	No mutagenic response
(+)	>2.0 ranging <2.0 or <2.0 ranging >2.0	Questionable mutagenic response
+	2.0–4.0	Weak mutagenic response
++	4.0–8.0	Moderate mutagenic response
+++	8.0–16.0	Strong mutagenic response
++++	16.0–32.0	Very strong mutagenic response
+++++	>32.0	Extremely strong mutagenic response

Table 2. Results of Ames Test Assay of Known Polycyclic Aromatic Hydrocarbons[a]

Compound Name	Relative Carcinogenic Activity[b]	TA Strain				
		1535	1537	1538	98	100
Acridine	none	–[c]	–	–	–	–
1-Aminopyrene	high	–	N.T.[d]	+	+++++	++
without S-9		–	N.T.	++	+++++	+
Anthracene	none	–	–	–	–	–
Benz(a)anthracene	weak	–	++	+	(+)	+
Benzo(g,h,i)perylene	moderate	–	+++	–	–	–
Benzo(a)pyrene	high	–	++++	+	+++	++
Carbazole	none	–	–	–	–	–
Dibenz(a,h)anthracene	moderate	–	+	–	–	(+)
13-H-Dibenzo(a,i)carbazole	weak	–	–	–	–	–
7, 12-Dimethylbenz(a)anthracene	high	–	+++	–	(+)	+
Fluoranthene	none	–	+	–	+	+
Perylene	none	–	+++++	–	(+)	–
Phenanthrene	none	–	–	–	–	–
Pyrene	none	–	–	–	–	–
Triphenylene	none	+	–	–	–	–

[a]All assays performed with the addition of rat liver homogenate (S-9) unless otherwise stated (without S-9).
[b]Data from Dipple [9].
[c]Symbols are used to indicate relative mutagenic strength. Refer to Table 2.
[d]Not tested.

or the other kind. In particular benzo(g,h,i) perylene appears to cause mutations in the specific sequence of bases found in the mutant TA 1537.

Mutagenicity Testing of Spent Shale Extracts

Four types of spent shale were used to obtain samples for mutagenicity testing. All were the result of early surface retorting research and therefore the data reported should not be considered reflective of commercial-scale processed shales.

The results of mutagenicity screening of extracts from these spent shales are presented in Table 3. The same format used in Table 2 to indicate relative mutational response for the known PAH is repeated for the unknown mutagens in the shale extracts. Also included in this table is a brief description of the procedures (extraction, concentration and separation) used in obtaining each sample.

The results of Ames test assays indicate that mutagens are present in each of the four types of spent shales and that mutagens can be obtained by a variety of different extraction methods. In general, these mutagens require metabolic activation (addition of enzyme-induced S-9) for their detection.

Table 3. Results of Ames Mutagenicity Testing of Spent Shale Extracts[a]

Spent Shale Type and Sample No.	TA Strain					Extraction Procedure[b]
	1535	1537	1538	98	100	
Type A						
#1	(+)[c]	+	(+)	(+)	−	Sox. Ext. one day with φH followed by one day with MeOH, KD concentration, in MeOH
#2	−	−	−	Δ[d]	(+)	Sox. Ext. four days with φH followed by four days with MeOH, RE concentration, in MeOH
#3	++	−	−	−	(+)	Sox. Ext. four days with φH followed by four days with MeOH, KD concentration, in MeOH
#4	N.T.[e]	+	+	−	−	Samples #2 and #3 combined
#5	N.T.	Δ	Δ	N.T.	Δ	Sox. Ext. four days with φH-MeOH solution 1:5 by volume, KD concentration, in φH-MeOH
Type B						
#1	+	++	+	+	−	Sox. Ext. three days with MeOH, KD concentration, in MeOH
#2	N.T.	++	+	N.T.	(+)	Same as #1
#3	−	(+)	−	−	−	TLC on silica gel of sample #1, eluted with MeOH: p-Dioxane solution 1:1 by vol.
#4	+	++	−	−	−	Same as #3 except eluted with MeOH:p-Dioxane solution 1:4 by vol.
#5 w/S-9	N.T.	++Δ	+Δ	+Δ	Δ	Sox. Ext. three days with φH:MeOH solution 1:5 by vol., KD concentration in φH:MeOH
w/o S-9[f]	N.T.	++Δ	++Δ	Δ	Δ	
#6	N.T.	Δ	Δ	Δ	Δ	Sox. Ext. one day with φH followed by one day with MeOH, RE concentration, in MeOH
#7	N.T.	(+)	(+)	−	−	Sox. Ext. one day with φH followed by two days with MeOH, in MeOH
#8[g]	N.T.	(+)	−	−	−	Sox. Ext. one day with Pentane, KD concentration, TLC on silica gel eluted 5 fractions w/MeOH

						Description
Type C						
#1	N.T.	++	+++	++	−	Sox. Ext. three days with φH followed by five days with MeOH, RE concentration, in MeOH
#2	N.T.	+++	+++	++	(+)	Same as #1 except KD concentration
#3	−	−	−	−	−	TLC on silica gel of sample #2, eluted fractions with MeOH[g]
#4	N.T.	△	△	△	△	Separation of sample #2 on Al_2O_3, eluted with φH
Type D						
#1	N.T.	++	+++	+	(+)	Sox. Ext. one day with MeOH, KD concentration, in MeOH
#2	N.T.	−	−	−	−	TLC on silica gel of sample #1 eluted fractions with MeOH
#3	N.T.	(+)	−	△	−	AQ. Leach. filtered and passed through Sephadex gel, eluted with MeOH
#4	N.T.	−	−	−	−	AQ. Leach. used in agar media preparation
#5	N.T.	(+)	(+)	−	−	AQ. Leach. filtered and passed through XAD-2 resin eluted with MeOH

[a] Results of plate incorporation assays with rat liver homogenate (S-9) except where indicated, from at least two replicates.

[b] Sox. Ext. = Soxhlet Extraction   KD = Kuderna Danish   Vol. = Volume
φH = Benzene   RE = Roto-evaporation   AQ. Leach. = aqueous leachate
MeOH = Methanol   TLC = Thin layer chromatography

[c] Symbols are used to indicate relative mutagenic strength (refer to Table 1).

[d] △ indicates toxicity at high concentrations of sample.

[e] N.T. = not tested.

[f] w/o S-9 = assay performed without rat liver homogenate.

[g] Of 5 fractions eluted from silica gel, only one $R_f = 0.18$ showed any mutagenicity.

The repeated response of TA strains 1537, 1538 and 98 indicate that many of these mutagens cause frame-shift substitutions of bases in bacterial DNA. There were fewer base-pair substitution mutagens detected in these samples. In addition, a toxic response was common and usually was associated with the solvent benzene in the sample, but not always (A-2, B-6, D-3). This toxic response may have obscured the detection of mutagens in these and other samples (A-5, B-5, C-4). Attempts to separate the complex mixtures for the purposes of removing the toxic agents and allowing more accurate detection and identification of chemical mutagens were unsuccessful. When thin layer chromatography was applied to certain mixtures which had high mutagenic activity (B-1, C-2, D-1), the mutagenic activity of the fractions (B-3, C-3, D-2, respectively) was weak or nonexistent. The weak response may indicate that the compounds are present in low concentrations (i.e., poor recovery and dilution of sample).

Table 3 also summarizes some of the attempts to obtain mutagens in aqueous leachates of shale. The various methods of concentration using sephadex and XAD resins worked poorly in this regard, resulting in questionable mutagenic activity (D-3, D-5). The attempts to use the aqueous sample in the agar media were also unsuccessful (D-4).

Mutagenicity of Chemical Mixtures

Assays of simple two-component mixtures were performed to assess the capability of the Ames test to integrate the total mutagenic potential of environmental samples. The chemicals paired in these mixtures were chosen from those in the list of knowns (Table 2) that gave a unique mutagenic response when assayed with a particular mutant strain. The ideal mutagenic response to this mixture would be one that was additive.

In the first set of experiments, ten pairs of PAH compounds were assayed (1:1 by weight mixture) in a total of eighteen experiments. In none of these experiments did the simple addition of the response to the compounds assayed separately approximate the mutagenic response when the compounds were assayed together. In many cases the outcome for the mixture paralleled that of one of the components assayed separately. Chi-square analysis for interaction was applied to the resulting dose-response curves to determine whether the set of three curves (component A alone, component B alone, and both A and B) were distinguishable. In Table 4 the outcome of this analysis is reported. The dose-response curve that results when two mutagens are assayed together was found to be one of the following: (1) indistinguishable from the dose-response curves of either of the individual components, which are themselves indistinguishable (a,a,a) (4 of 18 pairs); (2) distinguishable from the dose-response curve of one of the individual

Table 4. Ames Testing of Solutions of Two Mutagens*

		TA Strain		
Pair No.	Paired Compounds	1537	98	100
1	Benzo(ghi) perylene, BP	a[†] +++**		
	7,12-Dimethylbenz(a) anthracene, DMBA	a +++		
	BP/DMBA	a		
2	Benzo(a)pyrene, BaP	a ++++	a +++	
	Perylene, P	b +++++	b (+)	
	BaP/P	a	a	
3	Benzo(a)pyrene, BaP	a ++++	a +++	a ++
	7,12-Dimethylbenz(a) anthracene, DMBA	b +++	b (+)	b +
	BaP/DMBA	a	a	a
4	Benz(a)anthracene, BA		a (+)	a +
	7,12-Dimethylbenz(a) anthracene, DMBA		a (+)	b +
	BA/DMBA		a	c
5	Benz(a)anthracene, BA	a ++	a (+)	
	Fluoranthene, F	b +	a (+)	
	BA/F	a, b	a	
6	7,12-Dimethylbenz(a) anthracene, DMBA	a +++	a (+)	
	Perylene, P	b ++++	b (+)	
	DMBA/P	c	a,b	
7	Benzo(ghi) perylene, BP	a +++		
	Perylene, P	b ++++		
	BP/P	c		
8	Fluoranthene, F		a (+)	
	Perylene, P		b (+)	
	F/P		a	
9	7,12-Dimethylbenz(a) anthracene, DMBA		a (+)	a +
	Fluoranthene, F		a (+)	b +
	DMBA/F		a	a
10	Benzo(a)pyrene, BaP		a +++	a ++
	Fluoranthene, F		b +	b +
	BaP/F		a	c

*Results of plate incorporation assays with rat liver homogenate (S-9).

†Small case letters (a, b, c), refer to the results of chi-square analysis which was used to test whether the dose-response curves of the set were distinguishable (different letters) or not (same letters) at $\alpha \leqslant 0.01$.

**Symbols are used to indicate relative mutagenic strength (refer to Table 1).

components but not from the other (a,b,a) (8 of 18 pairs) or (3) distinguishable from the dose-response curves of both of the individual components (a,b,c or a,b,ab) (6 of 18 pairs). Twelve of the outcomes (cases 1 and 2) support the hypothesis that some degree of masking occurs in mixed samples, resulting in the underestimation of mutagenic potential. Little can be said of the six pairs which exhibited the case 3 response except that possibly some interaction was occurring in the mixture, causing a response that was intermediate but never additive.

The assaying of chemical mixtures was carried one step further in a set of experiments designed to test whether a mutagen which appeared to dominate the response in the mixture, thus masking the presence of a subordinate mutagen, could maintain this dominance when it composed only 10% of the two-component mixture. The paired components chosen to test this were those that had demonstrated a clearly defined dominance interaction (BAP/ DMBA, TA 100 and 98, and BAP/P on TA 1537). One component, A, of the pair was assayed at three concentrations comprising 10, 50 and 90% of the mixture, and the concentration of the other component, B, was held constant. Then the process was repeated for B with A constant. The results of these tests are shown in Table 5. The chi-square analysis indicates that in two of the three experiments (BAP/DMBA, TA 98 and 100) the dose-response curve of the dominant mutagen, BAP, is indistinguishable from that of BAP with constant DMBA (Table 5). The response curve is again not demonstrative of an additive mutagenic response and even when BAP comprises 10% of the mixture the presence of DMBA remains undetected. In the other experiment

Table 5. Results of Ames Test Assay of Non-One-to-One, Two-Component Mixture*

Compound Pairs	TA Strain		
	1537	98	100
BAP		a[†]+++**	a ++
BAP w/const DMBA		a	a
DMBA		b (+)	b +
DMBA w/const BAP		c	c
BAP	a ++++		
BAP w/const P	b		
P	c +++++		
P w/const BAP	a		

*Results of plate incorporation assays with rat liver homogenate (S-9).
[†]Small case letters (a, b, c), refer to the results of chi-square analysis, which was used to test whether the dose-response curves of the set were distinguishable (different letters) or not (same letters) at $\alpha \leqslant 0.01$.
**Symbols are used to indicate relative mutagenic strength (refer to Table 1).

(BAP/P, TA 1537) the weaker mutagen, BAP, is the masking factor that dominates over the normally stronger response of TA 1537.

DISCUSSION

Much controversy exists concerning the proper use of short-term muta-genic assays in screening environmental samples for potential carcinogens. Therefore this study made a special effort to characterize one particular assay, the Ames test, in terms of its strengths and limitations regarding its application to extracts from spent oil shale and petroleum-related effluents in general.

Several PAH compounds were obtained for Ames testing initially to deter-mine whether these procedures were correct and later to validate the assay. The consistency with which the magnitude and shape of the dose-response curves could be reproduced using this assay, allowing a unique characteriza-tion of each known compound, indicated that the procedures were reliable. Generally, when comparisons could be made, our data correlated well with those reported in the literature [7, 10, 11].

The Ames test was used with mixed success in detecting mutagens in extracts from spent oil shale. The best results were obtained when Soxhlet extraction was performed using organic solvents. The first few samples ob-tained in methanol following benzene extraction (Table 3: A1-A4 and C1-C2) were shown to produce dose-response curves similar to those obtained for the known standards. Then, in order to get a more detailed description of the chemical composition of these extracts and the mutagenically active fractions, separation techniques were employed. For the most part, these procedures did not enable further characterization of the mutagenic potency of these extracts. Dilution and solvent incompatibility appeared to be major problems. Subsequently, it was necessary to determine whether the muta-genic response to the complex mixtures being obtained prior to separation and purification was indicative of the total response contributed by each mutagenic component.

The results of experiments using two-component mixtures indicated that the response of the bacterial strains to two mutagens assayed together was nonadditive. In the majority of cases, the response to the mixture was in-distinguishable from the response to one of the components assayed separ-ately (masking of one component in the presence of another, more dominant compound). In a few instances, the response to the mixture was intermediate to that of the two separate components. A review of the literature indicates that researchers are finding an array of responses when the Ames test is applied to environmental samples. Mutagenic activity was found to be additive for petroleum fractions [12, 13] and cigarette smoke condensate fractions [14]. The phenomenon of comutagenesis (i.e., the enhancement of the mutagenicity of one compound by the addition of another compound which may or may not be a mutagen) also has been reported [15]. Several other researchers have confirmed our finding that mutagenic potential of a mixed sample may be underestimated in the Ames test [16-18].

While the spent shale extracts obtained using Soxhlet extraction and organic solvents were assayed with fairly good success, with results indicating the potential for groundwater pollution from mutagens present in spent shale, the assays of aqueous leachates did not support the observation by others [4] that they could be washed out of the shale. If mutagens were present in these leachates they were either in very small concentrations or were lost during the procedures of extraction purification and handling. Neither the standard Ames plate incorporation assay nor the alternate method for assaying aqueous samples indicated a sample having positive mutagenic activity. Another obstacle encountered was that one of the solvents, methylene chloride, which is recommended for use in sample separation, is a mutagen to the bacterial strains, thus preventing many fractions from being assayed.

In conclusion, it can be said that mutagens, many of which are certainly carcinogens, are present or at least can be formed from precursors in processed shale and that they can be extracted in organic solvents through Soxhlet extraction. The failure to obtain positive evidence of mutagens in the aqueous leachates does not negate the possibility that they were extracted. Due to limitations in the quantity of spent shale with which to experiment and in the procedures used to concentrate and prepare the extracts for testing, the compounds may have been present but in concentrations below the sensitivity level of the assay. This speculation is corroborated by the GC/MS data [8], indicating that many of the PAH previously determined to be mutagenic are present in the aqueous extracts below the limit of their solubility (ng/L). It is likely, given this information, that further biological testing of concentrated leachates from processed shale would successfully demonstrate that the disposal of spent oil shale can be a source of mutagens to the environment through leaching.

ACKNOWLEDGMENTS

The authors wish to acknowledge the Office of Water Research and Technology (Project no. B-154-UTAH, Contract no. 14-34-0001-8123), United States Department of the Interior, Washington, DC, which provided funds for research and publication.

REFERENCES

1. Ward, J. C. "Water Pollution Potential of Spent Oil Shale Residues," prepared for the U.S. Environmental Protection Agency, Grant No. 14030EDB (1971).
2. Dassler, G. L. "Assessment of Possible Carcinogenic Hazards Created in Surrounding Ecosystems by Oil Shale Development," MS thesis, Utah State University, Logan, UT (1976), 99 pp.
3. Cleave, M. L. "Effects of Oil Shale Development on Freshwater Phytoplankton," PhD dissertation, Utah State University, Logan, UT (1979).
4. Schmidt-Collerus, J. J. "The Disposal and Environmental Effects of Carbonaceous Solid Wastes from Commercial Oil Shale Operations," First Annual Report to NSF, NSFGI 34282 x 1, Denver Research Institute (1974), 169 pp.
5. Slawson, Jr., G. C., Ed. "Groundwater Quality Monitoring of Western Oil Shale Development: Identification and Priority Ranking of Potential Pollution Sources," U.S. EPA-600/7-79-023 (January 1979).
6. Ames, B. N., W. E. Durston, E. Yamasaki and F. D. Lee. "Carcinogens are Mutagens: A Simple Test System Combining Liver Homogenates for Activation and Bacteria for Detection," *Proc. Nat. Acad. Sci., U.S.* 70(8):2281–2285, (1973).
7. McCann, J., N. B. Spingarn, J. Kobori and B. N. Ames. "Detection of Carcinogens as Mutagens: Bacterial Tester Strains with R Factor Plasmids," *Proc. Nat. Acad. Sci., U.S.* 72(3):979–983 (1975).
8. Maase, D. L. "An Evaluation of Polycyclic Aromatic Hydrocarbons from Processed Oil Shale," PhD dissertation, Utah State University, Logan, UT (1980), 202 pp.
9. Searle, C. E., Ed. *Chemical Carcinogens* (Washington, DC: American Chemical Society, 1976).
10. Kraybill, H. F., C. T. Helms and C. C. Sigman. "Biomedical Aspects of Biorefractories in Water," paper presented at the Second International Symposium on Aquatic Pollutants, Amsterdam, The Netherlands, September 26–28, 1977. In *Aquatic Pollutants Transformation and Biological Effects*, O. Hutzinger, L. H. Van Leylveld and B. C. J. Zacteman, Eds. (New York: Pergamon Press 1978), pp. 419–459.
11. Belser, W. Personal communication (1978).
12. Epler, J. L. et al. "Analytical and Biological Analyses of Test Materials from the Synthetic Fuel Technologies. I. Mutagenicity of Crude Oils Determined by the *Salmonella typhimurium*/Microsomal Activation System," *Mutat. Res.* 57(3):265–276 (1978).
13. Epler, J. L. "Evaluation of Mutagenicity Testing of Shale Oil Products and Effluents," *Environ. Health Pers.* 30:179–184 (1979).
14. Hutton, J. J., and C. Hackney. "Metabolism of Cigarette Smoke Condensates by Human and Rat Homogenates to Form Mutagens Detectable by *Salmonella typhimurium* TA 1538," *Cancer Res.* 35:82461–2468 (1975).

15. Sugimura, T., T. Kawachi, T. Matsushi, M. Nagao, S. Sato and T. Yahagi. "Critical Review of Sub-Mammalian Systems for Mutagen Detection," in *Progress in Genetic Toxicology,* D. Scott, B. A. Bridges and F. H. Sobels, Eds. (Amsterdam: Elsevier/North Holland Biomedical Press, 1977), pp. 125-140.

16. Pelroy, R. A., and M. R. Petersen. "Use of Ames Test in Evaluation of Shale Oil Fractions," *Environ. Health Pers.* 30:191–203 (1979).

17. Loper, J. C., et al. "Residue Organic Mixtures from Drinking-Water Show in Vitro Mutagenic and Transforming Activity," *J. Toxicol. Environ. Health* 4(5–6):919–938 (1978).

18. Chrisp, C. E., G. L. Fisher and J. E. Lammert. "Mutagenicity of Filtrates from Respirable Coal Fly Ash," *Science* 199(4324): 73–75 (1978).

PART 2

IMPACT OF OIL SPILLS AT SEA

CHAPTER 6

BASELINE STUDIES FOR HYDROCARBONS AND ORGANIC CARBON ISOTOPE RATIOS OF RECENT SEDIMENTS IN THE BANK OF CAMPECHE BEFORE THE IXTOC-I OIL SPILL

A. V. Botello, S. A. Castro and R. Guerrero

Instituto de Ciencias del Mar y Limnología, UNAM
México 20, D. F. México

INTRODUCTION

To determine the presence of petroleum hydrocarbons in the components of a marine ecosystem one must first evaluate the relative quality of the hydrocarbons biosynthesized in the system. In this way it is possible to distinguish some of the biogenic hydrocarbon series from those of fossil origin occurring in marine organisms and sediments.

Organisms possess specific biosynthetic pathways which favor the production of hydrocarbons in preferred size ranges. Crude oils and oil products, on the other hand, are wide-range mixtures that contain molecules of different sizes in fairly even distribution. Indeed the measurement of the organic carbon isotope ratios in recent sediments is a helpful tool to understand the input of organic carbon in coastal areas from natural or anthropogenic sources. Most of the studies on the distribution of hydrocarbons and the isotopic composition of organic carbon in sediments and marine organisms of near-shore regions of the Gulf of Mexico have been conducted mainly along the Gulf coast of the United States.

The primary purpose of this study is to establish the distribution of biogenic hydrocarbons and the organic carbon isotope ratios in recent marine sediments of the continental shelf adjacent to the Terminos Lagoon in the

Bank of Campeche. Taking into consideration the unpolluted nature of this coastal region before the Ixtoc-I oil spill, this study will provide a basis for the assessment of man-induced alterations.

MATERIALS AND METHODS

Surface sediments of Campeche Bank (Figures 1 and 2) were collected with a Van Veen Grab sampler and representative uncontaminated samples were placed in specially cleaned glass jars and immediately frozen.

To determine the carbon isotope ratios the thawed sediments were treated following the procedure of Parker et al [1], and aliquot samples of sediments were burned, using a modified LECO radio frequency furnace.

The δ^{13}C values of the resulting gases were determined with a dual collector isotope ratio mass spectrometer (Nucleide 6-60 RMS). The working standard used was a lubricating oil with a δ^{13}C of -27.32 $^{o}/oo$ versus the Chicago PDB standard. Based on the analysis of 46 paired replicate samples, the average overall error in combustion and measurement was ± 0.3 $^{o}/oo$.

The hydrocarbons were extracted from sediments with toluene–methanol (2:1 volume) and concentrated. This concentrate was separated by column chromatography into a hexane and benzene eluate on a mixed bed alumina-silica gel column (2:1 volume).

The fractions were analyzed and quantified by gas chromatography on SP-1000 or OV-101 glass capillary columns using a Perkin Elmer Model 910 gas chromatograph equipped with a flame ionization detector (FID).

The oven temperature was programmed from 80°C to 240°C at 4°C/min and the carrier gas was nitrogen flowing at 25 mL/min under 50 psi.

Each hydrocarbon fraction was coinjected with a standard hydrocarbon mixture containing nine normal alkanes from C_{14} to C_{30} plus pristane and phytane. The concentration of each component was determined by its corresponding peak area in the chromatograms.

RESULTS

The distribution of δ^{13}C values and the concentration of hydrocarbons for sediments obtained during the Oplac-I cruise in July 1978 are shown in Table 1. The range of δ^{13}C values found in the sediments studied varied from -19.8 $^{o}/oo$ to -23.2 $^{o}/oo$. These values are very similar to those reported by Gearing et al [2] for sediments of the near-shore northern and western shelf of the Gulf of Mexico with δ^{13}C values ranging from δ = -21 to -24 $^{o}/oo$.

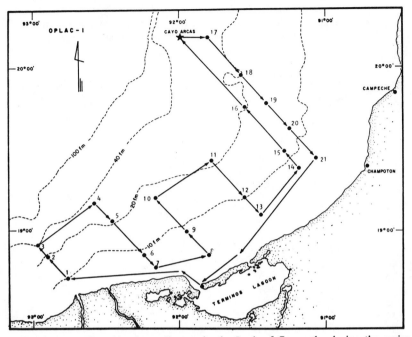

Figure 1. Sampling sites for sediments in the Bank of Campeche during the cruise OPLAC-I, July 1978.

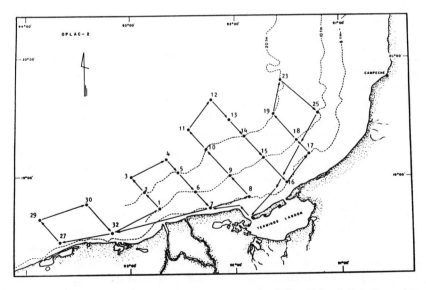

Figure 2. Sampling sites for sediments in the Bank of Campeche during the cruise OPLAC-II, March 1980.

Table 1. Summary of the δ^{13}C Values and Total Hydrocarbons Concentration in Marine Recent Sediments from the Bank of Campeche, Mexico

Cruise OPLAC-1, July 1978

| Station No. | Location | | δ^{13}C$^0/^{00}$ | Total Hydrocarbons (ppm dry wt) |
	Latitude	Longitude		
1	18°42'	92°46'	-23.0	56
2	18°50'	92°54'	-23.2	48
3	18°55'	92°58'	-22.4	51
4	19°10'	92°34'	-21.9	46
5	19°04'	92°28'	-21.9	43
6	18°50'	92°14'	-21.1	42
7	18°44'	92°08'	-21.4	51
8	18°50'	91°38'	-21.1	37
9	18°59'	91°57'	-23.0	48
10	19°12'	92°14'	-21.2	35
11	19°26'	91°45'	-20.5	23
12	19°12'	91°32'	-19.8	26
13	19°05'	91°26'	-21.9	25
14	19°23'	91°10'	-19.9	18
15	19°29'	91°15'	-19.8	21
16	19°45'	91°32'	-21.7	15
17	20°11'	91°47'	-21.0	12
18	19°57'	91°33'	-20.9	16
19	19°47'	91°23'	-20.7	14
20	19°38'	91°14'	-20.3	16
21	19°28'	91°04'	-21.2	18

The values of the carbon isotope ratios show three different sedimentary organic carbon environments: one zone with values δ^{13}C = -22.4 to -23.2 $^0/oo$ which is under the runoff influence of the Usumacinta and Grijalva rivers, another zone of typical marine influence with δ^{13}C values between -20.5 to -21.9 $^0/oo$, and finally, one zone with lower values due to the influence of carbonates in the Bank of Campeche (δ^{13}C = -19.8 $^0/oo$).

According to Sackett [3] and Parker et al [1] the isotopic composition of the Gulf coast sediments is usually fairly constant and any drift from the normal could indicate a major change in the source and isotopic composition of the organic carbon contributing to the sediments.

The more negative δ^{13}C values of the organic matter in the sediments of the studied area before the oil spill (δ^{13}C = -23.2 $^0/oo$) can be attributed to organic matter with lighter carbon isotope ratios originating from terrestrial industrial and urban sources [4]. However, we must take into consideration that biologically derived, isotopically light CO_2 (from bacteria acting on sewage and oil sludges) may be partially responsible for a shift in the isotopic matter generated in coastal environments [5].

Compared with the results of the Oplac-I cruise, the δ^{13}C values shown by the Oplac-II cruise in March 1980 remain very constant and show the same pattern of distribution (Table 2). The only exception is the value recorded at sampling site No. 12 (δ^{13}C = -27.2 ^{o}/oo) which is a clear indication that hydrocarbons originating from the oil activities have settled in the column sediment, causing a considerable change in the prerecorded δ^{13}C value [6].

The analysis of hydrocarbons in sediments and organisms of the Bank of Campeche confirmed the predominance of odd-carbon number n-paraffins in an area for which no analysis had been available. The odd-carbon predominance was well within the range reported by Cooper and Bray [7]. The predominance of odd-carbon number paraffins in recent sediments of the continental shelf adjacent to Terminos Lagoon suggests that the source of a large proportion of the paraffins may be organisms indigenous to the sediments or terrestrial plant detritus [8-10].

Table 2. Summary of the δ^{13}C Values and Total Hydrocarbons Concentration in Marine Recent Sediments from the Bank of Campeche, Mexico

Cruise OPLAC-2, March 1980

Station No.	Location		δ^{13}C^o/00	Total Hydrocarbons (ppm dry wt)
	Latitude	Longitude		
01	18°43'30"	92°44'	-22.7	42
02	18°42'	92°53'	-23.3	92
03	19°00'	93°00'	-22.1	73
04	19°09'	92°39'	-19.7	76
05	19°02'30"	92°34'	-19.9	55
06	18°52'	92°23'	-21.0	68
07	18°44'	92°15'	-21.2	27
08	18°49'	91°52'	-20.2	51
09	19°00'30"	92°03'	-19.5	67
10	19°14'	92°17'	-20.3	75
11	19°24'	92°27'	-20.9	81
12	19°30'	92°14'	-27.0	715
13	19°23'	92°03'	-26.6	117
14	19°21'	91°55'	-19.7	44
15	19°09'	91°43'	-19.5	43
16	18°56'	91°30'	-18.3	9
17	19°11'	91°17'	-18.2	51
18	19°19'	91°24'	-19.0	47
19	19°32'	91°38'	-19.9	40
23	19°49'	91°33'	-10.5	52
25	19°32'	91°12'	-20.2	27
27	18°26'30"	93°42'	-20.2	32
29	18°38'30"	93°53'	-21.4	35
30	18°47'	93°26'	-21.8	52
32	18°31'	93°11'	-22.2	35

During the Oplac-I cruise (July 1978) the concentration of n-paraffins in the surface sediments of the studied area ranged from 13 to 56 ppm dry weight. These figures are within the values recorded (<70 ppm) for un-polluted coastal areas and deep marginal seas and basins [11].

The chromatograms for n-paraffins of sampling sites No. 1 and 12 are represented in Figures 3 and 4. There is a clear predominance of odd n-paraffins, mainly C_{15}, C_{17}, C_{19}, C_{25}, C_{27}, C_{29}, C_{31} and C_{33}. However, this predominance is lost during March 1980 at site No. 12 because of the introduction of hydrocarbons in the sediments by the oil activities on the area. The chromatogram shows the saturated fraction containing normal paraffins from those with a low molecular weight like the n-tetradecane (C_{14}) to n-pentatriacontane (C_{35}); the isoprenes pristane and phytane are also present.

The pristane/phytane ratio was 1.2 and the odd-even predominance (OEP) value was 1.0, which agrees with previously reported values for crude oils that have not had a remarkable weathering [12, 4].

DISCUSSION

Our present knowledge of the concentrations and lethal and sublethal effects of petroleum hydrocarbons on the marine environment has greatly benefited from the results of a series of important scientific contributions which have shed light on the physiological response of a good number of marine organisms exposed to oil, as well as on the impact caused by spill accidents on both coastal and open ocean ecosystems.

Because the Campeche Bank has a paramount importance as a fishery area, there have been few attempts to evaluate the impact of the offshore petro-leum industry on fishery resources.

It has generally been conceded in the literature that chronic exposure to oil can be more harmful to the environment than the effects derived from the always dramatic oil well blow-outs. In this respect Hall et al. [13] have pointed out that day-to-day offshore-rig operations introduce into the local ecosystem an amount of pollutants whose consequences can be quite adverse, particularly to the coastal environment.

Since 1979, the offshore oil operations conducted by PEMEX (Mexican Petroleum Company) in the Campeche Bank have maintained an accelerated pace of exploration and exploitation which very likely will extend in the future over other areas where important fishing grounds are located. Under these circumstances petroleum hydrocarbons in or on seawater can become associated with suspended sediment particles, and be transported downwards and incorporated into the sediments.

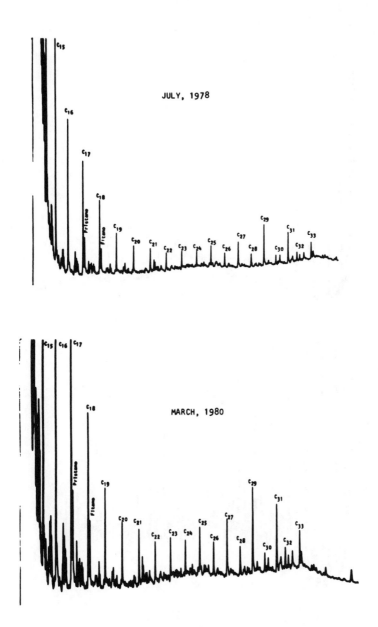

Figure 3. Distribution of n-paraffins in sediments from sampling site No. 1 during the cruises OPLAC-I and OPLAC-II.

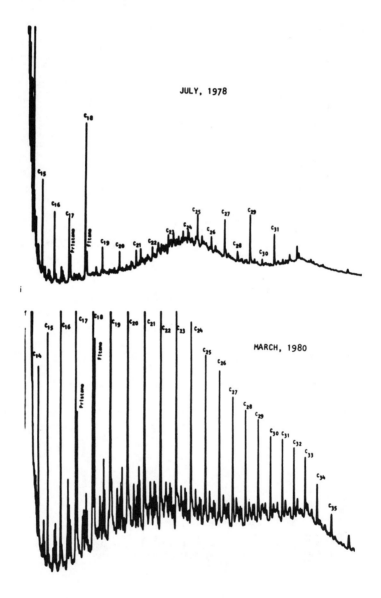

Figure 4. Distribution of n-paraffins in sediments from sampling site No. 12 during the cruises OPLAC-I and OPLAC-II.

The degree of toxicity will greatly depend on the type of oil introduced into the ecosystem, and the changes produced by weathering factors of the different oil components.

Recent findings by Lee et al. [14] indicate, for instance, that spilled Mexican oil failed to induce acute effects on marine organisms due to the evaporation of the most toxic components such as benzenes and naphthalenes.

To minimize the undesired consequences that this development may cause on a renewable resource, it is imperative to gain accurate data on the concentration of hydrocarbons and on the behavior of the stocks.

Even though the information presented here is by no means a thorough analysis of the interaction between petroleum hydrocarbons and fisheries, it does provide "base-line" data that can be used to detect large changes in the ecosystem. In addition, it establishes the critical levels of hydrocarbon concentrations in marine organisms.

ACKNOWLEDGMENTS

We want to thank L. Celis, R. Salal and J. M. Cortés for their technical assistance, and Miss Sharon Washtien for the translation of this paper. Contribution No. 269.

REFERENCES

1. Parker, P. L., W. E. Behrens and D. J. Shultz. "Stable Carbon Isotope Ratio Variations in the Organic Carbon from Gulf of Mexico Sediments," *Mar. Sci.* 16:139–147 (1972).
2. Gearing, P., J. N. Gearing, T. R. Lyttle and J. S. Lyttle. "Hydrocarbons in 60 Northeast of Mexico Shelf Sediments: A Preliminary Survey," *Geochim. Cosmochim. Acta* 40(9):1005–1017 (1977).
3. Sackett, W. M. "The Depositional History and Isotopic Organic Carbon Composition of Marine Sediments," *Mar. Geol.* 2:173–185 (1964).
4. Botello, A. V., E. F. Mandelli, S. A. Macko and P. L. Parker. "Organic Carbon Stable Isotope Ratios of Recent Sediments from Coastal Lagoons of the Gulf of Mexico, Mexico," *Geochim. Cosmochim. Acta* 44:557–563 (1980).
5. Lloyd, R. M. "Variations in the Oxygen and Carbon Isotope Ratios of Florida Bay Mollusks and Their Environmental Significance," *J. Geol.* 72:84–111 (1964).
6. Botello, A. V., and S. A. Castro. "Chemistry and Natural Weathering of Various Crude Oil Fractions from the Ixtoc-I Oil Spill," in Proceedings on Preliminary Results of the Researcher Pierce Cruise Ixtoc-I Workshop, Miami, FL (June 1980).
7. Cooper, J. E., and E. E. Bray. "A Postulated Role of Fatty Acids in Petroleum Formation," *Geochim. Cosmochim. Acta* 27:1113–1127 (1963).

8. Clark, R. C. and M. Blumer. "Distribution of n-Paraffins in Marine Organisms and Sediments," *Limnol. Oceanog.* 12:79–87 (1967).
9. Blumer, M., M. M. Mullin and D. W. Thomas. "Pristane in the Marine Environment," *Helgolaender Wiss. Meeresuntersuch* 10:187–201 (1964).
10. Botello, A. V., and E. F. Mandelli. "Distribution of n-Paraffins in Seagrasses, Benthic Algae, Oysters and Recent Sediments from Terminos Lagoon, Campeche, Mexico," *Bull. Environ. Contam. Toxicol.* 19(2): 162–170 (1978).
11. National Academy of Sciences. "Petroleum in the Marine Environment," National Academy of Science Workshop on Inputs, Fates and the Effects of Petroleum in the Marine Environment, Airlie House, VA (1975), 107 pp.
12. Scalan, R. S., and J. Smith. "An Improved Measure of the Odd-Even Predominance in the Normal Alkanes of Sediment Extracts and Petroleum," *Geochim. Cosmochim. Acta* 34:611–720 (1970).
13. Hall, C. A. S., R. Howarth, B. Moore and C. J. Vorosmarty. "Environmental Impacts of Industrial Energy Systems in the Coastal Zone," *Ann. Rev. Energy* 3:395–475 (1978).
14. Lee, W. Y., A. Morris and D. W. Boatwright. "Mexican Oil Spill: A Toxicity Study of Oil Accommodated in Seawater on Marine Invertebrates," *Mar. Poll. Bull.* 11:231–234 (1980).

CHAPTER 7

INVESTIGATION OF THE TRANSPORT AND FATE OF PETROLEUM HYDROCARBONS FROM THE IXTOC-I BLOWOUT IN THE BAY OF CAMPECHE— SAMPLING AND ANALYTICAL APPROACHES

Paul D. Boehm, David L. Fiest, Keith Hausknecht,
Jack Barbash and George Perry

(ERCO) Energy Resources Co. Inc.
Environmental Sciences Division
Cambridge, Massachusetts 02138

INTRODUCTION

The fate of components of spilled oil, especially the subsurface transport and physical and chemical fractionations and reactions that are associated with interactions of petroleum hydrocarbons (PHC) and the marine environment, are closely linked to the biological impact of these pollutants. During an oil spill caused by a tanker mishap or an offshore blowout, scientists seek information not only on the concentrations of bulk petroleum in the water column and other compartments (e.g., sediments) but on the distribution of toxic petroleum constituents in these compartments. Some of this information is needed on a semi-real-time basis (i.e., aboard ship), and other chemical information can be generated on a longer time scale and using sophisticated instrumentation in the laboratory. An assessment of the concentrations, compositions and three-dimensional distribution of petroleum's organic compounds is needed during the acute phases of the spillage event to determine environmental exposure levels and concentration gradients. Predictions of the

129

"potential environmental impact" of petroleum spillages must link the actual spillage data to a wealth of information from laboratory exposure studies through realistic dilution factors based on actual field measurements. By necessity, samples of seawater, dispersed oil, dissolved and particulate hydrocarbons, and actively sedimenting oil must be obtained under heavy surface oil concentrations to examine subsurface plumes and dispersions of oil. Such samples, by their very nature, are subject to contamination and cross-contamination.

Although difficult to accomplish, it is extremely important to sample both particulate and dissolved PHC during a spillage/blowout event, and to measure levels of individual hydrocarbon components in each phase. Obtaining and analyzing only whole seawater for PHC analysis without fractionation may obscure important physical chemical fractionations [1] and, hence, the physical/chemical behavior [2, 3], and biological effects or bioavailability [4] of pollutant compounds. Sampling beneath an oil-covered (oil, emulsified oil (mousse), surface films and sheens and tar flakes, chips, tar balls, etc.) sea surface without contaminating the seawater being collected has been a challenge to chemical oceanographers for years. Sampling devices themselves are notorious for either contaminating the sample or sorbing components of interest, thus altering the samples obtained [5-7].

Longer-term assessments of the ambient levels of petroleum transported to the benthos are required in conjunction with biological studies to determine the extent of long-lived contamination in the region of spillage and along the path of movement of the spilled oil.

A continuous point discharge of oil into the marine environment (i.e., a seabed blowout) offers unique opportunities to study many of the myriad of processes—chemical, physical, and microbiological—that act on the complex mixture of hydrocarbons and polar nitrogen/sulfur/oxygen (NSO) compounds that comprise petroleum once it is released to the marine environment [8]. Equally as important as the analytical techniques ultimately utilized are the sampling strategies and devices employed to obtain samples which enable the behavior of hydrocarbon constituents to be studied on a detailed basis.

The continuous seabed blowout event of the Ixtoc I drilling platform in the Bay of Campeche introduced at least 140 million gallons (3.5 million bbl $\cong$ 500 thousand metric tons) of oil into the offshore waters of the Bay of Campeche over a 10-month period [9].

Our studies were undertaken as part of a larger effort to examine the transport processes affecting petroleum introduced from the Ixtoc I blowout in the Bay of Campeche into the offshore marine environment. Our goals were: (1) to quantify the high-molecular-weight hydrocarbons in surface oil, the surface microlayer and in particulate, dissolved and whole water column samples [10, 11]; (2) to determine the detailed changes in hydrocarbon chemistry and hence "weathering" of the hydrocarbons in the samples [10-12]; (3) to examine vertical and horizontal concentration gradients and compositional changes; (4) to examine the compositional relationships and, hence, the relative rates of weathering between petroleum hydrocarbons in the water column and oil/mousse and sheen (microlayer) at the sea surface [10, 13]; and (5) to examine the transport of petroleum to the benthos [14].

This chapter focuses on our approaches to obtaining information on four major investigative areas, all dealing with aspects of dynamic oil spill chemistry. Extracting information from suites of environmental samples requires not only specialized sampling techniques and strategies, but an appropriate mix of analytical techniques as well. We discuss sampling methods and the analytical techniques, and summarize the findings of each segment of the investigation to illustrate the types of information obtainable by various combined sampling/analytical techniques.

GENERAL DESCRIPTION OF SAMPLING AREA

Two research vessels, the R/V Researcher (NOAA) and the C/V Pierce (Tracor Marine), operated in the vicinity of the Ixtoc I blowout between September 14 and 23, 1979. Samples of surface oil, subsurface water, sedimenting material and surface sediment were collected at various stations (Figure 1) along a 100-km transect oriented to the northeast of the blowout along the axis of surface plume movement and at a series of stations more distant from the blowout along the previously observed path of the oil as it moved toward the Texas coast.

STUDY AREA I: Determination of Concentrations, Compositions, Physical-Chemical State, and Geographic Distributions of PHC Beneath Surface Oil Layers

Sampling and Processing Methods

The dynamic state of seawater organic chemistry during petroleum spillage or blowout events dictates that chemical information be obtained in the water column beneath surface oil concentrations. Sampling devices and logistics associated with shipboard sampling of the water column must pre-

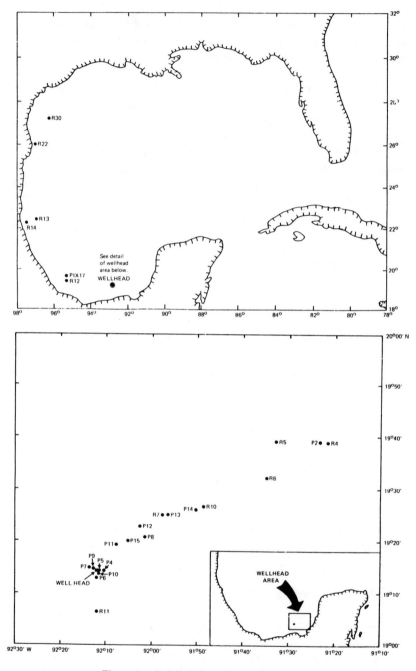

Figure 1. Detailed maps of sampling areas.

clude sample "contamination" by surface oil under which samples are being obtained. In addition, cross-contamination between samples taken at different depths must be avoided if subtle depth-dependent chemical variations are to be revealed.

There are two major choices as to type of sampling gear: (1) discreet samplers which enter through the surface in a closed mode and are opened at depth, (2) continuous pumping systems to obtain discreet samples for detailed analyses, or continuous information to "map" subsurface oil distributions. Samplers in the first category have been used successfully (e.g., Bodman-type samplers [1, 15-17], bacteriological "butterfly" samplers [18], drop samplers [19, 20], but are limited to spot-sampling and limited by time-consuming retrieval and redeployment operations.

A continuous sampler was chosen as the primary sampling device for this study to maximize the flexibility of the sampling operation and to enable both continuous monitoring (fluorometry) and discreet sampling of whole water or filtered water/particulate combinations to occur concurrently. Continuous towed fluorometers have been used in recent years to obtain qualitative maps of subsurface petroleum distributions [18]. However, only continuous sampling devices which bring water to the surface are useful for detailed chemical studies [20].

A combination of discreet and continuous samplers was actually employed. Discreet water samples from type 1 samplers [a 10-liter Teflon®* lined GO-FLO sampler (General Oceanics), a 30-liter glass Bodman sampler and a 90-liter aluminum Bodega-Bodman sampler] were obtained outside the areas of heaviest surface oil coverage, to supplement samples drawn under heavy oil slicks from the continuous pumping system. Our pumping system consisted of a modified submersible pump (Cole Parker Model 7111 with nylon impeller, silicone rubber gaskets, viton seals) connected to an in-line filtration system with alternate 2-m sections of stainless steel tubing (1.3 cm o.d. Type 903) and 0.5-m sections of flexible stainless steel couplings (Figure 2). The sections were alternately coupled with stainless steel unions and stainless steel quick connects (Swagelock with Viton seals) to allow water to be pumped from depths of 1 to 20 m.

The pumping system was introduced in the water column in a clear area of sea surface and remained submerged until pumping was completed at a given station. The system was flushed at each sampling depth for several minutes before measurements and samples were taken. The outflow from the pumping system was split into two streams, one for collecting whole water samples and the other for filtering and measuring fluorescence. The filtration system

---

*Registered trademark of E. I. du Pont de Nemours & Company, Inc., Wilmington, Delaware.

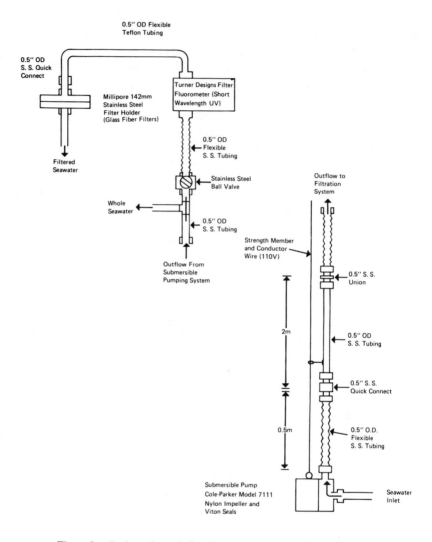

Figure 2. Design schematic for the submersible pumping system.

stream flowed through an in-line stainless steel ball valve which controlled the flow, a short-wavelength filter fluorometer (Turner Designs) which measured concentrations of oil (254 nm excitation, 360 nm emission) and a 142-mm stainless steel filter holder (Millipore) with a glass fiber filter (Gelman A/E) which removed particulates. The typical flow rate through the system was 1 liter/min, which corresponded to a linear velocity of 8 cm/sec and a flushing time of 3 min when pumping from a depth of 20 m. Seawater contacted only nylon, viton, silicone rubber, stainless steel and Teflon®.

Sample Analysis

A large number of 1-liter water column (whole water) samples were obtained for analyses by synchronous scanning spectrofluorometry (SSS) and a smaller number of large-volume (30 liters) samples obtained for chemical analysis by glass capillary gas chromatography (GC^2) and GC^2/mass spectrometry (GC^2/MS).

The 1-liter samples were extracted and analyzed aboard the research vessel [11] and the larger samples were preserved with dichloromethane, transported to the land-based laboratory and extracted there.

Synchronous Scanning Spectrofluorometry

SSS measurements on whole water hexane extracts enable one to examine the relative concentrations of petroleum hydrocarbons in the water column while at sea. Once back in the laboratory, absolute concentrations can be assigned based on accurate calibration of the method. In addition, gross qualitative differences between samples, denoting physical/chemical differences and fractionation, can be ascertained using SSS. These differences are revealed by examining spectral maxima (see below).

The solvent extracts were dried over sodium sulfate, transferred to a Kuderna-Danish apparatus and concentrated to 1 mL. The dichloromethane was displaced by repeatedly adding hexane (Burdick and Jackson, UV grade) and evaporating under a stream of purified nitrogen. The hexane extracts were analyzed for PHC using a synchronous spectrofluorometric technique [21, 22]. In summary, a measured aliquot of the sample extract was dissolved in a known volume of hexane. The intensity of the fluorescence emission was measured from 250 to 500 nm while synchronously scanning at an excitation wavelength 25 nm shorter than the wavelength at which the emission was measured. This technique measures aromatic hydrocarbons with a 2- to 5-ring aromatic structure [23]. The analysis was done on board the R/V Researcher using a Farrand Mark I spectrofluorometer equipped with corrected excitation and emission modules. The instrument conditions were as follows: excitation slit, 2.5 nm; emission slit, 5.0 nm; scan speed, 50 nm/min; sample cell, single 10-mm nonfluorescing quartz cell.

The intensities of the fluorescence spectra were measured at several wavelengths which correspond to peak maxima present in an Ixtoc I reference oil sample. The fluorescence spectra were converted to relative concentration units by comparing the peak height at 312 nm to that of a No. 2 fuel oil standard (API Reference No. 2). The No. 2 fuel oil is composed primarily of 2-ring aromatics, which are responsible for the fluorescence in the 312-nm region of the spectra.

The concentrations of oil in No. 2 fuel oil equivalents were converted to absolute concentrations by multiplying by a factor of 2.30. This factor was determined from a linear regression of oil concentrations in No. 2 fuel oil equivalents versus concentrations measured by microgravimetry using a Cahn electrobalance back in the land-based laboratory. The samples used for this calibration had gravimetric concentrations of oil which ranged from 74 to 1700 μg/L by gravimetry. The fluorescent material in samples with low concentrations (<20 μg/L) differs chemically from the material in the samples used for the regression. Although a lower conversion factor should have been used because of this discrepancy, none is available from existing data and the same conversion factor was used for all calculations.

Microgravimetry

Hexane extracts of large-volume water samples were fractionated by silicic acid column chromatography prior to GC^2 and GC^2/MS analyses.

The sample extract was charged to a glass chromatography column (10.5 mm i.d.) wet-packed with 11 g of silica gel (Davison 923; 100–200 mesh, 100% activated), 1 g of alumina (Fisher; 80–200 mesh, 5% deactivated), and 2 g of activated copper powder (Matheson; activated by rinsing with dilute HCl, organic-free water, methanol, dichloromethane and hexane). Both alumina and silica were successively extracted for 24 hr with methanol and dichloromethane, air-dried, and activated for 16 hr at 140°C prior to use. The sample was charged to the column with 2 × 0.5 mL of hexane and successively eluted with 17 mL of hexane (f_1), 21 mL of 1:1 hexane:dichloromethane (f_2) and 25 mL of methanol (f_3). The f_1, f_2 and f_3 fractions contain saturated, aromatic and polar compounds, respectively. The solvent in each fraction was concentrated and displaced with dichloromethane. Microgravimetric determinations were made on each fraction to determine gross PHC levels.

GC^2 and GC^2/MS

The f_1 and f_2 fractions were then analyzed by a combination of GC^2 and GC^2/MC analysis. A 1.0-μL aliquot of a dichloromethane solution of the sample was injected onto a 0.25-mm-i.d., 30-m SE-30 fused silica glass capillary column (J&W Scientific; AA grade) installed in a Hewlett-Packard 5840A gas chromatograph equipped with a splitless capillary injection port and a flame ionization detector and interfaced to a PDP-10 computer. Column conditions were as follows: carrier gas flow, 1 mL/min of helium; injection port temperature, 250°C; detector temperature, 300°C; and column oven temperature program, 60–275°C at a rate of 3°C/min.

Retention indices were calculated by comparing retention times of the peaks in each sample with the retention times of n-alkanes in a standard mixture which was analyzed daily. Peaks in the chromatograms of samples were identified by comparing their retention indices with retention indices of known compounds. These indices were measured by analyzing solutions containing known compounds or by identifying unknown peaks by GC^2/MS. The concentration of each peak was determined by comparing the area of the peak with the area of the internal standard, which was androstane for the f_1 and deuterated d_{10}-anthracene for the f_2.

Most of the f_2 fractions and selected f_1 fractions were analyzed on a Hewlett-Packard 5985 GC/MS system, which consists of a 5840A gas chromatograph, a quadrapole mass spectrometer, and a computer data system. The GC^2/MS conditions were the same as for the GC^2 analysis. The GC column was directly interfaced to the mass spectrometer. The mass spectrometer conditions were as follows: ionization voltage, 70EV; scan 46–500 AME, scan rate, 1 scan/2 sec; and electron multiplier voltage 2200–2600 V. The mass spectrometer was tuned daily with PBTBA.

Aromatic hydrocarbon compounds were identified and quantified by a combination of selected ion searches for either the molecular ion (m+) or another characteristic ion of each compound, retention index matching and matching of compound spectra with library spectra. The total ion current recorded for each peak in the selected ion search was compared to the area of the internal standard, d_{10}-anthracene. Response factors relative to d_{10}-anthracene were calculated from a regular analysis of a mixture of authentic aromatic compounds. When no standard was available—as was the case for C_4–naphthalene; C_3, C_4–phenanthrene; C_2, C_3–fluorene and C_1 through C_3 dibenzothiophenes—the response factor was obtained by extrapolation.

Results

SSS

Spectrofluorometric results calibrated with microgravimetry allowed for the determination of concentrations of oil on a large number of samples, thus resulting in cross-sectional maps of subsurface oil concentrations.

The concentrations of oil in the water column ranged from values of < 5 $\mu g/L$ at a distance of 80 km from the blowout to peak values of 1700 $\mu g/L$ within a few hundred meters of the blowout. The highest concentrations were observed within 25 km of the blowout in the top 6 m of the water column. The higher values reported here may be an underestimate since some oil was visibly adsorbed onto the walls of the bottle samplers during sampling in the plume.

The concentration data are summarized in Figure 3, which is a contoured vertical cross section of the oil concentrations along the plume axis. Elevated concentrations of oil occurred 40 km to the northeast of the well. At distances >40 km to the northeast and 2 km to the south and west, concentrations were, with a few exceptions, <5 μg/L. The northeastern orientation of the oil-contaminated seawater plume coincided with the observed direction of movement of the surface plume of oil. However, emulsified oil was observed floating on the ocean surface at distances greater than the apparent extent of the oil-contaminated seawater plume. Although surface oil was found 80 km or more to the northeast of the well, elevated concentrations of oil in the water column were limited to within 40 km. It is apparent that dissimilar processes are controlling the transport of surface and subsurface oil.

Several distinct spectral patterns were observed among the samples collected (Figure 4). Samples containing low concentrations ($<5\mu$g/L) had a spectrum with a single fluorescence peak at 308 nm (Type A). This spectrum results from either background fluorescent material in seawater or low level contaminants from the sample workup. Samples with concentrations from 5 to 20 μg/L had a single peak spectrum with a peak maximum at 312 nm (Type B). This peak results from a predominance of petroleum-derived 2-ring aromatics which fluoresce from 310 to 330 nm [23]. As discussed below, this spectral type reflects the selective dissolution by seawater of 2-ring aromatics from the whole oil released from the blowout. Spectral Type D is characterized by a series of fluorescent peaks at 312, 328, 355 and 405 nm. This spectrum was predominant for samples with concentrations >20 μg/L. The series of peaks results from 2-, 3-, 4-, 5- and larger-ring polycyclic aromatic compounds [23]. Type D spectra were similar to spectra of the whole oil collected from surface mousse samples.

At a few stations, samples of both whole seawater and filtered seawater through a 0.45-μm glass fiber filter from the same depth were analyzed. At stations with low concentrations of oil, no systematic differences between the filtered and unfiltered samples were found. At stations with moderate to high PHC levels, the spectra of the filtered water sample were depleted in the 3- to 5-ring region compared to the unfiltered sample.

The distinction between the three spectral types was confirmed by glass capillary gas chromatography analysis. The saturated (f_1) and unsaturated (f_2) fractions of samples with Type D, whole oil, spectra contain petroleum hydrocarbons in a boiling range equivalent to that of $<$n-C_{10} to n-C_{30}. The glass capillary gas chromatogram of the f_2 confirms the presence of polycyclic aromatic hydrocarbons (PAH) with 2-5 rings. Normal alkanes from $<$n-C_{10} to n-C_{34} and a low-boiling unresolved complex mixture predominate in the f_1. Samples with Type B, dissolved oil, spectra contain

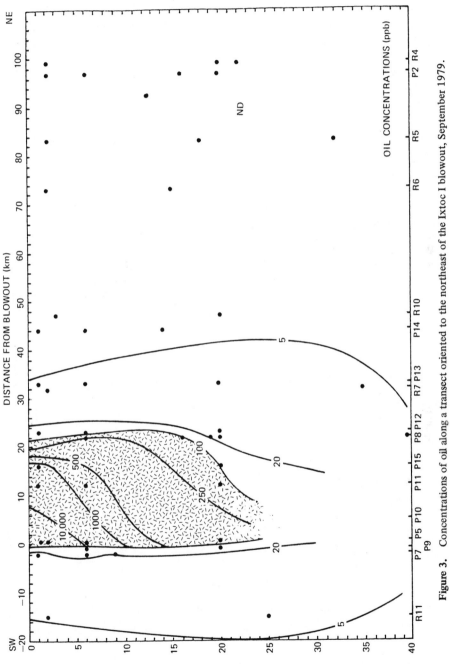

Figure 3. Concentrations of oil along a transect oriented to the northeast of the Ixtoc I blowout, September 1979.

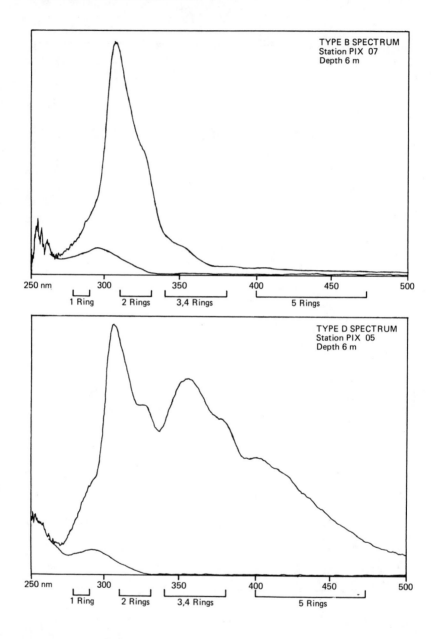

Figure 4. Representative synchronous fluorescence spectra of seawater samples collected near the Ixtoc I blowout.

predominantly substituted 1- and 2-ring aromatic hydrocarbons. Relatively small amounts of PAH with more than 2 rings and saturated hydrocarbons are present. Samples with Type A spectra contained very low amounts of material.

GC^2 and GC^2/MS

The detailed nature of data generated by these techniques allowed for the determination of levels of individual hydrocarbon components and, hence, much information on the physical/chemical fractionation between particulate and dissolved fractions, and the physical/chemical and microbiologically induced compositional changes (i.e., weathering).

Three ratios of saturated and aromatic hydrocarbons were calculated from the analytical results of each sample. Comparison of these ratios between samples revealed the dynamic time-dependent changes in the composition of the PHC assemblage.

The alkane/isoprenoid ratio (ALK/ISO) was used to measure the extent of microbial degradation through the preferential depletion of n-alkanes relative to isoprenoids [24]. The numbers in parentheses refer to the Kovats retention index of the compound on a 0.25-mm $\times$ 30-m SE-30 column.

$$\text{ALK/ISO}_{14-18} = \frac{(1400) + (1500) + (1600) + (1700) + (1800)}{(1380) + (1470) + (1650) + (1708) + (1810)}$$

The ALK/ISO ratio approaches 0 as the n-alkanes are depleted.

The saturated hydrocarbon weathering ratio (SHWR) was used to measure the relative abundance of low-boiling and high-boiling n-alkanes.

$$\text{SHWR} = \frac{\text{(Sum of n-alkanes from n-}C_{10}\text{ to n-}C_{25}\text{)}}{\text{(Sum of n-alkanes from n-}C_{17}\text{ to n-}C_{25}\text{)}}$$

The SHWR approaches 1.0 as low-boiling saturated hydrocarbons (n-C_{10} to n-C_{17}) are lost by evaporation.

The aromatic weathering ratio (AWR) was used to measure the relative abundance of low- and mid-boiling aromatics.

$$\text{AWR} = \frac{\text{Total naphthalenes + fluorenes + phenanthrenes + dibenzothiophenes}}{\text{Total phenanthrenes + dibenzothiophenes}}$$

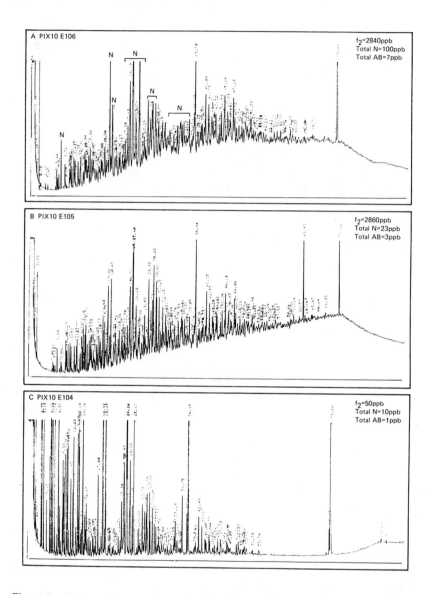

Figure 5. Glass capillary gas chromatograms of aromatic hydrocarbons in whole water (A), particulate (B), and dissolved (C) fractions from station PIX10 at 2 meters depth. (N = Naphthalenes; AB = Alkyl benzenes).

The AWR approaches 1.0 as low-boiling aromatics are lost by evaporation and/or dissolution.

Variations in the PHC composition of the different physical/chemical forms (i.e., particulate, dissolved) revealed that the dissolved fraction was much lower in overall concentration of PHC but much enriched in light aromatic hydrocarbons (alkyl benzenes, naphthalenes) (Figure 5). The saturated hydrocarbons showed little compositional variation between dissolved and particulate fractions, but were much reduced in concentration in the "dissolved" fraction. Both of these observations were predicted by laboratory studies [3, 25] but have previously eluded field detection.

Detailed compositional information revealed that—unlike spills in temperate and subarctic regions (*Amoco Cadiz* [18, 26]; *Tsesis* [1, 27])–PHC from the Ixtoc I were very slow to degrade through microbial processes. The ALK/ISO ratio, which would decrease as biodegradation proceeded, remained close to the wellhead oil value in water column samples from the study region.

GC^2/MS and GC^2 analyses also revealed compositional variations within the top 20 m of the water column. Closer to the wellhead ($\sim$10 km), the PHC in the top 20 m were compositionally invariant with the key aromatic (AWR) and saturated (SHWR) ratios fairly uniform in the water column (AWR = 2.9; SHWR = 1.8). However, at PIX 08 ($\sim$20 km from the wellhead), a discreet PHC layer located at 15-m depth, similar in composition (AWR = 2.5; SHWR = 1.8) to the wellhead oil (AWR = 2.7; SHWR = 2.5) and more similar to water column PHC compositions at PIX 11 ($\sim$10 km from wellhead) was observed. This evidence and that from the direct measurements of subsurface particulate concentrations from the acoustical reflectance data obtained by Walter and Proni (unpublished data) and the fact that this material could not be related to surface oil from the same station (AWR = 1.2; SHWR = 1.1), strongly suggest the existence of a subsurface, horizontally advected plume of petroleum. This material was effectively cut off from the surface and hence was not weathered evaporatively as was the oil transported to the same location on the sea surface. Detailed GC^2 and GC^2/MS were invaluable for revealing this important phenomenon.

STUDY AREA II: Determination of Weathering of Surface Oil Transported Away From the Wellhead

Sampling Methods

Unlike difficult water column sampling, obtaining samples of surface oil requires nothing more than a bucket if one wants to grab a gross surface oil sample. This method was employed here. This crude method, however, was supplemented by sampling of the fine structure of thick surface water-in-oil emulsions (mousse), often 6-12 in. thick, from a zodiac boat. Sampling from the water's surface, as opposed to sampling from a large ship, allowed us to obtain small tar particles and to examine the composition of sheens (i.e., the microlayer [10]) and mousse "flakes and skin" [12] relative to bulk surface mousse. Microlayer samples were obtained with a stainless steel, 16-mesh (1-mm) screen [28].

Sample Analysis

In order to examine detailed and often subtle chemical differences between samples, GC^2 and GC^2/MS proved essential.

Oil/mousse samples were extracted with dichloromethane dried over sodium sulfate and an aliquot of the dried solution was fractionated and analyzed by GC^2 and GC^2 MS as previously described.

Findings

These detailed analyses generated a wealth of data and not only enabled us to follow the weathering history of the Ixtoc I oil as it journeyed toward United States coastal waters, but enabled us to compare the different compositions of surface oil and subsurface water column PHC, and to mathematically model these changes as well [13].

The saturated and aromatic compositions changed mainly by physical/chemical weathering including photooxidation with loss of n-alkanes from n-C_{10} to n-C_{22} (Figure 6A) and volatile/soluble aromatics (Figure 6B). The high values of the ALK/ISO ratio, usually > 3.0, indicate that little microbial degradation of the oil had taken place.

An exception was a sample obtained from a tar ball collected in windrows of plant debris at Station RIX 13. The tar ball was extensively weathered by microbial and evaporative processes (Figure 7). N-alkanes were virtually absent (ALK/ISO = 0.2), and isoprenoid peaks and an unresolved envelope dominated the f_1. The aromatics in the f_2 were similar to those of mousse samples in which no microbial degradation had occurred.

The surface microlayer samples were chemically distinct from any of the oil/mousse forms. The low SHWR and AWR values (~ 1.0) for all of the microlayer PHC indicated that these sheens represent an advanced weathering state, compositionally depleted in lower-boiling alkanes and aromatics. The composition of the "freshest" microlayer PHC resembled that of the most weathered bulk oil/mousse.

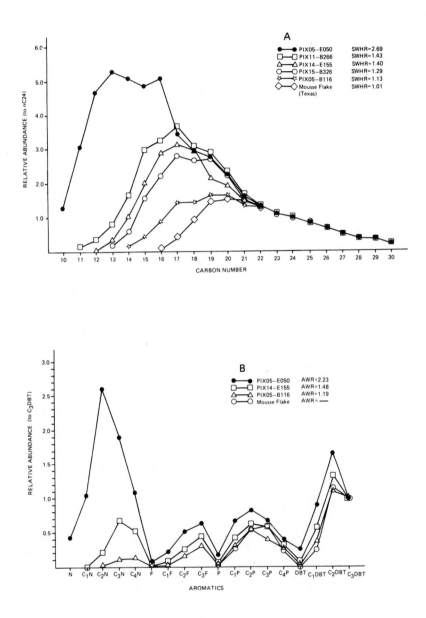

Figure 6. Composition of saturated (A) and aromatic (B) hydrocarbons from surface oil samples illustrating weathering histories of samples.

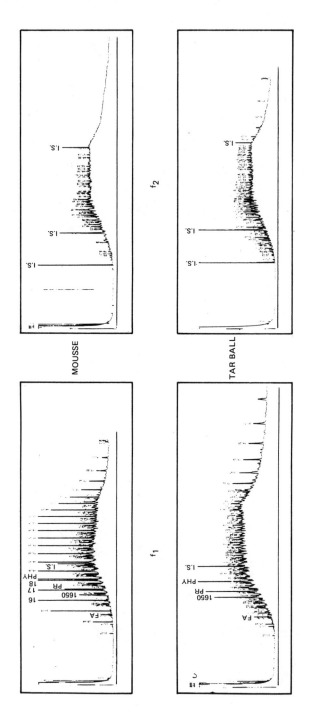

Figure 7. Comparison of mousse and tar ball samples collected at Stations PIX17 and RIX13, respectively, showing extensive biodegradation of the latter sample (FA, 1650, PR, PHY = Isoprenoid hydrocarbons).

STUDY AREA III: Determination of Transport of Oil
to the Benthos via Sedimentation Processes

Prior to the present study, there had been few studies directly pertaining to the transport of oil to the offshore continental shelf benthos via the important phenomena of adsorption of oil on living or detrital particulate matter followed by sedimentation to the benthos. An evaluation of the extent of this process is extremely important in order to predict the exposure of important benthic resources to petroleum hydrocarbons released from offshore blowouts. This process is dependent on the availability and concentration of suspended particulates and their surface area [8, 29-31]. Another possible route of transport to the benthos is by ingestion of oil by zooplankters followed by fecal pellet transport [32]. These two processes are those most likely to result in water column-to-benthos transport of petroleum hydrocarbons in continental shelf environments.

McAuliffe et al. [33] have associated the spilled oil in sediments in the vicinity of the Chevron platform blowout at the mouth of the Mississippi River with sorption and sedimentation processes. Boehm et al. [24] have examined the detailed chemistry of sedimenting oil captured in sediment traps deployed during the Tsesis tanker spill in Sweden. They found that microbial degradation caused rapid alteration of the chemical composition of the spilled cargo, and that the hydrocarbon composition of benthic deposit feeders *(Macoma balthica)* echoed this composition. Johansson et al. [27] estimated that 15-20% of the oil spilled during the Tsesis event was transported to the benthos by sorption and sedimentation and/or by ingestion and zooplankton fecal pellet transport.

Sampling

Although several types of sophisticated sediment traps or particle interceptors have been designed, we required a series of inexpensive, rapidly deployed and retrieved traps constructed of easily obtainable material to capture representative, actively sedimenting material in a semiquantitative manner. Part of the sampling rationale was to test the usefulness of rapidly deployable traps for determining surface oil-to-benthos coupling. Consequently, sediment traps were constructed quite simply by attaching wide-mouthed jars to a weighted Kevlar-coated hydrowire at several depths (2.5, 5 and 15 m) along the wire. Duplicate traps were deployed from the C/V Pierce at three stations near the wellhead along the observed direction of movement of the surface oil as well as at one control station. The traps remained deployed for approximately 8 hr, at which time they were retrieved. During retrieval the trap strings were slowly brought on board and each jar capped before breaking the sea surface. The jars' contents were preserved with methylene chloride and frozen at $-10°C$ until analyses were begun.

Sample Analyses

The overall analytical strategy was geared to determining the comparative detailed hydrocarbon chemistries of samples by a combination of GC^2 and GC^2/MS. The small amount of material (most often 10–100 mg) trapped precluded sacrificing any material to other analytical techniques (e.g., SSS).

Sediment trap samples consisted of a 500-mL jar of seawater with 0 to ~50 mg of sediment at the bottom. To lower the analytical detection limit and to avoid handling losses, the three to four jars comprising each trap array string were combined and treated as water samples (i.e., added to a separatory funnel and extracted vigorously with three 100-mL portions of methylene chloride). The extracts were treated as were the other PHC extracts (fractionation, GC^2, GC^2/MS).

However, unlike the oil or high-level water column samples, the low levels of PHC expected in trapped material dictated a vigorous routine of laboratory cleanliness and careful monitoring of procedural blank levels. The striking differences in the nature of these two sample sets point to the heart of both the shipboard and laboratory quality control program which must allow for determination of trace levels of PHC in samples in close proximity to PHC gross contamination.

Findings

Several of the sediment trap composite samples analyzed contained significant traces of what is believed to be actively sedimenting petroleum hydrocarbon-bearing particulates. Both trap arrays deployed in the surface slick plume contained detectable saturate hydrocarbon material primarily of a petroleum origin and likely associated with Ixtoc I oil. The nature of the captured material, as revealed by GC^2, appears to consist of undegraded, physically/chemically weathered (primarily evaporation) Ixtoc I oil. The n-alkane distribution appears quite similar to the filtered particulate oil and to the oil in sediments from the immediate area. Two noteworthy differences are apparent. The sedimentary petroleum hydrocarbon distribution (see below) appears to be richer in naphthenic (cyclic saturated) material in the unresolved complex mixture than the trapped particulates, presumably due to the lack of observed biodegradation in the water column and its increased role in the benthos. Secondly, the presence of phytoplanktonic material in the traps is signified by the prominence of n-C_{15} (pentadecane) and n-C_{17} (heptadecane) in the saturate capillary GC trace [34]. Filtered particulates in control areas contain phytoplanktonic hydrocarbons as the primary hydrocarbon input to the suspended particulate fraction. However, the sediments containing petroleum appear to be influenced by a combination of petroleum and terrigenous biogenic inputs ($n-C_{23}$, $n-C_{25}$, $n-C_{29}$, $n-C_{31}$).

That the actively settling material is undegraded seems remarkable in light of findings on *Tsesis* [24] and Amoco Cadiz spills [18] which both indicated that rapid biodegradation alters the composition of oil shortly after introduction into the marine system. However, findings of this work indicate that rapid biodegradation in some environments, perhaps limited by available nutrients, cannot be assumed to occur. Results from the sediment trap samples as revealed by GC^2 indicate the mechanism of transport of oil to the benthos appears to depend on the association of dispersed oil with phytoplanktonic material by adsorptive processes.

The absolute concentration of petroleum on the trapped hydrocarbon-bearing particulates is difficult to determine. The weight of trapped material was not measured due to the fear of contamination and losses. However, we estimate that particulates in the traps ranged from 10 to 50 mg. Absolute quantities of oil in the traps ranged from 50 to 250 μg (8 hr capture). Therefore, in those traps containing petroleum, the particulates may contain on the order of 1–25 μg petroleum hydrocarbons/mg. Assuming a 150-cm^2 cross-sectional area of capture (the area of the openings of three jars/traps), and assuming ideal capture, a vertical flux of 1–5 μg/cm^2/day of oil is estimated. Admittedly, much uncertainty is associated with these calculations.

STUDY AREA IV: Determination of Levels and Composition of Oil Transported to the Benthos (Surface Sediment)

Obtaining samples of surface sediment for PHC analysis to determine the extent and nature of contamination of the substrate is of a fundamentally different nature than water column samplings described previously. The goal of sediment sampling is to determine the levels, areal extent and detailed chemical nature of the oil transported to the geochemical sink. We are interested in a chemical distribution over a larger time scale, so the urgency of sampling is somewhat mitigated. Samples need not be obtained through heavy surface oil contamination because the surface water-column-benthos coupling is not necessarily precisely related to the direction of movement of surface oil. Indeed, the problem in offshore environments of locating oil in the benthos outside of the acute impact zone (~ 1 km) is that the oil in the benthos is initially distributed and redistributed by resuspension in unknown directions. Patchiness is likely to result. Furthermore, the newly deposited oil resides in a mobile flocculent layer [24, 35] of millimeter dimensions. Most conventional samples acquire samples of a sediment layer 5-10 cm in thickness.

Thus the problems are twofold: locating and quantifying the oil chemically in a relatively inexpensive manner (i.e., screening), and sampling in such a manner so as to preserve the flocculent layer.

Sampling

Ideally, sampling should be performed by divers who can preferentially obtain true surface sediment. In the depths of water encountered this was not feasible. A modified Box-corer (Soutar) was used, but sample penetration was poor due to the sandy nature of sediment texture. Finally, a conventional Smith-MacIntyre grab sampler was used successfully. Replicate sampling and sample pooling were performed at all stations to minimize potential effects of patchiness. The top several centimeters of sediment were isolated from each grab and the replicates composited back in the laboratory prior to analysis. We felt that a 2-3 cm sample would represent a reproducible subsample, allow for any bioturbation that could mix oil down into the sediment, and not significantly dilute any millimeter-thick surface PHC concentrations.

Sample Analyses

The overall analytical strategy was geared to selecting those samples containing Ixtoc I oil and then determining the comparative detailed hydrocarbon chemistries (saturates: $n\text{-}C_{10}$ to $n\text{-}C_{34}$; aromatics: alkylbenzenes to benzopyrenes) of samples by a combination of analytical techniques: GC^2 and GC^2/MS.

SSS

The hydrocarbon compositions of the surface sediments were screened by SSS [11, 21, 36]. Approximately 3 g dry weight of sediment was transferred to a centrifuge tube, solvent was added and the tube agitated using a vibrating mixer for several minutes. Different solvent mixtures (hexane; methylene chloride/methanol; hexane/methylene chloride) were tried to compare extraction methods. Hexane compared quite favorably to methylene chloride mixtures in its ability to qualitatively extract fluorescent compounds from wet sediment and was subsequently used in all screenings to address the qualitative sedimentary hydrocarbon distributions. Hexane extracts were analyzed directly on a Farrand MK-1 scanning spectrofluorometer by offsetting the excitation and emission monochronometers 25 nm, scanning both simultaneously and monitoring the resultant emission spectrum of the sample. Families of aromatics [23] are revealed by this method. Twelve samples were chosen on the basis of their fluorescence spectra for more detailed GC and GC/MS analyses. Those samples whose spectra revealed the probable presence of Ixtoc I oil or those whose spectra illustrated a "typical background" distribution of fluorescent polycyclic aromatic hydrocarbons were chosen for further scrutiny.

GC^2 and GC^2/MS

The solvent extraction, silica gel column chromatographic fractionation, GC^2, GC^2/MS and quantification procedures are described in detail elsewhere [26, 37, 38]. Briefly, 50-100 g of wet sediment is added to a Teflon® canister and internal standards are spiked to the sediment. The sediment is dried by shaking with methanol, extracted by a methylene chloride: methanol azeotrope (9:1); the methanol is back extracted with methylene chloride; all extracts are combined, reduced in volume, displaced with hexane and charged to a silica gel/alumina column; two fractions (f_1 = hexane eluate = saturated hydrocarbons; f_2 = hexane/methylene chloride eluate = unsaturated = aromatic and olefinic hydrocarbons) are eluted.

The weights of each of the fractions (f_1 and f_2) as determined by microgravimetry yields the absolute hydrocarbon levels. GC^2 analyses reveal the composition of the hydrocarbon assemblage.

Findings

SSS

The SSS determinations provided a rapid and clear picture as to the basic nature of the hydrocarbon composition of the surface sediments. The spectra fall in three nontrivial (nonzero) categories illustrated in comparison to a reference oil-in-water dispersion sampled at the wellhead (Figures 8A-E). The spectra of Figure 8B is characteristic of all sediment samples examined from Stations RIX 07 and RIX 10, ~30 and 50 km from the wellhead, respectively, and RIX 30, in deep water off Texas. These spectra compare quite favorably to the reference oil with wavelength maxima at 312 and 350 nm and significant shoulders at 322 and 400 nm. The 312- and 322-nm responses are attributable to families of 2-ringed aromatic hydrocarbons; the 350- and 400-nm maxima are due to 3-ringed compounds and minor amounts of 4-ringed aromatic structures [23]. The 312- and 322-nm peaks are depleted relative to the presumed source material (Figure 8A), undoubtedly due to weathering of the PHC mixture.

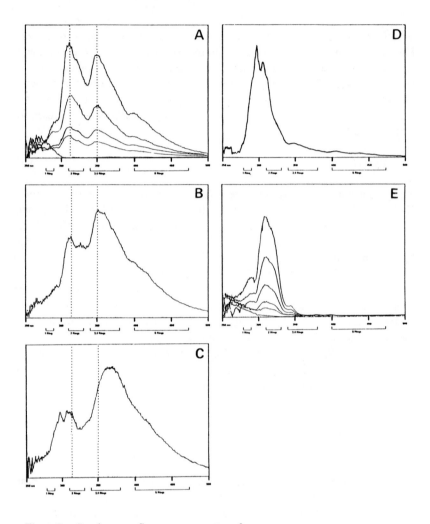

Figure 8. Synchronous fluorescence spectra of:
  A. Reference Ixtoc oil
  B. Sediment hydrocarbons resembling A
  C. Other anthropogenic polycyclic aromatic sources in sediments
  D. Sediment hydrocarbons containing primarily two-ringed aromatic compounds
  E. No. 2 Fuel Oil

The second type of spectrum was observed in sediment samples from Station RIX 12 off Veracruz (Figure 8C). The prominent multiple maxima between 290 and 315 nm are quite distinct and represent several biaromatic responses. The broad band and shoulders from 345 to 450 nm with a maximum at 370 nm are attributable to a set of 3- and 4-ringed compounds other than that associated with the spilled oil and hence another pollutant source.

The third spectral type, illustrated in Figure 8D, is characteristic of a prominent set of 2-ringed aromatic compounds as revealed by two spectral maxima at 295 nm and 305 nm. A small degree of triaromaticity is revealed by a small band centered at 345 nm. These spectra common to RIX 14 sediments north of Tampico are quite unlike those from other stations examined from the study area and are most similar to a light distillate oil (e.g., No. 2; see Figure 8E). However, GC^2 and GC^2/MS data do not reveal the presence of any marked distillate pollution; therefore, the origin of this distribution is unknown. Sample contamination is possible but unlikely in view of the fact that two of the three samples from RIX 14 examined, and no others, contain this unique distribution of fluorescent compounds.

Microgravimetry and GC^2/MS–Quantitative Aspects

The hydrocarbon content (f_1 and f_2) of sediments from the wide geographic region examined ranged from 15.1 μg/g at Station RIX 14 to 143.6

Table 1. Surface Sediment Data Summary

	Hydrocarbon Content (μg/g) (Gravimetric)			Composition as
Station	f_1	f_2	Total	Determined by GC[a]
RIX 04	5.4	19.2	25.2	B
RIX 07–1	11.4	18.8	30.2	I/B
RIX 07–2	26.3	54.0	80.3	I/B
RIX 10–1	33.6	110.0	143.6	I/B
RIX 10–2	23.5	38.0	61.5	I/B
RIX 12	8.3	16.4	24.7	T
	9.6	17.3	28.9	T
RIX 14–1	6.7	12.3	19.0	T/C
RIX 14–2	5.1	10.0	15.1	T/C
RIX 22	17.9	20.5	38.4	C/T
RIX 30	16.8	30.0	46.8	T/B/C/I
Blank	2	2	4	–

[a]I = predominantly *Ixtoc-I* oil in f_1; weathered aromatics and biogenics (olefins, sterenes) in f_2

B = biogenic inputs in f_1 and f_2

T = terrigenous biogenic inputs, (i.e., n-C_{23}, n-C_{25}, n-C_{29}, n-C_{31}, n-alkanes = vascular plant waxes)

C = chronic pollution; unresolved hump in f_1; pyrogenic polynuclear aromatic hydrocarbon inputs in f_2.

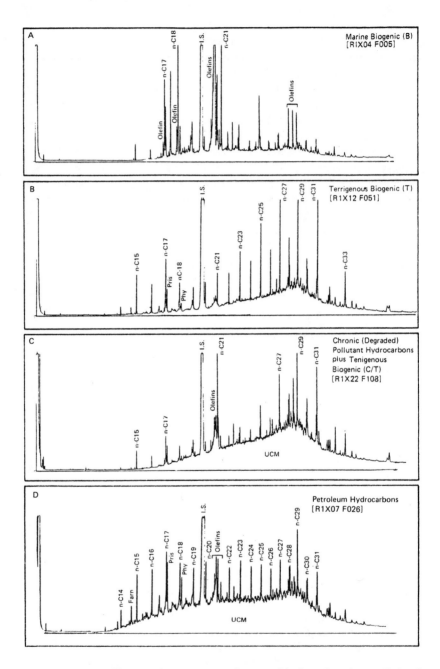

Figure 9. Glass capillary gas chromatograms of saturated hydrocarbons in nonoiled and oiled surface sediments.

μg/g at RIX 10 (Table 1). A direct relationship between absolute hydro-carbon levels and the presence of Ixtoc I oil does not seem to exist although the presence of petroleum hydrocarbons attributable to Ixtoc I was noted at Stations RIX 07 and RIX 10. The background distributions of hydrocarbons are related to the organic content of the sediments as determined by non-spill-related geochemical considerations (e.g. riverine influences).

The capillary GC traces (Figure 9) of hydrocarbons in the sediments illustrate the major compositional features referred to in Table 1. Background saturated hydrocarbon distributions are characterized by marine biogenic distributions (Figure 9A), terrigenous biogenic distributions (Figure 9B) and a combination of chronic pollution and biogenic inputs (Figure 9C). The unsaturated (f_2) fractions of most samples are composed of prominent sets of normal and branched olefinic compounds and compounds of the triterpene series. Several of the compounds reveal mass spectral fragments characteristic of this series (m/e = 191; MW 368, 380 [39, 40]) which probably originate in the marine microbiota [39]. Those sediment samples which do contain signi-ficant quantities of petroleum hydrocarbons (RIX 07, RIX 10) (Figure 9D) exhibit a moderately weathered n-alkane distribution overriding an unre-solved complex mixture (UCM) resulting from weathering of the saturate fraction, which visually enhances the UCM on the GC traces, and possibly results in the production of UCM compounds (cyclic saturated compounds) by marine bacteria [26]. That biodegradation is proceeding is evidenced not only by the UCM prominence, but also by the decreasing ALK/ISO ratios (1.1-1.9) for these samples.

The unsaturated fraction (f_2) of spill-impacted surface sediment contains major biogenic components as do the background samples. The petrogenic aromatic hydrocarbons are revealed only through GC^2/MS-generated re-constructed mass chromatograms. The quantitative results from the GC^2/MS examinations of the f_2 fractions are detailed elsewhere [14]. The prominent families of phenanthrene and dibenzothiophene homologues dominate the aromatic hydrocarbon distributions of RIX 07, RIX 10 and RIX 30. No naphthalene compounds were detected in the oil-impacted sediments. The phenanthrenes present when dibenzothiophenes and hence oil are absent are the unsubstituted parent compound with lesser amounts of alkylated phenan-threnes. The small amounts of phenanthrenes (1.0-3.0 ng/g) can be attributed along with the polynuclear aromatic hydrocarbon (m/e 202, 228, and 252) components, to pyrogenic sources (i.e., combustion of fossil fuels) [41] introduced to the sediment probably through fallout and riverine runoff. The sediments of RIX 22 and RIX 30 contain greater quantities of these indicators of pyrogenic sources most likely due to their proximity to the influence of the Rio Grande. RIX 30 also contains dibenzothiophene and alkyl phenanthrene PAH compounds which strongly suggest a petrogenic source. In addition, the synchronous spectrofluorometry spectrum of the RIX 30 sample suggests a similarity to Ixtoc I oil. However, other petroleum sources prevalent in the Texas Gulf Coast region (platform drilling discharges, tanker discharges) may result in similar distributions. Therefore, assignment of the trace levels of petroleum hydrocarbons found in RIX 30 sediments to any particular source is quite ambiguous.

A mass budget of oil found in the benthos indicates that a small percentage of oil is accountable for in the offshore sediments ($\sim$1-2%). This value includes material directly transported to the offshore benthos and does not include that material in the nearshore subtidal/offshore bar system transported as a result of landfall followed by beach erosion and offshore deposition. This latter sink for Ixtoc I oil is probably an order-of-magnitude higher than for the offshore direct water column-to-benthic transport.

ACKNOWLEDGMENTS

The authors would like to thank the scientists and crews of the R/V Researcher and C/V Pierce for their invaluable help in collecting and processing samples under the most difficult of sampling conditions.

We also acknowledge the valuable assistance of Dr. Ed Overton, Mr. Larry McCarthy, and associates at the University of New Orleans, Center for Bio-Organic Studies, for performing extractions on large-volume water samples and for acting as a reference/distribution library for mousse samples. Adria Elskus, John Yarko, Neil Mosesman, and Ann Jefferies assisted with phases of the analytical work.

Finally and perhaps most importantly we thank the support and efforts of Dr. Donald Atwood and Dr. George Harvey of NOAA/AOML, Dr. John Farrington of Woods Hole Oceanographic Institutiion, and Ms. Judith Roales and Dr. Lou Butler of NOAA/Office of Marine Pollution Assessment (OMPA), on behalf of the entire Ixtoc I research team. Our research was supported by a research contract No. NA80RAC00054 from NOAA/OMPA.

REFERENCES

1. Boehm, P. D. "The Decoupling of Dissolved, Particulate and Surface Microlayer Hydrocarbons in Northwestern Atlantic Continental Shelf Waters," *Mar. Chem.* 9:255–281 (1980).
2. Shaw, D. G., and S. K. Reidy. "Chemical and Size Fractionation of Aqueous Petroleum Dispersions," *Environ. Sci. Technol.* 13:1259–1263 (1979).
3. Zürcher, F., and M. Thüer. "Rapid Weathering Processes of Fuel Oil in Natural Waters: Analyses and Interpretations," *Environ. Sci. Technol.* 12:838–843 (1978).

4. Anderson, J. W., et al. "Characteristics of Dispersions and Water Soluble Extracts of Crude and Refined Oils and Their Toxicity to Estuarine Crustaceans and Fish," *Mar. Biol.* 27:75–81 (1974).

5. Gordon, Jr., D. C., and P. D. Keizer. "Hydrocarbons in Seawater Along the Halifax-Bermuda Section: Lessons Learned Regarding Sampling and Some Results," in *"Marine Pollution Monitoring (Petroleum),"* NBS Special Publication 409, Washington, DC (1974), pp. 113–116.

6. Boehm, P. D. and D. L. Fiest. "Analyses of Water Samples from the *Tsesis* Oil Spill and Laboratory Experiments on the Use of the Niskin Bacteriological Sterile Bag Sampler," Final Report, NOAA Contract No. 03–A01–8–4718, Boulder, CO (1978).

7. Zsolnay, A. "Caution in the Use of Niskin Bottles for Hydrocarbon Samples," *Mar. Poll. Bull.* 9:23–24 (1978).

8. National Academy of Sciences. "Petroleum in the Marine Environment." NAS, Washington, DC (1975), 107 pp.

9. Oil Spill Intelligence Report (OSIR). "Special Report: *Ixtoc I*," January 4, 1980 Newsletter, Center for Short-Lived Phenomena, Cambridge, MA (1980).

10. Boehm, P. D., and D. L. Fiest. "Subsurface Water Column Transport and Weathering of Petroleum Hydrocarbons During the *Ixtoc I* Blowout in the Bay of Campeche and Their Relation to Surface Oil and Microlayer Compositions," in *Proceedings of the Symposium on the Preliminary Findings of the Researcher Cruise to the Ixtoc I Blowout* (Rockville, MD (NOAA/OMPA, 1980).

11. Fiest, D. L., and P. D. Boehm. "Subsurface Distributions of Petroleum from an Offshore Well Blowout: the *Ixtoc I* Blowout, Bay of Campeche," *Environ. Sci. Technol.* (in press).

12. Patton, J. S. et al. "The *Ixtoc I* Oil Spill: Flaking of Surface Mousse in the Gulf of Mexico," *Nature* 290:235–238 (1981).

13. Boehm, P. D. et al. "Physical Chemical Weathering of Petroleum Hydrocarbons from the *Ixtoc I* Blowout. Chemical Measurements and a Weathering Model," in *Proceedings of the 1981 Oil Spill Conference* (Washington, DC: American Petroleum Institute, 1981).

14. Boehm, P. D., and D. L. Fiest. "Aspects of the Transport of Petroleum Hydrycarbons to the Offshore Benthos During the *Ixtoc I* Blowout in the Bay of Campeche," in *Proceedings of the Symposium on the Preliminary Findings of the Researcher Cruise to the Ixtoc I Blowout* (Rockville, MD: NOAA/OMPA 1980).

15. Bodman, R. H., L. W. Slabough and V. T. Bowen. "A Multipurpose Large Volume Seawater Sampler," *J. Mar. Res.* 19:141–148 (1961).

16. Gagosian, R. B., et al. "A Versatile Interchangeable Chamber Seawater Sampler," *Limnol. Oceanog.* 24:583–588 (1979).

17. Payne, J. R., et al. "Hydrocarbons in the Water Column. Southern California Baseline Study, Vol. III, Report 3.2.3., pp. 1–207." Final Report, submitted to the Bureau of Land Management, Washington, DC (1979).

18. Calder, J. A., J. Lake and J. Laseter. "Chemical Composition of Selected Environmental and Petroleum Samples from the *Amoco Cadiz* Oil Spill," in "The *Amoco Cadiz* Oil Spill: A Preliminary Scientific Report," W. N. Hess, Ed., NOAA/EPA Special Report (1978).

19. Gump, B. H., et al. "Drop Sampler for Obtaining Fresh and Seawater Samples for Organic Compound Analysis," *Anal. Chem.* 47:1223-1224 (1975).
20. McAuliffe, C. D. et al. "The Dispersion and Weathering of Chemically Treated Crude Oils on the Ocean." *Environ. Sci. Technol.* 14:1509-1518 (1980).
21. Wakeman, S. G. "Synchronous Fluorescence Spectroscopy and its Application to Indigenous and Petroleum-Derived Hydrocarbons in Lacustrine Sediments," *Environ. Sci. Technol.* 11:272-276 (1977).
22. Gordon, D. C., Jr., and P. D. Keizer. "Estimation of Petroleum Hydrocarbons in Seawater by Fluorescence Spectroscopy: Improved Sampling and Analytical Methods," J. Fish Res. Bd. Can. Technical Report No. 481 (1974).
23. Lloyd, J. B. F. "The Nature and Evidential Value of the Luminescence of Automobile Engine Oil and Related Materials. I. Synchronous Excitation of Fluorescence Emission," *J. Forensic Sci. Soc.* 11:83-84 (1971).
24. Boehm, P. D., et al. "A Chemical Investigation of the Transport and Fate of Petroleum Hydrocarbons in Littoral and Benthic Environments: the *Tsesis* Oil Spill," *Mar. Environ. Res.* (in press).
25. Boehm, P. D., and J. G. Quinn. "The Solubility Behavior of No. 2 Fuel Oil in Seawater," *Mar. Poll. Bull.* 5:101-105 (1975).
26. Atlas, R. M., P. D. Boehm and J. A. Calder. "Chemical and Biological Weathering of Oil from the *Amoco Cadiz* Spillage within the Littoral Zone," *Estaur. Coastal Mar. Sci.* 12:589-608.
27. Johannson, S., U. Larsson and P. Boehm. "The *Tsesis* Oil Spill: Its Impact on the Pelagic Ecosystem," *Mar. Poll. Bull.* 11:284-293 (1980).
28. Garrett, W. D. "Collection of Slick-Forming Materials from the Sea Surface," *Limnol. Oceanog.* 24:602-605 (1965).
29. Poirier, O. A., and G. A. Thiel. "Deposition of Free Oil by Sediments Settling in Seawater," *Bull. Am. Assoc. Petrol. Geol.* 25:2170-2180 (1941).
30. Mattson, J. S., and P. L. Grose. "Modeling Algorithms for the Weathering of Oil in the Marine Environments," Final Report, Research Unit No. 499, Outer Continental Shelf Environmental Assessment Program, NOAA, Boulder, CO (1979).
31. Thuer, M., and W. Stumm. "Sedimentation of Dispersed Oil in Surface Waters," *Prog. Water Technol.* 9:183-194 (1977).
32. Conover, R. J. "Some Relations between Zooplankton and Bunker C Oil in Chedabucto Bay Following the Wreck of the Tanker *Arrow*," *J. Fish. Res. Bd. Can.* 28:1327-1330 (1971).
33. McAuliffe, C. D., et al. "Chevron Main Pass Block 41 Oil Spill: Chemical and Biological Investigations," in *Proceedings of the Joint Conference on Prevention and Control of Oil Spills, San Francisco, CA (1975)*, pp. 555-566.
34. Clark, Jr., R. C., and M. Blumer. "Distribution of Paraffins in Marine Organisms and Sediments," *Limnol. Oceanog.* 12:79-87 (1967).
35. Gearing, P. J., et al. "Partitioning of No. 2 Fuel Oil in Controlled Estuarine Ecosystems. Sediments and Suspended Particulate Matter," *Environ. Sci. Technol.* 14:1129-1136 (1980).

36. Gordon, Jr., D. C., et al. "Fate of Crude Oil Spilled on Seawater Contained in Outdoor Tanks," *Environ. Sci. Technol.* 10:580–585 (1976).
37. Boehm, P. D., D. L. Fiest and A. Elskus. "Comparative Weathering Patterns of Hydrocarbons from the *Amoco Cadiz* Oil Spill Observed at a Variety of Coastal Environments," in *Proceedings of the International Symposium on the Amoco Cadiz: Fates and Effects of the Oil Spill, November 19–22, 1979* Brest, France: CNEXO (in press)).
38. Brown, D. W., et al. "Analysis of Trace Levels of Petroleum Hydrocarbons in Marine Sediments Using a Solvent/Slurry Extraction Procedure," in "Trace Organic Analysis: A New Frontier in Analytical Chemistry," NBS Special Publication 519, Washington, DC (1979), pp. 161–167.
39. Ensminger, A., et al. "Pentacyclic Triterpanes of the Hopane Type as Ubiquitous Geochemical Markers: Origin and Significance," in *Advances in Organic Geochemistry*, B. Tissot and F. Bienner, Eds. (1973), pp. 245–260.
40. Bieri, R. H., et al. "Polynuclear Aromatic and Polycyclic Aliphatic Hydrocarbons in Sediments from the Outer Continental Shelf," *Int. J. Environ. Anal. Chem.* (1979).
41. Youngblood, W. W., and M. Blumer. "Polycyclic Aromatic Hydrocarbons in the Environment: Homologous Series in Soil and Recent Marine Sediment," *Geochim. Cosmochima. Acta.* 39:1303–1314 (1975).

CHAPTER 8

DISTRIBUTION OF HYDROCARBONS IN WATER AND MARINE SEDIMENTS AFTER THE AMOCO CADIZ AND IXTOC-I OIL SPILLS

Michel Marchand

Centre Océanologique de Bretagne, CNEXO
29273 Brest, France

Jean-Pierre Monfort and Amanda Cortés-Rubio

Instituto Mexicano del Petróleo
Mexico 14, DF

INTRODUCTION

The shipwreck of the Amoco Cadiz supertanker on the rocks of the Brittany coast in France (March 1978) and the blowout of Ixtoc-I well in the Gulf of Mexico (June 1979) were the most important oil spills ever recorded. The crude oils discharged in the marine environment from both accidents were light petroleums and their chemical compositions were similar (Table 1). After these two oil spills we examined the hydrocarbon pollution in the marine environment. Our chemical studies were limited to an overall estimate of the oil content to assess the importance of and the extent of the pollution at the seawater surface and into the water column plus the sediment contamination. The analytical techniques used were UV spectrofluorometry for the seawater samples and IR spectrophotometry for the sediment samples. The Ixtoc-I study was less important than the Amoco Cadiz one, and was limited to the analysis of samples collected during only one oceanographic cruise.

161

Table 1. Characteristics of Amoco Cadiz and Ixtoc-I Oil Spills

Oil Spill	Amoco Cadiz	Ixtoc-I
Nature of incident	stranding of a tanker	blowout (drilling well)
Location	Brittany coast, Portsall (France)	Gulf of Mexico (Mexico) $19°24.5'$ N $- 92°12.5'$ W
Date	March 16, 1978	June 3, 1979 to March 20, 1980
Crude oils	"Arabian Light" and "Iranian Light" (about 1:1)	Ixtoc-I
Quantities discharged	223,000 ton	~600,000 ton
Density	0.85	0.84
Chemical composition		
Saturated	39% (Ref. 1)	50% (Ref. 3)
Aromatics	34%	32%
Resins and Asphaltens	27%	18%
Nickel	14 ppm (Ref. 2)	10.5 ppm
Vanadium	45 ppm	55 ppm

EXPERIMENTAL

Seawater

Subsurface seawater samples (2 liters) were collected at 1-m depth with a glass bottle held in a metal frame. The bottle was closed with a Teflon*® cylindro-conical stopper, the opening and the closing of which were controlled by the manipulator when the sampler was at the sampling depth. The collection of seawater samples at different depths was carried out with sterile plastic bags fixed on a "Butterfly" model sampler. The bags also were opened and closed at the sampling depths. Comparative analysis of two water samples collected with the glass bottle and plastic bag did not show any significant difference [4].

Immediately after collection, the seawater sample was extracted successively with 100 mL of chloroform and 100 mL of hexane. The two organic extracts were set in a glass flask and kept for analysis at the laboratory. The

---

*Registered trademark of E. I. Du Pont de Nemours & Company, Inc., Wilmington, Delaware.

organic phase (chloroform-hexane mixture) was then reduced by evaporation to about 5 mL. The traces of water were removed with anhydrous sodium sulfate. The extract was then concentrated to dryness under a nitrogen stream and finally rediluted to 5 mL with hexane.

Hydrocarbons were measured by UV spectrofluorometry (Turner apparatus, model 430), according to the method previously described [5-8]. The excitation wavelength was fixed at 310 nm. The emission spectra were set from 500 to 320 nm and maxima of fluorescence appeared between 390 and 360 nm. The spectrofluorometer was calibrated with a solution of emulsified crude oil in hexane, collected respectively near the Amoco Cadiz shipwreck* and Ixtoc-I well.** For comparison, we give the responses of the spectrofluorometer with the two calibration solutions (Table 2). Although only the aromatic compounds are detected by spectrofluorometry, the data are expressed in total hydrocarbons related to calibration crude oil solution. Nevertheless, the data reported are not necessarily representative of absolute hydrocarbon concentrations in seawater.

Immediate Analysis of Seawater on Board

During oceanographic cruises we collected duplicate seawater samples (1 liter) which were extracted with only 20 mL of hexane. The organic extracts were immediately measured on board ship by UV spectrofluorometry under identical conditions. Thus we were able to obtain during these cruises preliminary and immediate information about oil pollution in seawater. A comparison of the results of duplicate water samples analyzed either on board ship or later in the laboratory is illustrated by Figures 1 (Amoco Cadiz) and 2 (Ixtoc-I). For most of the data compared, the correlation is good (85%). So, the spectrofluorometry technique can be used as an operational analytical method to rapidly observe, while still on board the ship, the extent of oil pollution in the marine environment.

Table 2. Spectrofluorometer Responses, Expressed in Arbitrary Units
$\lambda_{Excitation}$: 310 nm; $\lambda_{Emission}$: 360 nm

Solution at 1 /μg/mL in Hexane	Spectrofluorometer Responses (arbitrary units)
Amoco Cadiz emulsion	760
Ixtoc-I emulsion	465

*Emulsion sample given by Dr. Calder (NOAA-USA).
**Emulsion sample given by Ing. Teyssier (IMP-Mexico).

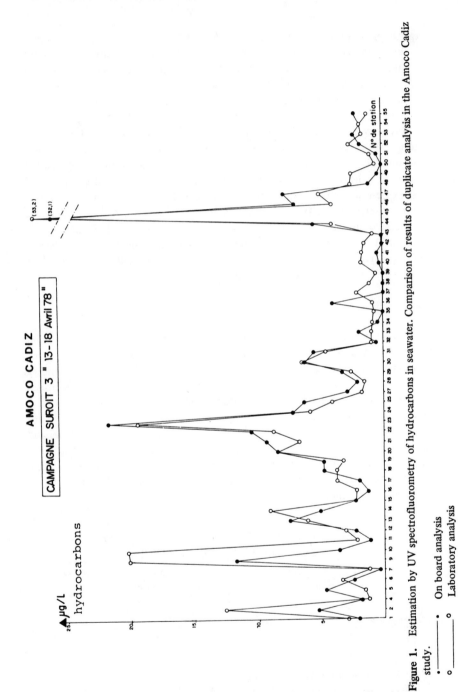

Figure 1. Estimation by UV spectrofluorometry of hydrocarbons in seawater. Comparison of results of duplicate analysis in the Amoco Cadiz study.

• ———— On board analysis
o ———— Laboratory analysis

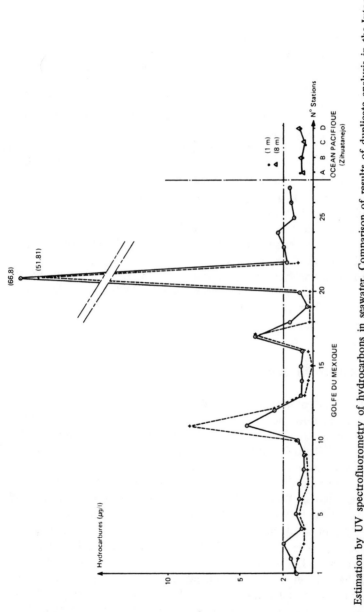

Figure 2. Estimation by UV spectrofluorometry of hydrocarbons in seawater. Comparison of results of duplicate analysis in the Ixtoc-I study.

+ ———— + On board analysis
• ———— • Laboratory analysis

Sediment

Surface marine sediments were collected with a Shipeck grab device. In coastal areas for the Amoco Cadiz study, we also used the little Ekman grab and Hamon grab devices. The samples, stored in a congelator, were either dried in an oven (70°C) or freeze-dried before analysis. The dried sample (100-200 g) was extracted in a Soxhlet apparatus or by stirring with chloroform or carbon tetrachloride. The organic extract was concentrated to dryness, then dissolved back in 10 mL of carbon tetrachloride.

A first indication of petroleum pollution in sediments was obtained with direct analysis of nonpurified extracts by IR spectrophotometry (Perkin Elmer Model 397). Quantitative measurements were carried out at 2920 cm^{-1} and correspond to the presence of hydrocarbons and polar compounds. The results are overestimated because of the response of coextracted natural substances (fats, fatty acids, etc.) from sediments. The IR spectrophotometer was calibrated with either a mixture of Arabian and Iranian light crude oils or Ixtoc-I crude oil.

Hydrocarbon analysis was performed by IR spectrophotometry after cleanup of organic extracts on activated alumina (200°C) in a glass column (i.d. = 0.6 cm, h = 15 cm). The hydrocarbons were eluted with 15 mL of carbon tetrachloride.

AMOCO CADIZ OIL SPILL

On the night of March 16, 1978 the Amoco Cadiz oil tanker became stranded on shallow rocks off Portsall (North Brittany), 1.5 miles from the coast. From March 17 to March 30, 223,000 tons of a mixture of Arabian light crude oil (100,000 ton) and Iranian light crude oil (123,000 ton) flowed into the sea without interruption. During this period the wind direction (W, NW, SW) induced a large drift of the spill eastward. The slicks successively reached the Aber-Wrac'h (March 19), Roscoff (March 20), the Bay of Lannion (March 21), the Sept-Iles (March 22) and the Sillon du Talbert (March 23). In April, the wind direction changed, thus reversing the slick drift that got to le Conquet and Ouessant Island (April 11), the Raz de Sein zone (April 13) and Douarnenez (April 22). A few hydrocarbon traces reached the coast in the Bay of Audierne at the beginning of May. The maximum extent of the oil slicks is presented in Figure 3. A chemical follow-up of the hydrocarbon pollution was made during several oceanographic cruises [4] (Figure 4) to assess the size and extension of the seawater pollution at the surface and in various water depths and in the sediment.

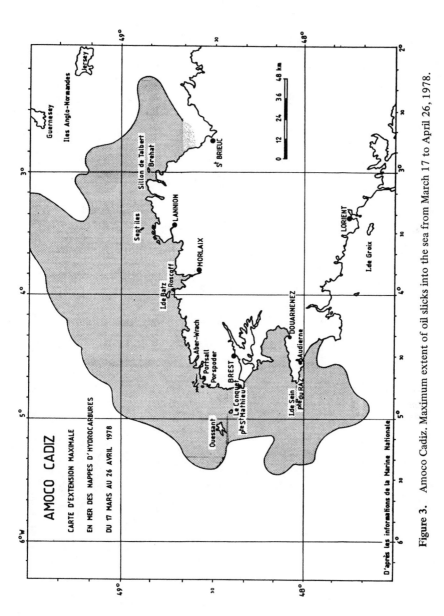

Figure 3. Amoco Cadiz. Maximum extent of oil slicks into the sea from March 17 to April 26, 1978.

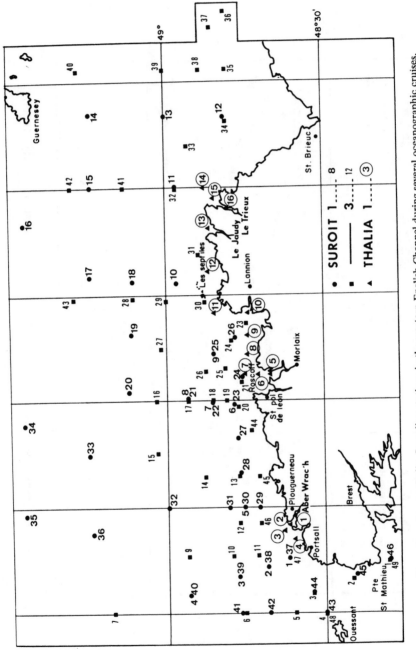

Figure 4. Amoco Cadiz. Sampling stations in the western English Channel during several oceanographic cruises.

Seawater

The chemical study by UV fluorescence geographically determined the spread limits of the oil pollution a fortnight and one month after the stranding. It also determined the diffusion of hydrocarbons into the water column and the evolution of the hydrocarbon contents. The first data showed that the Amoco Cadiz oil spill affected a very large section of the western English Channel. The range of hydrocarbon concentrations in seawater was found to be from 0.5 μg/L to more than 100 μg/L. The lowest concentrations were ranged from 0.5 to 1.0 μg/L and were very similar to those reported by other authors [9-10] in unpolluted areas from the NW Atlantic basin ($<$ 1.0 μg/L). Like Keizer and Gordon [6], we adopted the limit of 2.0 μg/L as a criterion of oil pollution in seawater. The main conclusions of this study were the following:

Spreading of the Pollution

A fortnight after the Amoco Cadiz stranding, the water pollution, under the action of winds from the West, spread eastward to the Bay of St. Brieuc. The western limit was found at the level of the 5°W meridian and the northern one along the 49°20'N parallel.

One month later (April 13-18) a change in wind direction reversed the drift of the oil spill. The average hydrocarbon concentrations in subsurface seawater, collected in different marine and coastal areas, are presented in Figure 5. The Bay of St. Brieuc did not show any significant pollution (1.2 ± 1.0 μg/L) and the eastern limit could be located approximately at the Sillon du Talbert. A slight increase of the hydrocarbon content was recorded west of Portsall (2.2 ± 0.9 μg/L). The 49°N parallel roughly constituted the northern limit, beyond which no more oil pollution was observed (1.6 ± 0.5 μg/L). The most polluted areas were located in the sheltered coastal zones such as the Abers area (38.9 ± 6.7 μg/L), Bay of Morlaix (11.5 ± 5.1 μg/L) and Bay of Lannion (10.7 ± 3.0 μg/L).

Diffusion of Hydrocarbons into the Water Column

The hydrocarbon concentrations were determined in seawater samples collected at various depths (1, 2, 5, 20 and 50 m above sea bottom), in the western English Channel, during several oceanographic cruises. The data, reported in Table 3, showed that the entire water column was contaminated by the oil slick drift. The in-depth diffusion of the hydrocarbons might be due to the dynamic mixing of water masses (hydrological conditions, sea state), the type of oil spilled (light crude oils facilitating the natural disso-

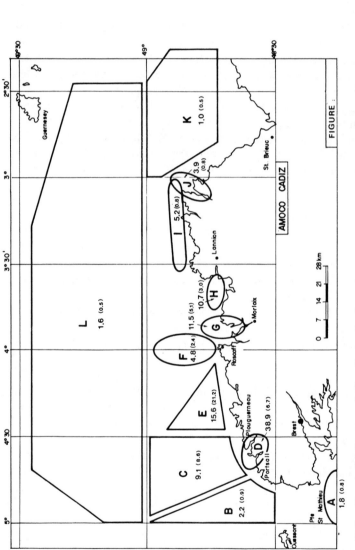

Figure 5. Amoco Cadiz. Chemical follow-up of oil pollution in the seawater. Average concentrations in subsurface seawater collected in different areas (April 13–18, 1978). Concentrations determined by UV spectrofluorometry and expressed in /µg/L. ( ) is standard deviation.

<1.0 /µg/L: no oil pollution
>2.0 /µg/L: oil pollution.

Table 3. Diffusion of Oil into the Water Column (Amoco Cadiz Study)
(Hydrocarbon concentrations in seawater expressed in $\mu g/L$.)

	Oceanographic Cruises												
	"Suroit 1" (March 30–April 4, 1978)								"Suroit 3" (April 13–18, 1978)				
Stations, N°	1	3	6	7	9	16	1	4	10	19	21	23	24
Latitude (N)	48°37'	48°45'6	48°46'6	48°49'2	48°52'5	49°27'4	48°17'6	48°29'7	48°47'	48°49'8	48°46'	48°44'8	48°44'1
Longitude (W)	04°42'5	04°49'2	04°00'7	03°57'7	03°49'3	03°10'	04°46'5	05°01'7	04°44'2	03°59'8	03°55'1	03°38'7	04°09'1
Depths (m)													
1	138.0	14.3	46.4	15.6	17.9	1.0	2.9	1.3	20.2	3.2	6.7	19.4	4.0
2	—	19.7	36.4	9.9	8.3	0.6	2.3	0.9	3.7	2.5	5.9	15.0	4.9
5	152.9	19.9	38.6	12.1	13.8	1.1	1.9	0.8	4.5	3.3	6.9	20.3	3.3
20	84.1	18.6	51.1	16.6	19.8	—	2.3	1.3	4.2	7.5	12.2	39.2	4.1
bottom + 5 m	102.7	42.3	27.7	18.3	26.6	—	10.7	1.8	9.6	4.9	11.0	—	4.5
(depth)	(44)	(95)	(40)	(70)	(70)		(28)	(80)	(95)	(75)	(45)		(65)
Average, $\bar{m}$	122.8	23.0	40.0	14.5	17.3	0.9	4.0	1.2	8.4	4.3	8.5	23.5	4.2
± σ	±28.3	±11.0	±9.1	±3.4	±6.8	±0.3	±3.7	±0.4	±7.0	±2.0	±2.8	±10.7	±0.6
(%)	(23%)	(48%)	(23%)	(24%)	(39%)	(29%)	(94%)	(33%)	(83%)	(46%)	(33%)	(46%)	(14%)

lution process), its physical and chemical evolution (emulsification, natural sinking) and the use of dispersant products. This dispersion process into the water column should have been quick because each water mass maintained some vertical homogeneity of the oil content observed from one site to the other.

Evolution of Surface Seawater Pollution

We observed a general and rapid decrease of the hydrocarbon concentrations in seawater. From March to June 1978, the half-time of hydrocarbons in subsurface water was given at 11 days in the oceanic zone east of Portsall, 14 days in the coastal area near the Abers, and 28 days in the bays of Morlaix and Lannion (Figure 6). Other data reported [1] gave a half-time of oil in water of 40 days in the sheltered estuary zone of Aber-Wrac'h.

Marine Sediments

Pollution of the Sea Bottom in April 1978

One month after the stranding of the Amoco Cadiz, we collected marine sediments samples during an oceanographic cruise (R/V Suroit) to assess the sea bottom contamination in the western English Channel. The sediments sampled were coarse to medium calcareous sands ($> 70\%$ $CaCO_3$). In coastal areas, the organogen calcareous content decreased (50–70% $CaCO_3$) in sands collected in the bays of Morlaix and Lannion and near the Abers. Organic carbon content generally was low, from 0.02 to 0.6% ($\overline{m} = 0.18 \pm 0.13\%$). The hydrocarbon concentrations in sediment samples ranged from 10 to 1100 ppm. Pollution of the sedimentary phase was observed in the coastal and offshore areas reached by the drifting slicks (Figure 7). The diffusion of the oil into the water column seems to show that the seawater had been a transfer agent of the oil pollution from the surface to the bottom. Off the Sept-Iles, a gradient was observed from the coast to the open sea (219, 52, 42 and 34 ppm). At the level of the 49°N parallel, from west to east, one could observe an increasing and then decreasing gradient: 21, 19, 48, 102, 54, 52 and 24 ppm. The high petroleum accumulations in marine sediments were located in the coastal zones, in the Abers (100 to $> 10,000$ ppm) and in the bays of Morlaix and Lannion (10 to > 1500 ppm) (Figure 8).

Evolution of Oil Pollution in Coastal Sediments

We have followed the oil pollution in sediments collected from the two Abers (Aber-Benoit, Aber-Wrac'h) which are small estuaries, 10 to 15 km

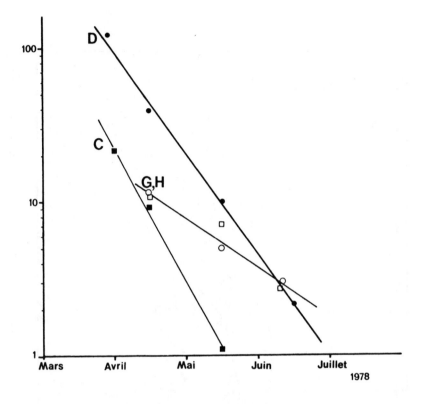

Figure 6. Evolution of hydrocarbon concentrations in subsurface seawater from March to June 1978 in different areas of the western English Channel (cf. Figure 5).

long and 1 km wide, including sandy and muddy areas. The study also was carried out in the bays of Morlaix and Lannion [4]. The chemical follow-up showed that the natural decontamination process was related to two essential factors: the type of sediment and the energy level of the geographic zone. Table 4 shows the main facts we observed during one year of study, and the results are briefly summarized here:

- Aber-Benoit: In the mud zone the sediments acted as an oil trap (oil contents >10,000 ppm) and the decontamination process was not observed. However, the medium to fine sands sampled from the downstream part of the estuary were well decontaminated after one year.
- Aber-Wrac'h: On the downstream part the decontamination process, activated by the sea nature of the environment, was considerably reduced by the muddy consistency of the polluted sands. On the upstream part, low energy zone, the decontamination of the muddy sands was not observed in any significant manner.

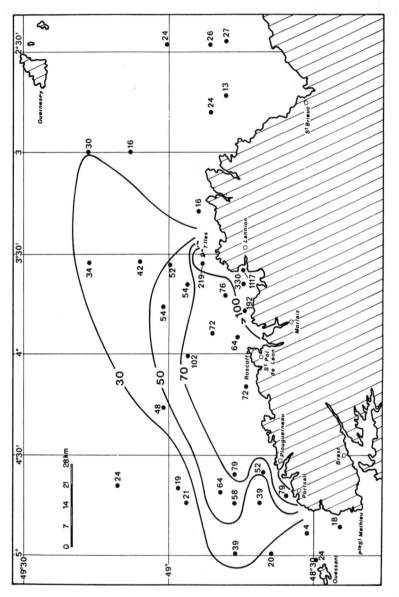

Figure 7.   Amoco Cadiz. Petroleum pollution of marine sediments in April 1978. Concentrations expressed in ppm.

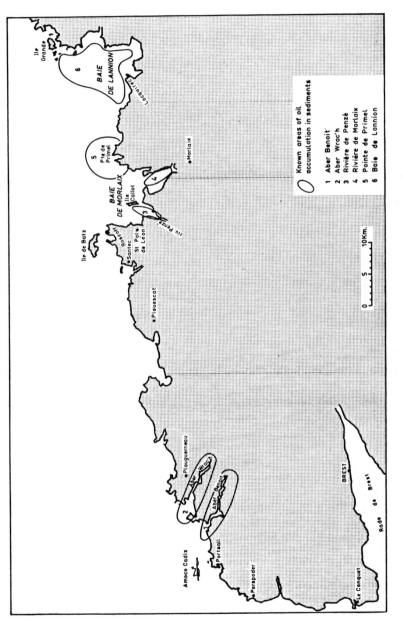

Figure 8.   Amoco Cadiz. Known coastal areas of oil accumulation in sediments.

Table 4. Evolution of the Pollution Caused by Oil on the Coastal Sea Floors

	Coastal Zone	Description	Type of Sediments	Hydrocarbon Content (mean values)		Decontamination Process
	1. The Two Abers			April 1978	March 1979	
Sheltered Area	Aber-Benoit	loc majan (mud)	muddy	>10,000	>10,000	no
	Aber-Wrac'h	upstream part	muddy sands	1,500	1,700	no
	Aber-Wrac'h	downstream part	muddy sands	4,200	1,700	low
	Aber-Benoit	upstream part	no muddy sands	700	27	yes
	2. Bays of Morlaix and Lannion			July 1978	February 1979	
Exposed Area	Morlaix River	bottom of the bay	sandy mud	311	172	low
	Bay of Lannion	bottom of the bay	fine sands	281	126	low
	Primel Area	sector exposed to winds and storms	fine to coarse sediments	600	19	yes

- Bays of Morlaix and Lannion: In the bottom of these bays, the decontamination process of muddy sands and fine sands was observed but remained low. In the eastern part of the bay of Morlaix, located around Primel, an area more exposed to winds and storms than the bottoms of bays, we observed that the coarse and fine sands were well decontaminated.

So, one year after the Amoco Cadiz stranding, the long-term evolution of oil pollution in the marine environment was focused in the sediments from low-energy level coastal zones, such as estuaries and bays.

IXTOC-I OIL SPILL

The blowout of Ixtoc-I in the Gulf of Mexico occurred on June 3, 1979 and the oil pollution continued until March 20, 1980. The exact quantities of oil spread into the sea will never be known, but according to some information the oil spill was ~4000 ton/day during the first weeks, then decreased to ~ 2000 ton/day. We can thus estimate an oil spill of more than 600,000 ton of light crude oil with the blowout on the Ixtoc-I well. A significant quantity of oil was burning at the sea surface in a flame field, but the remainder drifted and was dispersed in the marine environment.

During an oceanographic cruise with R/V Oceanográfico HO2 (Mexican Navy), August 16-21, 1979, from Vera Cruz to Ixtoc-I and Ciudad del Carmen (Figure 9), we collected seawater and sediment samples to determine the extent of the oil pollution from this source and to determine the dispersion of hydrocarbons in the ocean.

Seawater

During the cruise, we observed oil pollution on the surface of water as droplets, pancakes of emulsified petroleum, and oil slicks near stations 8-10, 12 and 21. The hydrocarbon concentrations, determined by UV spectrofluorometry in subsurface seawater (1 m), were on the whole low, generally <2.0 μg/L (Table 5). To compare these data, we also analyzed some seawater samples collected in unpolluted areas in the Pacific Ocean, off Zihuatanejo; the results ranged from 0.6 to 1.0 μg/L. In this study, we again kept the limit of 2.0 μg/L as a criterion of oil pollution in water. Pollution of the seawater was observed only at some stations but was not very important (2.0-4.6 μg/L). The highest level recorded (66.8 μg/L) was found at station 21, the nearest to the Ixtoc-I well.

In the *water column,* at stations 5, 9 and 20, where the oil pollution was not observed on the surface, the hydrocarbon concentrations determined along the depth profile were homogeneous and <2.0 μg/L (Table 6). The

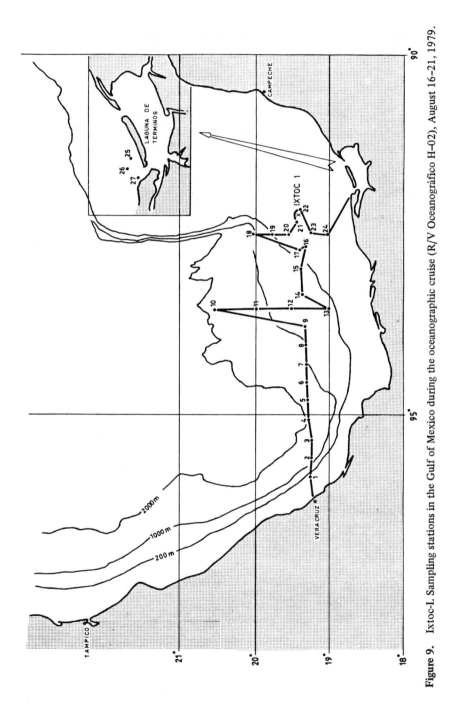

Figure 9. Ixtoc-I. Sampling stations in the Gulf of Mexico during the oceanographic cruise (R/V Oceanográfico H–02), August 16–21, 1979.

Table 5. Spectrofluorometric Determinations of Total Hydrocarbons ($/\mu g/L$) in
Seawater Collected in the Gulf of Mexico (August 16–21, 1979)
—Subsurface seawater (1 m)

| Station Number | Position | | Depth (m) | Hydrocarbon Concentration |
	Latitude (N)	Longitude (W)		
1	19°15.3'	95°51.9'	180	1.1
2	19°13.5'	95°36.4'	–	1.5
3	19°13.9'	95°20'	1640	2.0
4	19°15.1'	95°04'	2180	0.7
5	19°08'	94°45.9'	2180	1.1
6	19°16.3'	94°32.4'	2350	0.9
7	19°16.9'	94°17.1'	900	0.9
8	19°17.9'	94°01'	1115	0.6
9	19°19.9'	93°44.9'	820	0.6
10	20°30'	93°29.9'	1800	1.0
11	20°00'	93°30'	1200	4.6
12	19°30'	93°30'	600	2.6
13	19°00'	93°30'	485	0.8
14	19°21.2'	93°13.4'	540	0.7
15	19°21.8'	92°57.7'	–	0.8
16	19°18'	92°38.2'	130	0.7
17	19°22'	92°42.1'	176	4.0
18	20°00'	92°30'	1080	1.6
19	19°44.9'	92°30'	205	0.4
20	19°30.8'	92°31.2'	110	0.9
21	19°24.7'	92°18.8'	52	66.8
22	19°22.7'	92°08'	58	1.8
23	19°15.3'	92°30'	75	2.0
24	19°00.1'	92°30'	36	2.4
25			–	1.3
26 }	Ciudad del Carmen		–	1.5
27	Laguna de Terminos		–	1.6

Table 6. Spectrofluorometric Determinations of Total Hydrocarbons ($/\mu g/L$) in
Seawater Collected in the Gulf of Mexico (August 16–21, 1979)
–Diffusion of Oil into the Water Column

	Station Number					
	5	9	12	20	21	22
Depths (m)	Hydrocarbons ($/\mu g/L$)					
1	1.1	0.6	2.6	0.9		
5	1.3	1.3	3.5	1.1		
20	1.9	1.0	1.3	1.6		
100	1.2	1.0	1.7	1.8		
300	–	1.3	–	–		
$\bar{m}$	1.4	1.1	2.3	1.3		
$\pm\sigma$	±0.3	±0.3	±1.0	±0.4		
(%)	(26%)	(28%)	(43%)	(31%)		
1					66.8	1.8
5					2.9	–
20					3.9	10.9
50					73.1	–

water sampling at stations 12 and 21 was made under oil slicks. At station 21, near the Ixtoc-I well, the subsurface water was polluted (66.8 $\mu g/L$) but the diffusion of oil into the water column remained very low: 2.9 $\mu g/L$ at 5 m and 3.9 $\mu g/L$ at 20 m of depth. At station 12, located about 60 miles from Ixtoc-I, the oil content was low in the subsurface (2.6 $\mu g/L$) and was only observed down to a depth of 5 m (3.5 $\mu g/L$); beyond that the hydrocarbon concentrations were <2.0 $\mu g/L$. At station 8, where emulsified petroleum was observed in surface, the oil content in subsurface water remained at a background level (0.6 $\mu g/L$).

Two anomalies were observed in the vicinity of the well. At station 21, a very important oil content in water was found at a depth of 50 m (73.1 $\mu g/L$). At station 22, we noticed a significant contamination of water at a depth of 20 m (10.9 $\mu g/L$) but not in the subsurface water (<2.0 $\mu g/L$).

These observations showed that oil pollution in the seawater, at surface and in-depth, appeared especially in the vicinity of the Ixtoc-I well. These data contrast with those obtained during the Amoco Cadiz chemical study. The absence of diffusion of oil in the water can be explained by the unruffled state of the sea during the oceanographic cruise, the burning of volatile hydrocarbons (which are the more easily dissolved compounds in the water) above the well, and the local hydrological conditions. The mixing layer of waters was thin, at least during our observations, and there was a strong thermocline which restricted the diffusion process to the bottom.

Sediments

Sediments were collected on the continental shelf, near the Ixtoc-I well (calcareous muds) and near Ciudad del Carmen (muddy to fine sands). Results of these analyses are summarized in Table 7. The organic extracts ranged from 26 to 116 ppm ($\overline{m}$ = 59 ± 28 ppm). As a comparison, the average value of the organic extracts measured on coarse to muddy sediments collected in coastal Atlantic waters (Brittany, France) was 53 ± 39 ppm. The IR spectrophotometric analysis of nonpurified extracts gave values from 16 to 76 ppm ($\overline{m}$ = 36 ± 19 ppm). In a first approximation these results did not show any significant oil pollution in sediments. This information was confirmed with IR spectroscopic analysis of total hydrocarbons after clean-up of organic extracts on activated alumina. The hydrocarbon concentrations reported in muddy sediments sampled in the vicinity of the Ixtoc-I well were in a range of 9.7 to 20.9 ppm ($\overline{m}$ = 15.9 ± 3.6 ppm), and in sandy sediments collected near Ciudad del Carmen, in a range from 2.8 to 6.1 ppm ($\overline{m}$ = 5.8 ± 2.9 ppm). These data show that the sea bottom investigated was not polluted by the Oxtoc-I oil spill. As a comparison, total hydrocarbons in sandy sediments from the Brittany coastal area in the Atlantic Ocean (France) were found from 3.6 to 31.3 ppm ($\overline{m}$ = 11 ± 8 ppm) [11]. In the Northeast part of the Gulf of Mexico total hydrocarbon concentrations in shelf sediments ranged from 1.5 to 11.7 ppm [12]. The absence of contamination in the sediments sampled in the vicinity of the Ixtoc-I well indirectly confirmed the lack of diffusion of oil into the water column.

Table 7. IR Spectrophotometry Analysis of Sediments Collected in the Gulf of Mexico (Ixtoc-I)

Station Number	Type of Sediments	Water (%)	Extract[a] (ppm)	Index of Oil Pollution[a,b] (ppm)	Total Hydrocarbons[a] (ppm)
16	mud	60.4	82.7	61.5	16.6
17	mud	64.0	115.8	76.1	20.9
20	mud	62.4	29.8	23.0	9.7
21	mud	59.8	52.2	29.4	15.8
22	mud	58.7	57.0	32.5	15.6
23	mud	60.9	43.3	31.4	16.6
25	muddy sand	31.3	73.0	28.1	6.1
26	fine sand	21.0	26.4	16.3	8.5
27	sandy mud	43.5	50.6	28.8	2.8

[a]Concentrations expressed in dry weight.
[b]IR spectrophotometry analysis of nonpurified organic extracts.

CONCLUSION

These two chemical studies carried out after the Amoco Cadiz and Ixtoc-I oil spills show a difference in behavior of the crude oil spilled in the ocean, in regard to the oil diffusion into the water column and the pollution of the sediments. These differences can be related to the characteristics of the oceanic region polluted, the type of crude oil spilled and the conditions of the discharge of oil in the marine environment.

REFERENCES

1. Calder, J. A., and P. D. Boehm. "The Chemistry of "AMOCO CADIZ" Oil in the Aber-Wrac'h," in *Colloque: AMOCO CADIZ. Conséquences d'une pollution accidentelle par hydrocarbures* (Paris: CNEXO, 1981) pp. 143-158.
2. Ducreux, J., and M. Marchand. "Evolution des Hydrocarbures Présents dans les Sédiments de l'Aber-Wrac'h," in *Colloque AMOCO CADIZ: Conséquences d'une pollution accidentelle par hydrocarbures.* (Paris: CNEXO, 1981), pp. 175-216.
3. IFP. Personal communication from Institut Français du Pétrole, Service Analytique (1979).
4. Marchand, M., and M. P. Caprais. "Suivi de la pollution de l'AMOCO CADIZ dans l'eau de mer et les sédiments marins," in *Colloque AMOCO CADIZ: Conséquences d'une pollution accidentelle par hydrocarbures,* (Paris: CNEXO, 1981) pp. 23-54.
5. Levy, E. M. "The presence of petroleum residues off the east coast of Nova Scotia in the Gulf of St-Lawrence and the St-Lawrence river," *Water Res.* (1971), pp. 723-733.
6. Keizer, P. D., and D. C. Gordon. "Detection of Trace Amounts of Oil in Sea Water by Fluorescence Spectroscopy," *J. Fish Res. Bd. Can.* 30(8):1039-45 (1973).
7. Levy, E. M. "Fluorescence Spectroscopy: Principles and Practice as Related to the Determination of Dissolved and Dispersed Petroleum Residues in Sea Water," Bedford Institute of Oceanography, Report Series BI-R-77 (1977), 17 pp.
8. UNESCO. "Manuel sur la surveillance continue du pétrole et les hydrocarbures en mer et sur les plages," Supplément aux manuels et guides n°7. IOC-WMO-UNEP/MED-MRM/3, Supp. 2 (1977), 21 pp.
9. Gordon, D. C., P. D. Keizer and J. Dale. "Estimates Using Fluorescence Spectroscopy of the Present State of Petroleum Hydrocarbon Contamination in the Water Column of the Northwest Atlantic Ocean," *Mar. Chem.* 2:251-61 (1974).
10. Keizer, P. D., D. C. Gordon and J. Dale. "Hydrocarbons in Eastern Canadian Marine Waters Determined by Fluorescence Spectroscopy and Gas Liquid Chromatography," *J. Fish Res. Bd. Can.* 34:347-53 (1977).

11. Marchand, M. and J. Roucaché. "Critères de pollution par hydrocarbures dans les sédiments marins. Etude appliquée à la pollution du Böhlen," *Oceanol. Acta* 4(2): 171–183 (1981).
12. Gearing, P., et al. "Hydrocarbons in 60 North-East Gulf of Mexico Shelf Sediments: A Preliminary Survey," *Geochim. Cosmochim. Acta* 40:1005–17 (1976).

CHAPTER 9

BIODEGRADATION OF CRUDE OIL IN A MARINE ENVIRONMENT – GENERAL METHODOLOGY

D. Ballerini and J. P. Vandecasteele

Institut Français du Pétrole
Direction de Recherche Environnement et Biologie Pétrolière
92506 Rueil Malmaison
France

This study aims essentially at quantifying the processes of microbial degradation of crude oil in optimal conditions compatible with a marine environment. Our purpose, specifically, was to determine the proportion of crude which could undergo biological degradation, and to improve our knowledge of the biodegradation kinetics of the various hydrocarbon families which constitute a crude oil.

GENERAL METHODOLOGY

This work was performed in laboratory reactors where various physico-chemical parameters such as pH, temperature, stirring speed and air flow rate were strictly controlled. The study was conducted first in batch then in continuous cultures.

The composition of the mineral medium used for the culture is shown in Table 1. This composition was chosen by taking the average composition of the Atlantic Ocean as a basis [1] and enriching it in nitrogen, phosphorus

Table 1. Mineral Medium

Salts	Concentration (g/L)
NaCl	23.9
$MgCl_2, 6H_2O$	10.8
$CaCl_2$	1.15
KCl	0.74
Na_2SO_4	4.0
H_3BO_3	0.023
$NaBr, 2H_2O$	0.117
NaF	0.002
$NaHCO_3$	0.196
$SiO_2, Na_2O, 5H_2O$	0.022
$FeSO_4, 7H_2O$	0.002
NH_4Cl	0.9
K_2HPO_4	0.15

Table 2. N-Alkanes Composition of Saturated Hydrocarbons (%)

C Number	Arabian Light 240^+	C Number	Arabian Light 240^+
14	0.33	25	4.14
15	3.44	26	3.75
16	7.72	27	3.18
17	10.13	28	2.54
18	10.42	29	2.05
19	9.73	30	1.85
20	9.10	31	1.47
21	8.23	32	1.20
22	7.03	33	1.01
23	6.12	34	0.76
24	5.37	35	0.52

and iron. The concentrations of these three elements were respectively raised from 0.5 to 235 mg/L for nitrogen, from 0.07 to 26.7 mg/L for phosphorus and from 0.01 to 0.4 mg/L for iron and the pH of this medium was adjusted to 8.1, the average pH of seawater.

In order to approximate the conditions existing at sea, where evaporation is not negligible [2], we mainly used crude oil which had been topped at 240°C. In the present case, we used a residue of an Arabian light crude, obtained by distillation under reduced pressure, in which all the fractions distilling below 240°C had been removed.

Table 2 shows that in the alkane distribution of Arabian light crude oil topped at 240°C, n-paraffins of chain length below C_{14} have been completely removed. In oil spills at sea, all compounds with a number of carbons below 15 disappear after a residence time of about 10 days.

The reactor used in our studies is a glass cylinder equipped with two stainless steel endplates. Its total volume is approximately 3 liters. Stirring is achieved with a simple 4-bladed turbine, the speed of which is adjusted between 600 and 800 rpm. The reactor is equipped with automatic pH and temperature controls and with a dissolved oxygen analyzer. The pH of the culture is maintained at 8.1 by addition of 1N sodium hydroxide, the temperature is kept at $20°C \pm 0.2°C$ and the air flow rate is maintained at a value of 1 liter air/L of medium/hr. Experiments are conducted in nonsterile conditions.

The analytical methodology utilized to follow the processes of crude oil degradation is summarized in Figure 1. Gaseous effluent from the fermentor first goes through a flask containing carbon tetrachloride where the hydrocarbon vapors are trapped, then in a second flask containing a known amount of 1N potassium hydroxide, which traps carbon dioxide. Hydrocarbons are then determined by infrared spectrophotometry and the carbon dioxide is determined by titration.

Liquid samples are withdrawn at various times during the course of the fermentation. A sample is first centrifuged to separate the hydrocarbon from the aqueous phase. The latter is then filtered (diameter of the pores of the

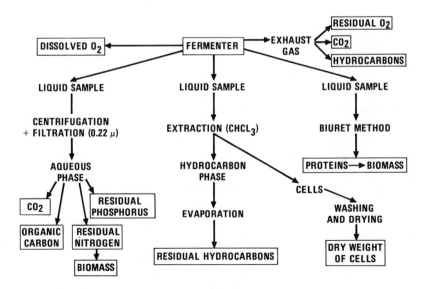

Figure 1. Analytical methodology.

filter: 0.22 μ) in order to remove the fine particles present in suspension. On the perfectly clear aqueous phase obtained, the following analyses are performed:

- total organic carbon (Dohrman DC 50 apparatus),
- dissolved carbon dioxide by manometry using a Warburg apparatus,
- residual phosphorus, and
- residual ammonia nitrogen.

The decrease in time of the concentration in ammonia nitrogen in the medium corresponds to an enrichment in cellular nitrogen (protein nitrogen) and thus to biomass formation during the considered time interval. Knowing the percentage of cellular nitrogen, which is easy to determine by the Kjeldahl method, the production of biomass during the course of the fermentation can be readily calculated. This indirect method for biomass determination is the one which we finally selected.

Another method for biomass determination, which is the direct measurement of the protein content of the cell on an aliquot of a culture sample, also has been used. The correspondence between this protein content and the cellular concentration has been established by measuring the protein content of dried cells prepared as described below and a value of 2 has been found for the ratio cell dry weight/protein content. However, the reproducibility and accuracy of the results obtained with this very simple and rapid method have not always been completely satisfactory.

On a third culture sample, extraction of residual hydrocarbons is performed. For this extraction, several solvents have been tried: in particular ethyl ether, carbon tetrachloride, benzene and chloroform, used pure or in mixtures. Finally, chloroform was selected for its extraction capacity and also for its relative convenience of handling. Three successive extractions are performed. Between each extraction, the emulsified organic and aqueous phases are separated by centrifugation and the organic phases are collected and evaporated at 50°C under a stream of nitrogen. The residue is then weighed. By using this procedure with reference samples, we measured the extraction yield of hydrocarbons and found that it was always in a range from 95 to 100%.

From the aqueous phase obtained after hydrocarbon extraction, the cellular mass is separated by centrifugation, washed with distilled water, dried and weighed. Nevertheless, owing to the complexity of the procedure and the very small mounts of biomass handled, the accuracy of this method of biomass determination is not very good.

Separation of asphaltenes is first performed on the hydrocarbon residue (Figure 2). Asphaltenes are precipitated by boiling with heptane under reflux for 1 hour. Insoluble materials are then separated by filtration, rinsed with hot heptane, then dried and weighed.

On the residue obtained after heptane evaporation, separation of the three main classes of compounds of crude oil, saturated hydrocarbons, aromatic hydrocarbons and resins is performed either by thin layer chromatography on silica gel or by liquid chromatography on a silica gel column using cyclohexane as the elution solvent.

The sum of the weights of the four fractions recovered in this procedure always represents 90 to 100% of the initial weight of the total oil sample. On the saturated and aromatic fractions, more detailed analyses using gas chromatography and mass spectrometry are then performed.

MICROBIOLOGY

The isolation and selection of the strains of microorganisms used were performed starting from marine silt and sand samples collected at places hit by crude oil spills.

The first enrichment cultures were performed by successive transfers in liquid medium to which 2 g/L of crude had been added, first in shaken flasks and, soon afterward, in laboratory fermentors in order to obtain perfectly defined culture conditions.

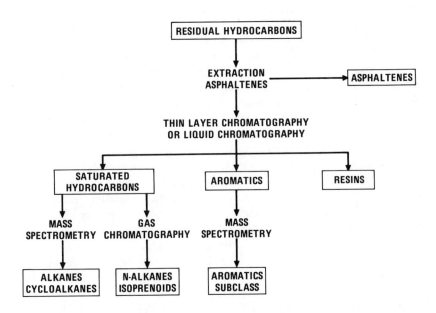

Figure 2. Hydrocarbons analysis.

In the course of these successive batch cultures we observed a very important decrease of the lag phase and an improvement of the final biodegradation performance. A large improvement of the dispersion of the hydrocarbon phase in the mineral medium was simultaneously observed. A very fine emulsion of the organic phase in the aqueous phase was thus obtained, leading to a marked increase of the reactional area.

The bacterial association present in our selected cultures consists of four dominant bacterial genera: *Pseudomonas, Moraxella, Acinetobacter* and *Flavobacterium.* During a batch culture we studied the changes in the bacterial flora. Although at the beginning the dominant strains belonged mainly to the genus *Moraxella*, the *Pseudomonas* strains predominated later on, as oil biodegradation advanced.

RESULTS AND DISCUSSION

Batch Cultures

The biodegradation process during the batch cultures can be illustrated by following the changes of various parameters. For example, on Figure 3, we have represented biomass and carbon dioxide production, the latter in both the aqueous and gaseous phases. Total production of CO_2 follows an evolution quite comparable to that of biomass production. At the end of the culture, whereas hydrocarbon consumption reaches 2 g/L, the total amount of carbon dioxide produced is 1.45 g/L and that of biomass is 1.4 g/L.

The changes in time of dissolved oxygen concentration in the medium, as well as consumption of ammonia nitrogen and hydrocarbons are shown in Figure 4. The amount of hydrocarbons lost by evaporation is quite negligible as it is measured on a 48 hour period and is always below 10 mg. A perfect correspondence between the evolution of the consumptions of ammonia nitrogen and hydrocarbon can be again noted.

In addition, during the acceleration period of the biodegradation process, a very distinct decrease in the oxygen dissolved in the medium occurs. Actually, dissolved oxygen concentration even becomes a limiting factor since values of 5% of the saturation level are read. Later, as the velocity of the biodegradation process decreases, the concentration of oxygen dissolved in the medium increases again, but without reaching the saturation value.

In Table 3, we present the hydrocarbon balance, performed on a batch culture which had lasted 48 hours. Of the 2.65 g/L of hydrocarbon initially present, 1.08 g/L has been consumed, corresponding to a degradation percentage of 41%. It appears clearly that the saturated fraction is the most susceptible to biodegradation since 67% of this fraction is consumed, whereas

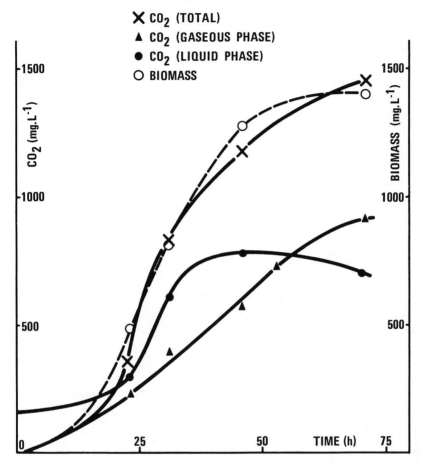

Figure 3. Batch culture: biomass and CO_2 production.

only 27% of the aromatic fraction is degraded. As for asphaltenes and resins, it can be noted that they are attacked little or not at all by microorganisms. After biodegradation, the composition of petroleum is considerably modified since it contains less saturated hydrocarbons but, as a counterpart, a very high content of asphaltenes and resins.

On Figure 5, we have represented the evolution in time of total hydrocarbons and of the saturated and aromatic fractions. It can be seen that after the first 20 hours of the batch culture 25% of the total hydrocarbons are consumed, but this is mainly due to the disappearance of the saturated fraction, the aromatic fraction undergoing little degradation. From 20 to 46 hours the pattern changes. On one hand, a slowing down of the overall degra-

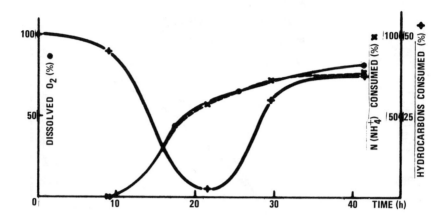

Figure 4.   Batch culture: dissolved oxygen concentration.

dation process takes place, since during this period only 16% of the total hydrocarbons and 18% of the saturated fraction are consumed. On the other hand, a speeding-up of aromatic biodegradation can be noted, along with a corresponding enrichment of the aqueous phase in organic carbon. In fact, the concentration in organic carbon in the aqueous phase, which is relatively stable during the first 20 hours, increases later from 55 mg/L to 258 mg/L. From the latter observation, it can be suggested that a large part of aromatic compounds are only partially oxidized, and then are solubilized in the aqueous phase.

On the saturated and aromatic fractions, obtained from samples taken at the beginning and at the end of a culture, gas chromatographic and mass spectrometric analyses were performed.

Analysis of the n-alkanes (C_{14}–C_{35}) and isoprenoids (C_{16}–C_{23}) detectable by gas chromatography showed that these compounds had practically disappeared at the end of the culture.

The results of mass spectrometric analysis of the saturated fraction are presented in Table 4. First they indicate that alkanes are the main target of biodegradation since 88.9% have disappeared at the end of the culture. This allows us to conclude that, in addition to n-alkanes and isoprenoids which represent only 14.8% of the saturated fraction, the great majority of iso-alkanes is consumed by microorganisms. Among naphthenic compounds, 1-ring and 2-ring naphthenes are the main targets of biodegradation, with respective percentages of biodegradation reaching 44 and 47%.

For the aromatic fraction results presented in Table 5 indicate that the microbial action is particularly evident at the level of mono- and diaromatic

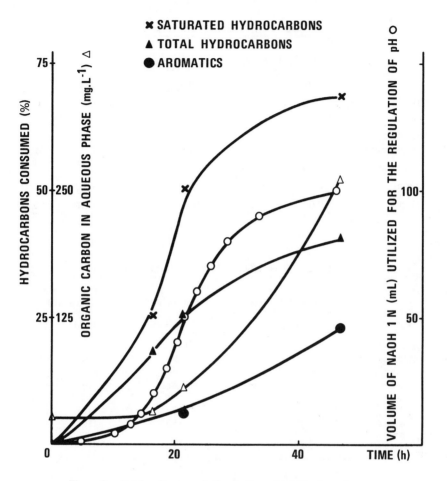

Figure 5.   Batch culture: evolution in time of total hydrocarbons.

compounds. At the end of the culture all mono- and diaromatic compounds with a carbon number below 16 have disappeared. Among monoaromatic compounds, it can be seen in Table 6 that the compounds most susceptible to microbial action are alkylbenzenes, 67.7% of which have disappeared at the end of the culture, and benzocycloparaffins, 46.2% of which have been consumed during this experiment.

The results presented in Table 7 illustrate the microbial action on di-aromatic compounds and, in particular, the effect of microbial attack on the residual content in naphthalene derivatives, 50% of which are consumed during the biodegradation process.

Table 3. Degradation of Hydrocarbons

	Start of the Batch		End of the Batch		Hydrocarbons Consumed	
	(g/L)	%	(g/L)	%	(g/L)	.%
Total Hydrocarbons	2.65		1.57		1.08	41
Saturated Hydrocarbons	1.17	44.2	0.39	24.9	0.78	67
Aromatics	1.00	37.5	0.73	46.5	0.27	27
Resins	0.42	15.9	0.39	24.2	0.03	7
Asphaltenes	0.06	2.4	0.06	4.4	0.00	0

Table 4. Degradation of Saturated Hydrocarbons

Hydrocarbons	Start of the Batch (mg/L)	End of the Batch (mg/L)	Hydrocarbons Consumed (mg/L)	(%)
Saturated Hydrocarbons	1170	390	780	67
Alkanes	739	82	657	88.9
1-ring	99	51	44	44.4
2-ring	136	89	47	34.5
3-ring	92	79	13	14.1
4-ring	55	50	(5)	—
5-ring	27	24	(3)	—
6-ring	22	15	(7)	—

Table 5. Degradation of Aromatic Hydrocarbons

Hydrocarbons	Start of the Batch (mg/L)	End of the Batch (mg/L)	Hydrocarbons Consumed (mg/L)	(%)
Monoaromatics	268	138	130	48.5
Diaromatics	251	168	83	33.0
Other Compounds	481	424	57	11.8

Table 6. Degradation of Monoaromatics

Hydrocarbons	Start of the Batch (mg/L)	End of the Batch (mg/L)	Hydrocarbons Consumed (mg/L)	(%)
Alkylbenzenes C_NH_{2N-6}	113.4	36.6	76.8	67.7
Benzocyclo Paraffins C_NH_{2N-8}	100.0	53.8	46.2	46.2
Benzodicyclo Paraffins C_NH_{2N-10}	54.9	47.4	7.5	13.7

Table 7. Degradation of Diaromatics

Hydrocarbons	Start of the Batch (mg/L)	End of the Batch (mg/L)	Hydrocarbons Consumed (mg/L)	(%)
Naphthalenes C_NH_{2N-12}	98.3	49.1	49.2	50.0
Acenaphtenes C_NH_{2N-14}	81.4	60.4	21.0	25.8
Fluorenes C_NH_{2N-16}	71.0	58.3	12.7	17.9

We will continue this work in batch cultures in the following directions:

- establishment of a more precise overall carbon balance,
- for the aromatic fraction, establishment of the extent of bacterial attack at the levels of side chains and of aromatic rings, and
- identification of the main compounds resulting from partial oxidation.

Continuous Cultures

In contrast to batch cultures, continuous cultures can be assimilated to "open systems" approximating the conditions existing at sea. In addition, they allow a perfect control of the concentration of nutrients and the adjustment of these concentrations to limiting values.

In continuous cultures, we studied the influence of the dilution rate (which is the reciprocal of residence time) and determined the requirements of the bacterial culture in nitrogen and in phosphorus.

The dilution rate was increased by steps from 0.04/hr to 0.12/hr. The results, presented in Table 8, show that an increase of the dilution rate D, that is, of growth velocity, leads to an improvement of the velocity of the consumption of hydrocarbons, which varies from 24.4 mg/L/hr for D = 0.04/hr to 69.6 mg/L/hr for D = 0.12/hr. At higher dilution rates, the microbial attack of saturated hydrocarbons is favored at the expense of aromatic compounds. This confirms that saturated hydrocarbons are the first compounds to undergo biodegradation.

During another experiment, performed at a dilution rate of 0.04/hr, we varied the residual content in nitrogen and phosphorus in the medium. We then determined the consumption of these two elements with respect to the corresponding consumption of hydrocarbons. The results of Table 9 show that biodegradation of 1 mg hydrocarbons requires assimilation by microorganisms of an average of 0.1 mg mitrogen and 0.012 mg of phosphorus.

As a conclusion, the results presented here allow us to hope that, on condition of making some improvements in the methodology presently used in the laboratory, it will be possible in the near future to obtain a reasonably complete and quantitative evaluation of the microbial processes of crude oil degradation.

Table 8. Continuous Culture

Dilution Rate (per hr)	Hydrocarbons Consumed (mg/L per hr)	Saturated Hydrocarbons (% degradation)	Aromatics (% degradation)
0.04	24.4	37.7	22.2
0.08	27.2	32.3	n.d.
0.12	69.6	30.6	8.6

Table 9. Continuous Culture

Residual Nitrogen (mg/L)	N Consumed (mg) Hydrocarbons Consumed (mg)	Residual Phosphorus (mg/L)	P Consumed (mg) Hydrocarbons Consumed (mg)
165	0.10	7.5	0.012
45	0.11	1.5	0.013
1	0.09		

REFERENCES

1. Ivanoff, A. *Introduction à l'océanographie,* Tome 1, Chapter 9, Vuibert, Ed. (1972).
2. Kreider, R. E. *Proceedings of the Joint Conference on Prevention and Control of Oil Spills,* (1971), pp. 119-124.

CHAPTER 10

BIODEGRADATION OF HYDROCARBONS IN MOUSSE FROM THE IXTOC-I WELL BLOWOUT

R. M. Atlas, G. E. Roubal, A. Bronner and J. R. Haines

Department of Biology
University of Louisville
Louisville, Kentucky 40292

INTRODUCTION

The Ixtoc-I well blowout in the Bay of Campeche created the world's largest oil spill to date. Before the flow of oil could be halted 10 months after the blowout occurred, an estimated three million barrels of oil had been spilled into the Gulf of Mexico. This oil spill has renewed questions about the safety of offshore drilling and the ecological effects of such spillages. The oil formed a stable emulsion or mousse. Mousse is a heterogenous mixture of hydrocarbons and water. The term generally is used to describe a water in oil emulsion with greater than 60% hydrocarbon.

Microorganisms play a major role in the removal of hydrocarbon pollutants from ecosystems. Important factors which influence the rates of biodegradation include the qualitative and quantitative composition of the microbial community, the chemical composition of the oil, the physical state of the oil, temperature, oxygen concentrations and nutrient concentrations, especially nitrogen and phosphorus. For a general review of the interactions of microorganisms and petroleum hydrocarbons see Bartha and Atlas [1] or Colwell and Walker [2].

Studies were initiated several months after the well blowout occurred to examine the microbial biodegradation of hydrocarbons in the mousse. These investigations included examining the microbial populations involved in hydrocarbon biodegradation, the rates of microbial hydrocarbon utilization and the changes in chemical composition of the mousse with time.

METHODS

Sample Collection

Surface water samples (top 1 m) were collected during the period September 14–23, 1979, at 13 sites shown in Figure 1. Samples were collected either with a Niskin butterfly sterile water collector (General Oceanics, Miami, Florida) or in areas of heavy oil accumulation with a clean bucket. Samples were visually examined and the presence of oil or mousse was recorded.

Enumeration of Microbial Populations

Total numbers of microorganisms per ml of surface water were determined by direct count procedures [3]. Portions of collected water samples were preserved with 2.5% formalin. Microorganisms in the preserved samples were collected on a 0.2-μm pore size Nuclepore filter which had been stained with irgalan black. The microorganisms were stained with acridine orange and viewed using an Olympus epifluorescence microscope. Cells staining orange or green were counted in 20 randomly selected fields and the mean concentration determined.

Hydrocarbon utilizing microorganisms were enumerated using a 3-tube most probable number (MPN) procedure [4]. Serial dilutions of water samples, prepared using Rila marine salts solutions, were inoculated into sealed serum vials containing 10 mL Bushnell Haas broth (Difco) and 20 μL of South Louisiana crude oil spiked with ^{14}C hexadecane (sp. act. 10 μCi/mL oil). After 14 days incubation at 25°C the [14]CO$_2$ (if any) in the headspace was collected by flushing and trapping in Oxifluor CO$_2$ (New England Nuclear) and quantitated by liquid scintillation counting. Vials showing [14]CO$_2$ production (counts significantly above background) were scored as positive and the most probably number of hydrocarbon utilizers per mL was calculated from standard MPN tables.

Hydrocarbon Biodegradation Potentials

Five-mL portions of water samples were placed in serum vials containing 20-μL-filter sterilized South Louisiana crude oil spiked with either ^{14}C hexa-

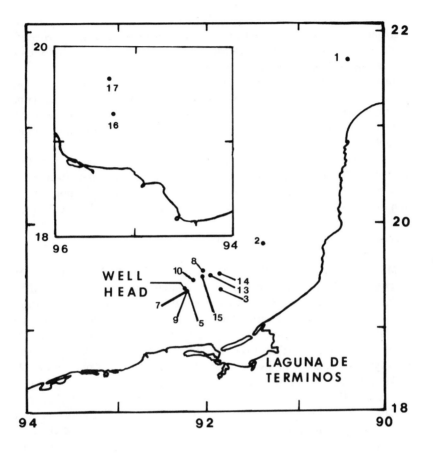

Figure 1. Chart showing P-series (Pierce Vessel) sampling sites.

decane, ^{14}C pristane, [14]C 9-methylanthracene or ^{14}C benzanthracene. (All specific activity $\cong 10$ μCi/mL oil.) The samples were incubated at 25°C. At 10 and 30 days further degradation was stopped by addition of KOH. All sample-substrate-time combinations were run in triplicate. The [14]CO$_2$ produced from mineralization of the radiolabeled hydrocarbon was determined by acidifying the solution, flushing the headspace, trapping the [14]CO$_2$ in 10 mL Oxifluor-CO$_2$ and quantitating by liquid scintillation counting. Filter sterilized controls were treated in a similar manner. The percent mineralization was calculated as [14]CO$_2$ produced (above sterile control) divided by ^{14}C hydrocarbon added.

The above procedures measure the potential for complete microbial degradation of oil in a slick under natural conditions. Any sparing or cometabolic degradation would be detected in the results of these procedures, as would any nutrient limitations. The influence of the physical state of the oil such as within "mousse" would not be determined in these tests.

In a separate series of tests five mL portions of water samples were supplemented with NH_4NO_3 (final concentration: 1 mM) and KH_2PO_4 (final concentration: 1 mM) to remove nutrient limitations. These supplemented samples were added to serum vials containing 20 μL of South Louisiana crude oil spiked with ^{14}C hexadecane (specific activity 10 μCi/mL oil). After 10 days incubation further degradation was stopped and the amount of $^{14}CO_2$ produced (^{14}C hexadecane mineralized) was determined as described above. Ratios of amounts of hydrocarbon mineralized with nutrient supplementation to amounts mineralized under natural conditions were calculated.

Oil and Mousse Biodegradation and Mineralization

Mousse samples were collected at stations P5, P8 and P17. Samples were weighed and "replicate" 5-g portions of mousse were placed in 250-mL flasks, each containing 75 mL of water collected at the same site. The quantity of oil in the mousse was determined by extracting with diethyl ether and weighing the residual water. The mousse contained 2.3, 4.0 and 4.3 g hydrocarbon/5-g mousse portion from stations P5, P8 and P17, respectively. Twenty-eight replicate flasks were established for each station. Four of the flasks from each station were fitted with sidearms and used to measure CO_2 production. Half of the flasks were treated with concentrated KOH to act as sterile controls. All flasks were incubated at 25°C on a rotary shaker at 50 rpm. CO_2 production was measured cumulatively by placing KOH in the sidearm flasks to trap CO_2. Periodically the KOH traps were recovered, $BaCl_2$ was added to precipitate $BaCO_3$, and the amount of trapped CO_2 was determined by titrating with 0.1 N HCl to a point of neutrality. Fresh KOH was added to the sidearms and the flasks were further incubated up to 6 months. During this time the mousse retained its physical integrity, at least by visual observation.

After 10, 40, 120 and 180 days incubation, oil was recovered from two active and two control flasks from each station. Extraction was with two sequential portions of methylene chloride. The analytical procedure described by Brown et al. [5] was used for extractions. The extracts were transferred to pentane solvent by distillation at 44°C under a Vigreaux reflux column. The volume was adjusted to 5 mL by evaporation under nitrogen. The extracts were subjected to column chromatography to split the extracts into aliphatic (f_1) and aromatic (f_2) fractions. Columns were prepared by suspending silica gel 100 (E. M. Reagents, Darmstadt, W. Germany) in CH_2Cl_2 and transferring the suspension into 25-mL burets with Teflon® stopcocks to attain a 15-mL silica gel bed. The CH_2Cl_2 was washed from the columns with 3 volumes pentane. Portions of the extracts in pentane were applied to the columns, drained into the column bed and allowed to stand for 3–5 minutes. The aliphatic fraction (f_1) was eluted from the column with 25 mL pentane. After 25 mL pentane had been added to the column, 5 mL of 20% (v/v) CH_2Cl_2 in pentane was added and allowed to drain into the column bed. Fraction f_1 was 30 mL. The aromatic fraction (f_2) was eluted from the column with 45 mL of 40% (v/v) CH_2Cl_2 in pentane.

The fractions of each extract were then concentrated to about 5 mL at 35°C and transferred quantitatively to clean glass vials. Fractions f_1 and f_2 were prepared for analysis by gas chromatography (GC) or gas chromatography/mass spectrometry (GC/MS). An internal standard, hexamethylbenzene (Aldrich Chemical Co., Milwaukee, Wisconsin), was added to each sample. In fraction f_1, hexamethylbenzene (HMB) was present at 12.6 ng/μL; in fraction f_2, HMB was present at 25.2 ng/μL.

Fraction f_1 was analyzed by GC on a Hewlett-Packard 5840 reporting GC with a flame ionization detector (FID). The column was a 30 m, SE54 grade AA glass capillary (Supelco, Bellefonte, Pennsylvania). Conditions for chromatography were: injector, 240°C; oven 70°C for 2 min to 270°C at 4°C/min and hold for 28 min; FID, 300°C; and carrier, He at 25 cm/sec. A valley-valley integration function was used for quantitative data acquisition. Response factors were calculated using n-alkanes, (C_{10}–C_{28}), pristane and phytane standards.

Fraction f_2 was analyzed with a Hewlett-Packard 5992A GC-MS. Conditions for chromatography were: injector, 240°C; oven 70°C for 2 min to 270°C at 4°C/min and hold for 18 min. Data were acquired using a selected ion monitor program. Thirteen ions were selected for representative aromatic compounds. The ions monitored were 128, 142, 147, 156, 170, 178, 184, 192, 198, 206, 212, 220, and 226. The representative compounds were naphthalene, methylnaphthalene, HMB as an internal standard, dimethylnaphthalene, trimethylnaphthalene, phenanthrene, dibenzothiophene, methylphenanthrene, methyldibenzothiophene, dimethylphenanthrene, dimethyldibenzothiophene, trimethylphenanthrene, and trimethyldibenzothiophene, respectively. The dwell time per ion was 10 msec. Instrument response factors were calculated by injecting known quantities of unsubstituted and C_1- and C_2-substituted authentic aromatic hydrocarbons and determining the integrated response for each compound. These values were used to extrapolate for quantitation of isomers and C_3 substituted compounds.

Statistical Tests

An analysis of variance procedure (SPSS Program) and the Duncan multiple range test were used to analyze the data. An α value $\leqslant 0.05$ was considered as necessary for establishing statistical significance.

RESULTS

Enumerations

The results of enumerations of microbial populations are shown in Table 1. There was no significant difference in total microbial biomass (direct count) between samples with visible oil and those from control sites. The highest direct counts did occur at sites P8 and P14 where mousse was present; direct counts at these sites were 2–4 times those at control sites P1 and P2. In other cases though, direct counts at sites with mousse (e.g., P9, P13 and P17) were not higher than at the control sites.

In contrast to the total microbial biomass there was a significant positive correlation between numbers of hydrocarbon utilizing microorganisms (MPN hydrocarbon) and the visible presence of mousse in the sample. At control sites P1 and P2 concentrations of hydrocarbon utilizers in the surface

Table 1. Enumeration of Microbial Populations

Site	Direct Count $(\times 10^5 \text{ mL}^{-1})$	MPN-Hydrocarbon (mL^{-1})
P1	1.5	0.3
P2	3.4	0.3
P3[a]	4.5	24
P5[b]	1.8	4
P7	2.8	4
P8	2.3	9
P8[a]	7.0	2400
P9[a]	2.2	2400
P11	2.5	4
P13[a]	1.7	24
P14	3.6	1
P14[a]	6.2	12000
P15[c]	4.2	46
P16	2.0	1
P17[a]	1.1	2400

[a]Heavy mousse water mixture.
[b]Oil but no mousse present in sample.
[c]Mousse-tar particles in sample.

waters were $< 1/mL$. The sample with oil and no mousse collected at site P5 had a concentration of 4 hydrocarbon utilizers/mL, which was an order of magnitude above the eastern control sites, but was not significantly different from other sites which did not show visible oil or mousse in the vicinity of the wellhead. Samples with visible mousse had significantly elevated counts of hydrocarbon utilizers. In samples with mousse-tar particles, counts of hydrocarbon utilizers were 10^1-10^2/mL. In samples with heavy mousse accumulations counts of hydrocarbon utilizers were 10^1-10^5/mL. The highest concentration of hydrocarbon utilizers occurred in a sample collected at site P14. High concentrations of hydrocarbon utilizers also were found at sites P8, P9 and P11. There appeared to be a relatively tight association between elevated numbers of hydrocarbon utilizers and water in direct contact with mousse. At both sites P8 and P14 surface water samples were collected with and without the presence of visible mousse. In both cases, the counts in the samples with mousse were several orders of magnitude higher than in the samples without the mousse.

Hydrocarbon Biodegradation Potentials

Representative hydrocarbon biodegradation potentials without added nutrients are shown in Table 2 for hexadecane, pristane, 9-methylanthracene and benzanthracene. In these experiments the radiolabeled hydrocarbon tracer had the same specific activity relative to the amount of oil added. The tracer was within an oil slick, not within mousse. The amount of hydrocarbon added (16 mg) was in excess of levels of hydrocarbons contained in the water samples. The amount of hydrocarbon degraded to CO_2 was very low for all of the hydrocarbons tested. For hexadecane, which is considered to be a very readily biodegraded hydrocarbon, generally less than 1% was mineralized during 30 days incubation. The greatest degradation of polynuclear aromatics measured was 0.2% mineralized during 30 days and in most cases no mineralization of either 9-methylanthracene or benzanthracene could be detected.

A possible cause for the limited hydrocarbon degradation was the limited availability of nitrogen- and phosphorus-containing nutrients in the water samples, which are necessary to support microbial oil degradation. This indeed appears to be the case. A comparison of the percent hexadecane mineralized during 10 days incubation with and without nutrient supplementation is shown in Table 3. The extent of mineralization was 1 to 2 orders of magnitude higher with nutrients added than under natural conditions. The extent of hydrocarbon mineralization with nutrients added was 9% at both control sites P1 and P2 and between 13 and 28% at sites nearer the wellhead. Unlike the enumeration results, there was no significant correlation between

Table 2. Natural Biodegradation Potentials—Percent Mineralization
During 30-Day Incubation

Site	Hexadecane	Pristane	9-Methylanthracene	Benzanthracene
P1	0.3	0.7	0.0	0.0
P2	0.5	0.5	0.0	0.0
P3[a]	0.0	0.0	0.0	0.0
P5[b]	0.1	1.0	0.0	0.0
P7	1.2	1.2	0.0	0.0
P8	1.0	0.4	0.2	0.0
P8[a]	2.5	2.0	0.2	0.2
P9[a]	0.8	1.0	0.2	0.0
P11	0.5	0.8	0.2	0.0
P13[a]	0.3	1.0	0.0	0.0
P14	1.0	0.4	0.1	0.0
P14[a]	1.2	1.3	0.0	0.0
P15[c]	0.1	1.0	0.0	0.0
P16	0.3	1.0	0.0	0.0
P17[a]	1.6	1.2	0.2	—

[a]Heavy mousse-water mixture.
[b]Oil but no mousse present in sample.
[c]Mousse-tar particles present in sample.

the biodegradation potentials with or without added nutrients and whether or not there was visible mousse present in the sample. The maximal rate of complete hydrocarbon degradation for the natural situation appears to be 3 mg/day/liter of surface water based on hexadecane and assuming 1% mineralization during 10 days for all hydrocarbons in the oil. For the nutrient-supplemented situation maximal rates could be 80 mg hydrocarbon mineralized/day/liter; this could occur if there was extensive mixing to prevent nutrient depletion in the vicinity of the oil in areas of the Gulf of Mexico where productivity is not severely limited by available concentrations of nitrogen and phosphorus.

Mousse Hydrocarbon Biodegradation

The evolution of CO_2 from the degradation of mousse hydrocarbons at sites P5, P8 and P17 is shown in Figure 2. Higher rates of CO_2 production were found for mousse collected at site P17 than at the other sites. Site P17 is relatively distant from the wellhead and the mousse from this site was presumed to be the oldest. Lower CO_2 production was found for site P8 compared to site P17 even though similar concentrations of hydrocarbon per gram mousse were found at the two sites. The lowest rates of CO_2 production were found for mousse samples from site 5, which had the lowest con-

Table 3. Natural and Nonnutrient Limited Biodegradation Potentials
for Hexadecane During 10-Day Incubation

Site	Nutrient Limited (Natural) (% mineralization)	Nonnutrient Limited (N, P-Supplemented) (% mineralization)
P1	0.2	9
P2	0.3	9
P5[a]	0.1	17
P7	0.3	15
P8	0.5	17
P8[b]	1.1	23
P9[b]	0.5	15
P11	0.5	18
P13[c]	0.2	26
P14	0.2	13
P14[b]	0.2	27
P15[c]	0.1	23
P16	0.2	13
P17[b]	0.3	28

[a]Oil only, no mousse present.
[b]Heavy mousse water mixture.
[c]Mousse-tar particles in sample.

centration of hydrocarbon within the mousse and the lowest amount of hydrocarbon added to each experimental unit. The rates of CO_2 production convert to mineralization rates of 1.8 mg hydrocarbon converted to CO_2/day/liter of water for site P5 and 2.5 mg hydrocarbon mineralized/day/liter of water for site P17. The rates of mineralization of hydrocarbons in the mousse, per volume of surrounding water, thus appear to be of the same order of magnitude as rates of hydrocarbon mineralization measured in the ^{14}C radiolabeled natural hydrocarbon degradation potential experiments. It should be noted, however, that only a maximum of $< 2\%$ of the mousse actually added in the flasks was mineralized during the 180-day incubation. Thus, the percent degradation of hydrocarbons in the mousse was lower than would be predicted from the biodegradation experiments.

While there was a continuous gradual evolution of CO_2 during the 180-day experiment, changes in the relative concentrations of hydrocarbons in the mousse, indicative of biodegradation in the alkane fractions, did not generally appear until the 120-day sampling (Figures 3-5). There was a high degree of variability in the relative concentrations of hydrocarbons in the mousse detected in replicate samples. In all cases, however, the concentrations of n-alkanes in freshly collected mousse were higher than those of pristane and phytane. Following 180 days incubation the ratios of n-alkanes to isoprenoid

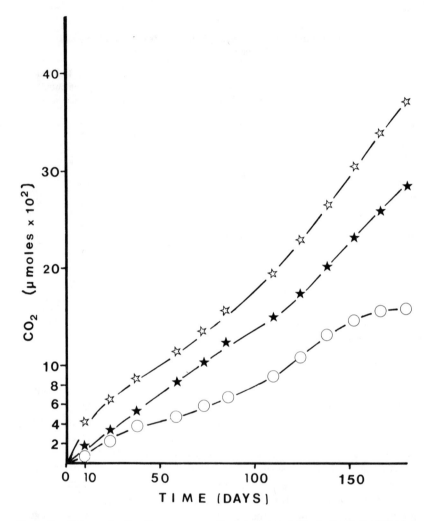

Figure 2. CO_2 production from mousse collected at several sites in the Gulf of Mexico. (Open circle = Site P5; closed star = Site P8; open star = Site P17).

hydrocarbons were reduced, but only for mousse from site P5 were the *n*-alkanes reduced to concentrations below those of the isoprenoid hydrocarbons. The changes in the overall composition of the oil indicate some preferential degradation of *n*-alkanes over isoprenoid hydrocarbons and reflect the overall low percentage of degradation of hydrocarbons in the mousse added to each flask. Analyses of mousse from site P5 after 180 days incubation showed an unusual complete disappearance of *n*-alkanes below

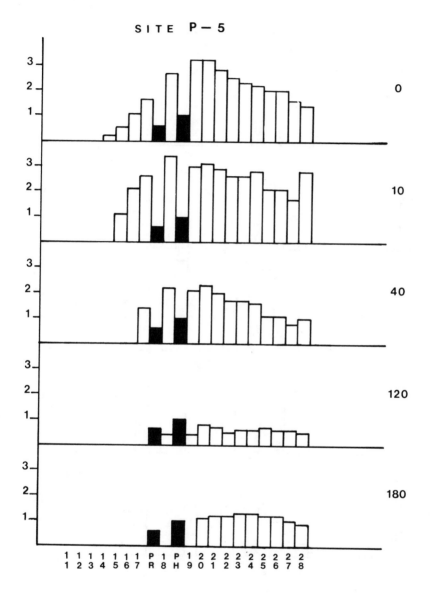

Figure 3. Changes in concentrations of selected hydrocarbons in the f_1 fraction relative to phytane during incubation of mousse collected at site P5. Y axis shows relative concentration; X axis shows identity of hydrocarbon. Numbers refer to chain length of n-alkanes; PR = pristane; PH = phytane. Time of incubation = 0, 10, 40, 120 and 180 days.

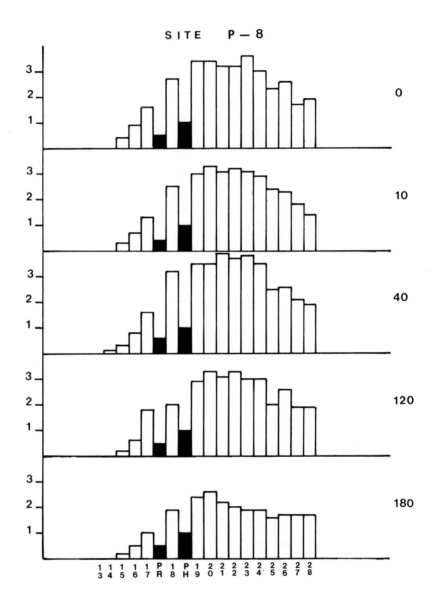

Figure 4. Changes in concentrations of selected hydrocarbons in the f_1 fraction relative to phytane during incubation of mousse collected at site P8. Y axis shows relative concentration; X axis shows identity of hydrocarbon. Numbers refer to chain length of n-alkanes; PR = pristane; PH = phytane. Time of incubation = 0, 10, 40, 120 and 180 days.

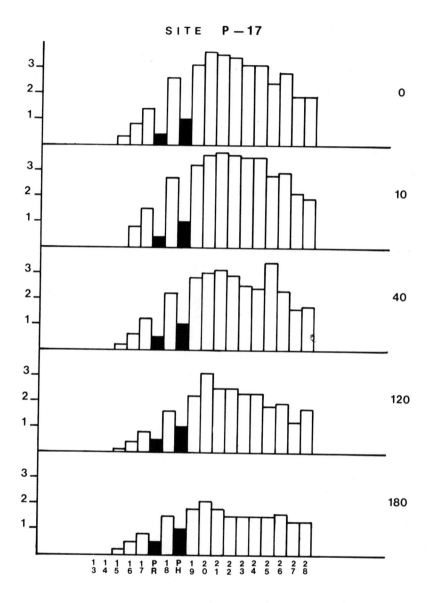

Figure 5. Changes in concentrations of selected hydrocarbons in the f_1 fraction relative to phytane during incubation of mousse collected at site P17. Y axis shows relative concentration; X axis shows identity of hydrocarbon. Numbers refer to chain length of n-alkanes; PR = pristane; PH = phytane. Time of incubation = 0, 10, 40, 120 and 180 days.

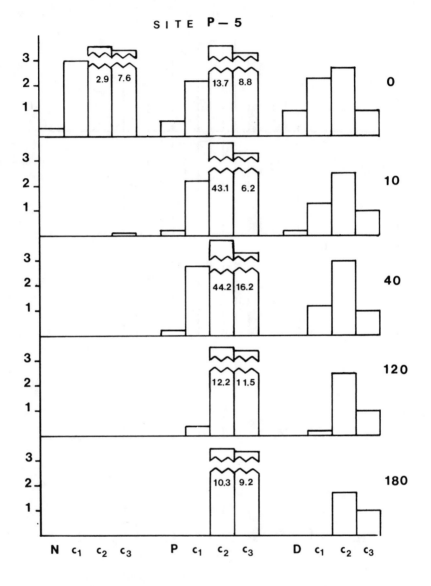

Figure 6. Changes in concentrations of selected hydrocarbons in the f_2 fraction relative to C_3 dibenzothiophene during incubation of mousse collected at site P5. Y axis shows relative concentration. X axis shows identity of hydrocarbon. N = naphthalene; P = phenanthrene; D = dibenzothiophene; C_1, C_2 and C_3 refer to the degree of substitution. Time of incubation = 0, 10, 40, 120 and 180 days.

n-C_{20} but not $\geqslant n$-C_{20}. This appears to indicate preferential biodegradation, but where such preferential degradation has been previously observed the split is usually at n-C_{25}–n-C_{28}, not at n-C_{20}.

Analyses of the aromatic fraction indicate that biodegradation of the aromatic fraction occurred during the incubation period (Figures 6-8). The dimethyl and trimethyl phenanthrenes appeared to be among the most persistent compounds in the aromatic fraction of the mousse and increased in importance relative to even substituted dibenzothiophenes during the 180-day incubation period. The naphthalene series and the C_1 and unsubstituted phenanthrenes were lost relatively rapidly; comparison with sterile controls indicated that losses of these compounds were due to both biodegradation and abiotic weathering; microbial degradation accelerated removal of these compounds from the mousse.

DISCUSSION

There was a significant elevation (several orders of magnitude) in counts of hydrocarbon-utilizing microorganisms in association with mousse formed from oil released from the Ixtoc I well in the Gulf of Mexico. This rise represented a shift within the microbial community; there was no significant elevation in total microbial biomass. Similar selective increases in concentrations of hydrocarbon utilizers have been reported in most cases following oil spills and have been noted by the authors of this chapter in previous studies of marine and inland oil spills, including in sediments impacted by the Amoco Cadiz oil spill.

Despite the elevated populations of hydrocarbon utilizers, rates of hydrocarbon degradation appear to be extremely slow. Unlike previous reports in the literature concerning the fate of crude oil in the Gulf of Mexico, e.g., Kator et al. [6], which would indicate extensive degradation within 24 hours, the complete biodegradation to CO_2 of hydrocarbons in mousse from Ixtoc-I was of the order of $< 5\%$/yr. There was evidence for a relatively severe nutrient limitation to extensive oil biodegradation. Although nutrient levels were not measured directly, the comparison of natural and nutrient-stimulated rates of hydrocarbon mineralization strongly suggests that concentrations of available nitrogen and phosphorus in the Bay of Campeche surface waters were very low.

The hydrocarbon composition of mousse collected near (site P5) and at some distance from the wellhead (site P17) did not show evidence of rapid microbial modification of the hydrocarbons in the mousse. Unlike analyses of hydrocarbons in sediments impacted by "mousse" from the Amoco Cadiz spill where n-alkane to isoprenoid hydrocarbon ratios shifted rapidly from

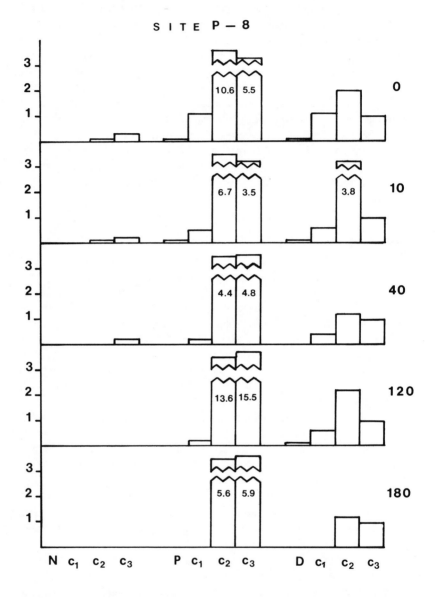

Figure 7. Changes in concentrations of selected hydrocarbons in the f_2 fraction relative to C_3 dibenzothiophene during incubation of mousse collected at site P8. Y axis shows relative concentration. X axis shows identity of hydrocarbon. N = naphthalene; P = phenanthrene; D = dibenzothiophene; C_1, C_2 and C_3 refer to the degree of substitution. Time of incubation = 0, 10, 40, 120 and 180 days.

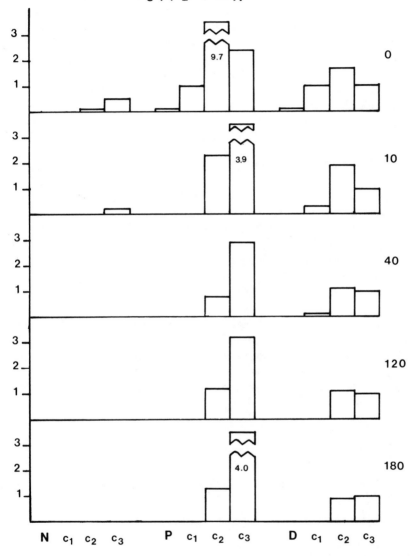

Figure 8. Changes in concentrations of selected hydrocarbons in the f_2 fraction relative to C_3 dibenzothiophene during incubation of mousse collected at site P17. Y axis shows relative concentration. X axis shows identity of hydrocarbon. N = naphthalene; P = phenanthrene; D = dibenzothiophene; C_1, C_2 and C_3 refer to the degree of substitution. Time of incubation = 0, 10, 40, 120 and 180 days.

4 to < 1 within a few weeks following the spill, the n-alkane to phytane ratios generally remained > 1 in mousse collected in the Gulf of Mexico even following extensive incubation times. A decrease in the n-alkane to isoprenoid hydrocarbon ratio is taken as evidence of biodegradation since microorganisms usually degrade straight chain alkanes more rapidly than branched alkanes.

There is some difficulty in establishing appropriate units for degradation rates of hydrocarbons in mousse. Compared to oil slicks, a lower percentage of hydrocarbon was found to be completely degraded to CO_2 in the mousse although absolute rates of CO_2 evolution extrapolated to comparable volumes of seawater were not significantly different between hydrocarbons in oil slicks and hydrocarbons in mousse. It is likely that hydrocarbons in large mousse accumulations are not as rapidly degraded as oil available in slicks or in fine emulsions within the water column due to unfavorable surface-area-to-volume relationships and poor transport of nutrients to hydrocarbons within large accumulations of mousse. It seems safe to conclude from the evidence that biodegradation of hydrocarbons in mousse floating on surface waters in the Gulf of Mexico was severely limited; it was occurring but at low rates. Mousse collected some distance from the wellhead did not show any evidence of extensive microbial degradation such as an alteration in the n-alkane to isoprenoid hydrocarbon ratios. It appears that sufficient time existed for transport of the mousse in relatively undegraded states to other systems such as benthic or intertidal sediments. Evidence for long-term mineralization of the hydrocarbons in the mousse was found but it is more likely that extensive degradation would occur in systems with greater availability of nutrients, such as in coastal lagoons and estuaries where the oil may be transported and deposited, than within the floating mousse accumulations.

ACKNOWLEDGMENTS

This project was supported by a contract with the Office of Marine Pollution of NOAA.

REFERENCES

1. Bartha, R., and R. M. Atlas. "The Microbiology of Aquatic Oil Spills," *Adv. Appl. Microbiol.* 22:225–266 (1977).
2. Colwell, R. R., and J. D. Walker. "Ecological Aspects of Microbial Degradation of Petroleum in the Marine Environment," *CRC Crit. Rev. Microbiol.,* 5:423–445 (1977).

3. Daley, R. J., and J. E. Hobbie. "Direct Counts of Aquatic Bacteria by a Modified Epifluorescence Technique," *Limnol. Oceanog.* 20:875-882 (1975).

4. Atlas, R. M. "Measurement of Hydrocarbon Biodegradation Potentials and Enumeration of Hydrocarbon Utilizing Microorganisms Using Carbon-14 Hydrocarbon-Spiked Crude Oil," in *Native Aquatic Bacteria: Enumeration, Activity and Ecology*, J. W. Costerton and R. R. Colwell, Eds. (Philadelphia, PA: ASTM, STP 695, 1979), pp. 196-204.

5. Brown, D. W., L. S. Ramos, A. J. Friedman and W. D. MacLeod. "Analysis of Trace Levels of Petroleum Hydrocarbons in Marine Sediments Using a Solvent/Slurry Extraction Procedure," in *Trace Organic Analysis: A New Frontier in Analytical Chemistry*, Spec. Pub. 519 (Washinton, DC: National Bureau of Standards, 1979), pp. 161-167.

6. Kator, H., C. H. Oppenheimer and R. J. Miget. "Microbial Degradation of a Louisiana Crude Oil in Closed Flasks and under Simulated Field Conditions," in *Prevention and Control of Oil Spills* (Washington, DC: American Petroleum Inst., 1971), pp. 287-296.

CHAPTER 11

CHEMICAL CHARACTERIZATION OF IXTOC-I
PETROLEUM IN TEXAS COASTAL WATERS

R. S. Scalan, J. K. Winters* and P. L. Parker

The University of Texas
Marine Science Institute
Port Aransas Marine Laborabory
Port Aransas, Texas 78373

INTRODUCTION

On June 3, 1979 the Ixtoc-I well in the Bay of Campeche, Gulf of Mexico, blew out with the subsequent release of approximately three million barrels of oil. Despite efforts to collect and disperse the spill, by late July it was obvious that floating oil would impact Texas coastal waters. On August 1, 1979, the Port Aransas Marine Laboratory, under contract to the U.S. Coast Guard, initiated studies to chemically characterize fresh and floating Ixtoc-I oil prior to its impacting United States waters. The studies included tests of acute toxicity of floating oil and of a "water soluble" fraction of the floating oil toward phytoplankton, larval fish, benthic animals and mature fish and invertebrates. The purpose of the study was to provide chemical and biological information that could be used to help estimate the duration and degree of impact of Ixtoc-I oil on coastal biota. This chapter describes the results of the chemical aspects of those studies, the biological results are reported elsewhere [1].

The U.S. Coast Guard cutter, Point Baker, stationed at Port Aransas, was sent to the area off Tampico, Mexico, 230 miles SSE of Brownsville, Texas to

---

*Present address: Coastal Science Laboratories, Port Aransas, Texas.

collect samples of floating oil. Large quantities of material derived from Ixtoc-I were being carried by winds and fast-moving surface currents toward Texas coastal waters. On July 28, 1979, 28 quarts of material subsequently known as the "Point Baker mousse" were delivered to our laboratory by Dr. Joseph Lafonara of the U.S. Environmental Protection Agency. Chemical characterization of this material was desirable because the mousse and a water soluble fraction prepared from it were to be used in acute toxicity tests.

As the oil spill progressed, the Texas and Mexican beaches were heavily oiled during the late summer and early fall of 1979. The need for chemical methods to identify Ixtoc-I oil became evident. Scientific surveys to determine the distribution of the oil and individual citizens bringing samples to various labs called for fairly rapid characterization techniques. The gas-liquid-chromatographic (GLC) methods developed earlier in our laboratory and those of others were the methods of first choice. However, GLC is slow and fairly expensive and the results are somewhat dependent on the weathering history of a particular sample. For these reasons it was decided to evaluate other identification methods including variations in the stable carbon, sulfur and nitrogen isotope ratios.

MATERIALS AND METHODS

Standard organic geochemical methods were used to characterize the Point Baker mousse and other samples which later came to our laboratory. These methods and results are described in this chapter.

Samples of Ixtoc-I oil, mousse and beach tar were obtained from several private and government sources as shown in Table 1. The R/V Longhorn of the University of Texas made several cruises which provided additional samples. Five samples of oil collected from a beach south of the United States-Mexico border were analyzed. Although the exact sampling location is not known, they were included because they represented samples arriving very early. A sample of what was reported to be the "first" Ixtoc-I oil to come ashore on South Padre Island, Texas was received from Craig Hooper of NOAA. A sample of mousse collected near the well site within forty hours of the blowout was received from Dr. Alfonso Botello of the National University of Mexico, Mexico City. This was compared to fresh samples obtained from PEMEX through Dr. Carl Oppenheimer and to a sample collected by Dr. John Robinson of NOAA.

Oil "accommodated" in seawater used for acute toxicity tests was prepared by shaking a 1% mixture of Point Baker mousse in seawater for one hour on an Eberbach shaker. The mixture was allowed to settle for one hour after shaking before the aqueous phase was siphoned off for chemical and

Table 1. Component Type Analyses of Ixtoc-I Samples

Sample Description	Source	% Saturate	% Aromatic	% NSO
Mousse, collected 40 hr after blowout	Dr. Botello Mexico	52.0	34.0	N.A.
Oil, unweathered	Dr. John Robinson NOAA	51.4	31.8	6.9
Mousse, unweathered	Dr. Oppenheimer	53.0	34.7	7.5
Mousse, Pt. Baker 22°59'N 96°26'W	U.S.C.G.	51.0	26.4	18.3
Pt. Baker mousse, accommodated particles	UTMSI/PAML	48.0	18.7	19.4
Beached oil, lower beach, Mexico	U.S.C.G.	34.7	28.5	34.1
Beached oil, upper beach, Mexico	U.S.C.G.	38.0	18.0	38.3
Beached oil, midbeach Mexico	U.S.C.G.	37.3	14.0	40.5
Tarballs in surf, Mexico	U.S.C.G.	29.4	17.3	40.2
Tar on beach, Mexico	U.S.C.G.	36.7	15.8	43.3
"First" Ixtoc-I tarball, South Padre	NOAA	34.4	15.4	39.4
Average of non-Ixtoc-I beach tars, 1978–1979 St. Joseph Island, Texas	Dr. Scalan UTMSI/PAML	46.8 (8–77)	32.3 (14–55)	20.8 (8–41)

biological studies. A water-soluble fraction (WSF) for chemical and biological studies was prepared from the oil-accommodated-seawater mixture produced on the shaker by filtering the mixture twice through glass fiber filters (Whatman GF/C) using gentle suction. Accommodated or WSF oil was back-extracted from seawater in separatory funnels with three aliquots of dichloromethane in order to determine its composition.

All samples of oil and oil extracts, except the WSF, were separated into four chemical fractions by column chromatography using silica gel. In this procedure a weighed sample is treated with n-pentane and the resulting asphaltene precipitate removed by filtration. The pentane filtrate was placed on a glass column (19 × 300 mm) packed with 200 mesh, activity 1, ICN Pharmaceuticals, silica gel. Care was taken not to overload the column. Three fractions were eluted: the saturated hydrocarbon fraction was eluted with two column volumes of hexane; the nonsaturated or aromatic hydrocarbon fraction was eluted with a hexane:benzene (1:1) mixture; and the nitrogen-, sulfur-, oxygen-containing compounds (NSO) were eluted with a volume of methanol sufficient to elute all color. All four fractions were suitable for stable carbon isotope analyses, $\delta^{13}C$, as described in a later section.

Gas chromatography was carried out on a Perkin-Elmer 910 gas chromatograph equipped with a flame ionization detector. Electronic integration of peak area was performed by a Hewlett-Packard 3352B Data System. The glass capillary column utilized for most analyses was OV-101, 11 m $\times$ 0.25 mm i.d. Oven temperature was programmed from 70 to 255°C at 3°/min.

Some analyses were performed on a 27-m OV-101 column programmed from 100 to 260°C at 2°/min. Combined gas chromatography/mass spectrometry (GC/MS) of selected samples was carried out on a Dupont 21-491 mass spectrometer with a Dupont 21-094 B data system to assist with compound identification.

The water content of the mousse was determined by adding an equal volume of hexane to the mousse, followed by centrifugation at 10,000 rpm for twenty minutes. The phases were separated and the volumes determined.

The methods of determining $\delta^{13}C$ and $\delta^{34}S$ are described in the literature and are only summarized here [2]. Organic matter is converted to CO_2 for carbon mass analysis by a method modified from Sofer [3]. Sulfur isotopes, $\delta^{34}S$, were measured on SO_2 which was prepared by a sealed tube method modified from that of Holt and Engelkemeir [4] and Bailey and Smith [5]. Samples of nitrogen gas suitable for isotope ratio mass spectrometry were prepared by a modified Dumas method developed by Macko [6].

The mass spectrometer used for isotope analysis was a VG-Micromass model 602 E, 90° sector field, double collector instrument, similar to that described by McKinney et al. [7]. The isotope analyses are expressed in terms of the parts per thousand (per mil) difference between the isotope ratio of the sample and that of an arbitrary standard.

$$\delta^{13}C = \frac{(^{13}C / ^{12}C) \text{ sample} - (^{13}C / ^{12}C) \text{ std}}{(^{13}C / ^{12}C) \text{ std}} \times 10^3$$

Analogous expressions are used for $\delta^{15}N$ and $\delta^{34}S$ definitions. The standard for carbon is the Peedee belemnite (PDB) carbonate, $\delta^{34}S$ variations are reported relative to the Canyon Diablo meteorite, and $\delta^{15}N$ variations are reported relative to air nitrogen. The experimental errors for sulfur, carbon and nitrogen are ± 0.5, ± 0.2 and ± 0.2, respectively.

RESULTS AND CONCLUSIONS

Results of Silica gel chromatographic separations are given in Table 1. The data indicate Ixtoc-I oil originally contained 52-53% saturates, 34-35% aromatics and 7-8% NSO compounds. The Point Baker mousse used for toxicity tests was found to contain a similar percentage of saturates (51%) but a considerably lower percentage of aromatics (26%). The NSO fraction increased to 18% in the Point Baker mousse. The Mexican beach samples collected by the Coast Guard and the first Ixtoc-I tarball collected on South Padre Island show similar percentages of silica gel fractions. These percentages are lower in aromatics and higher in NSO compounds than most tarballs collected in a study conducted on St. Joseph and Padre Islands for two years preceding the Ixtoc-I spill [8]. Results of studies of this oil spill, probably the world's largest, confirm many small-scale observations and guesses as to the chemical and physical changes that spilled oil undergoes in the marine environment.

Gas chromatography indicated the saturate fraction of the Point Baker mousse (Figure 1B) contained n-alkanes from C_{13} to greater than C_{36} with highest concentrations in the C_{18} to C_{20} range. The mousse sample collected forty hours after the blowout (Figure 1A) contained n-alkanes lower than C_{10} with a maximum at C_{13}.

Analyses of the aromatic fraction from the Point Baker mousse (Figure 2B) revealed the presence of alkyl naphthalenes, primarily C_2-(dimethyl + ethyl) and C_3 homologs. C_3-naphthalenes were present at a concentration which was about one-third of that present in the 40-hr sample (Figure 2A). Three-ring aromatic compounds such as phenanthrenes and dibenzothiophenes are present in both the Point Baker mousse and the 40-hr sample in similar relative concentrations. The Ixtoc-I oil contains a relatively high concentration of alkyl dibenzothiophenes with concentrations similar to those of alkyl phenanthrenes.

The WSF was chemically characterized for two purposes. As mentioned, the biology team at the laboratory was using it in acute toxicity studies. A second reason was that many scientists felt that the major biological impact of Ixtoc-I oil on the biota would be on organisms living near the surface of the water column near the well site. In this case the WSF and the chemical dispersing agents being used at the well site would be among the more toxic substances.

We later found total dissolved alkane values as high as 650 ng/L at offshore stations remote from the well site which were highly impacted by floating oil. One expects that the aromatic values may have been even higher.

The "water soluble" material prepared from the accommodated oil-seawater mixture was not fractionated on silica gel. A typical chromatogram

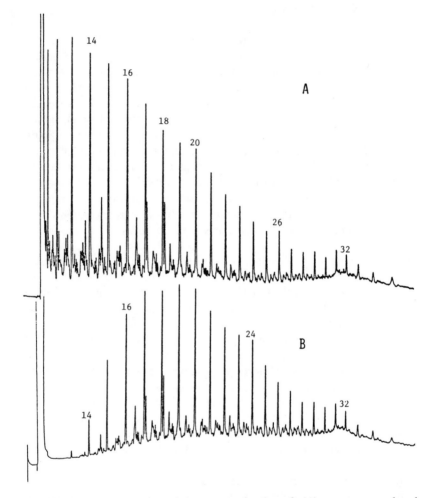

Figure 1. Gas chromatograms of the saturate fraction of: (A) a mousse sample collected 40 hr after the Ixtoc-I blowout and (B) the Pt. Baker mousse.

of the WSF is shown in Figure 3. The increased concentration of more soluble aromatic compounds relative to the Point Baker mousse was evident. The presence of n-alkanes in the WSF probably indicates that some oil particles were sufficiently small to pass through the glass fiber filters. The n-alkane distribution in the soluble fraction was from about C_{19} to greater than C_{37} with a maximum at about C_{31}. This distribution was significantly different from the Point Baker mousse which had a maximum at about C_{19}.

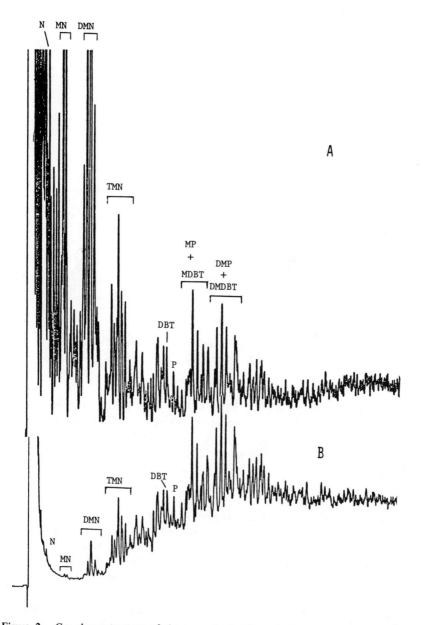

Figure 2. Gas chromatograms of the aromatic fraction of: (A) a mousse sample collected 40 hr after the Ixtoc-I blowout; and (B) the Pt. Baker mousse. N = naphthalene, MN = methylnaphthalene, DMN = dimethylnaphthalenes (includes ethylnaphthalenes), TMN = trimethylnaphthalenes (includes all methylethyl- and propylnaphthalenes), DBT = dibenzothiophene, P = phenanthrene, MP = methylphenanthrenes, MDBT = methyldibenzothiophenes, DMP = dimethylphenanthrenes (see DMN), DMDBT = dimethyldibenzothiophenes (see DMN).

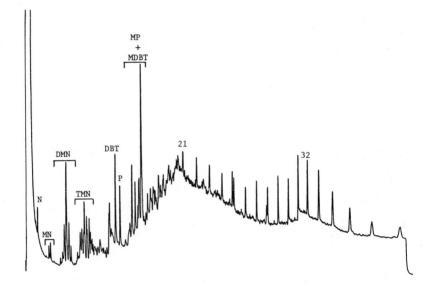

Figure 3. Gas chromatogram of the water-soluble fraction prepared from the Pt. Baker mousse (see Figure 2 for peak identification).

The samples of oil from a Mexican beach gave similar results, with some samples showing slightly more weathering. Typical chromatograms of saturate and aromatic fractions are given in Figure 4. Normal alkanes began at C_{15} with a maximum at C_{21}. The aromatic fraction indicated that alkyl naphthalenes had been almost completely lost. Phenanthrene, dibenzothiophene and methyl homologs were also significantly lower in concentration relative to the C_2 and C_3 homologs.

A chromatogram of the saturate fraction of a tarball from South Padre Island is shown in Figure 5A. Maximum n-alkane concentration occurred at C_{20}. The aromatic fraction (Figure 5B) was found to contain a large concentration of alkyl benzenes and naphthalenes relative to all other aromatic fractions analyzed other than samples of very fresh mousse. A chromatogram of the aromatic fraction of Point Baker mousse has been included in Figure 5C for comparison. The Point Baker mousse proved to be similar in composition, but not identical, to oil that later landed on the Texas coast. For this reason it may have been a poor choice as a sample with which to perform the toxicity tests.

Stable isotope ratios, $\delta^{13}C$, $\delta^{34}S$ and $\delta^{15}N$, have proved to be very powerful tools for characterizing Ixtoc-I oil and for distinguishing it from

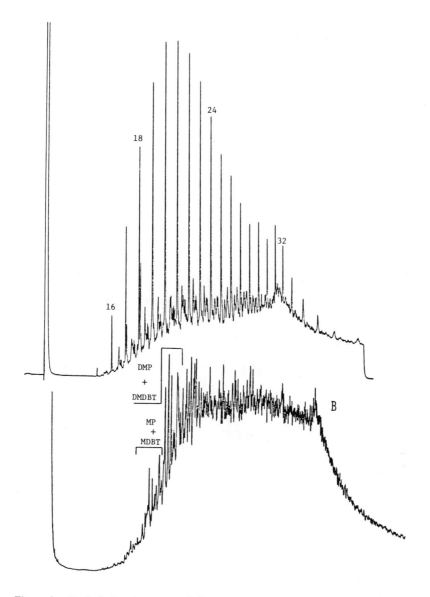

Figure 4. Typical chromatograms of the saturate (A), and aromatic (B), fractions of Ixtoc-I oil from a Mexican beach (see Figure 2 for peak identification).

other oils present in the local environment. Our laboratory was doing a two-year study of $\delta^{13}C$ of oils and tar on South Texas beaches started prior to the Ixtoc-I spill. These data provided a baseline for comparison of the Ixtoc-I

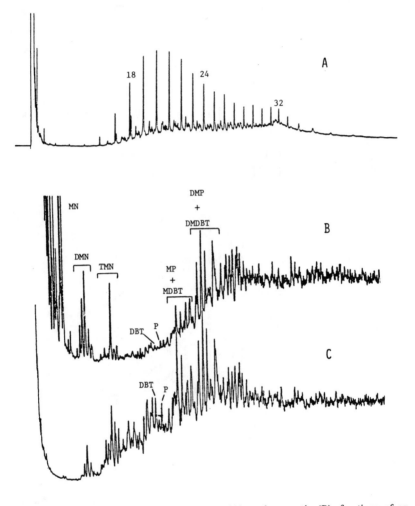

Figure 5. Gas chromatogram of the saturate (A), and aromatic (B), fractions of an Ixtoc-I tarball from South Padre Island, Texas. Pt. Baker mousse aromatic fraction (C) shows much lower volatile aromatics (see Figure 2 for peak identification).

oil. In order to enhance the utility of the comparison the oil was resolved into three fractions—saturates, aromatics, and nitrogen/sulfur/oxygen compounds (NSO)—by silica gel chromatography.

Values for $\delta^{13}C_{PDB}$ in the first two lines of Table 2 fall in a relatively narrow range. These are samples which are known to be authentic Ixtoc-I oil. They are compared in the bottom two lines with samples collected a year

Table 2. Ranges of Values of $\delta^{13}C_{PDB}$ of Ixtoc-I Oil and Beached Oil for Separated Fractions

Sample	Saturate	Aromatic	NSO
Fresh mousse (7) summer 1979	−27.6±.31	−27.0±.24	−26.8±.24
Weathered mousse (16) summer 1979	−27.4±.18	−26.9±.12	−26.8±.22
Padre Island National Seashore (4) summer 1980	−27.5±.30	−27.3±.30	−27.0±.24
South Padre Island (4) summer 1980	−27.5±.12	−27.2±.15	−26.9±.22

Table 3. $\delta^{13}C_{PDB}$ of Fractions of Tarballs Collected from Mustang and St. Joseph Islands, Texas, Prior to the Ixtoc-I Spill

Sample	Saturate	Aromatic
1	−29.3	−26.9
2	−25.4	−24.6
3	−27.9	−27.1
4	−26.6	−25.9
5	−28.0	−25.9
6	−29.4	−28.5
7	−28.1	−26.9
8	−25.4	−24.2
9	−27.1	−26.4
10	−27.1	−26.4

following the spill. The latter samples have stable carbon isotopic values which suggest an Ixtoc-I origin.

For comparison with the Ixtoc-I samples some values for ten tarballs collected on one day in 1978 from St. Joseph and Mustang Islands are given in Table 3. In some instances carbon isotope values of a fraction may fall within the Ixtoc-I range, but seldom are both saturate and aromatic values within range. This comparison serves as a test of the method which can be used for post-spill identification of beach tars and oils.

In the fall of 1979 Dr. Henry Hildebrand made an extensive collection of oil on Mexican beaches from Matamoros south to Santa Clara. The data in Table 4 were derived from part of these samples. They provide an application of the $\delta^{13}C$ method for establishing the probability that samples are derived from Ixtoc-I oil. Only the sample from La Pesca is clearly from a non-Ixtoc-I source. The others are probably derived from Ixtoc-I oil. These represent

Table 4. $\delta^{13}C_{PDB}$ of Tarballs Collected from Mexican Beaches in the Fall of 1979

Sample Location	Saturate	Aromatic	NSO
Matamoros	−27.8	−26.9	−26.7
La Pesca	−25.9	−24.8	−25.1
Tampico	−27.1	−26.8	−26.7
Tuxpan	−27.4	−27.4	−27.1
Veracruz	−27.3	−27.0	−26.9
Alvarado	−27.3	−26.8	−26.8
Coatzacoalcos	−27.1	−26.9	−26.7
Barra Tupilco	−27.4	−	−26.9
Isla Del Carmen	−27.4	−27.0	−26.7
Celestum	−27.9	−26.9	−26.7
Santa Clara	−	−26.8	−26.9

Table 5. $\delta^{34}S_{CDM}$ and $\delta^{15}N_{Air}$ for Oil from Various Sources

	Whole oil $\delta^{34}S$	NSO fraction $\delta^{15}N$
Ixtoc I	−5.1 (7)	0.7 (2)
Burma Agate	16.4* (1)	5.6 (1)
Beach Tar July 1980	−4.2 (8)	0.7 (5)

*This sample was very low in total sulfur and may be appreciably contaminated with seawater sulfate.

material deposited on Mexican beaches after the fall coastal current reversal took most of the floating mousse back south.

We made limited but interesting observations on values of $\delta^{34}S$ and $\delta^{15}N$ of Ixtoc-I oil, post-Ixtoc-I beach tar and Burma Agate oil (Table 5). The Burma Agate oil was derived from a tanker by that name which was grounded off Galveston, Texas during the Ixtoc-I spill. It appeared for a while that the two oils might mix. Values of $\delta^{15}N$ for the two spills are well resolved: 0.7 vs 5.7. The $\delta^{34}S$ values are also well resolved but because the Burma Agate sample was very low in total sulfur content, the single value given may be in doubt. Possible contamination of such samples with seawater sulfate is always a problem.

In summary, it has been demonstrated that a combination of high-resolution gas chromatography and stable carbon, nitrogen and sulfur isotope ratio variations were adequate to characterize and distinguish the Ixtoc-I oil from other oil randomly found in the coastal waters of South Texas.

ACKNOWLEDGMENTS

Several graduate students helped with the field and laboratory work which Ixtoc brought us. We thank Mark A. Northam and Lee Enzeroth for leaving their thesis research to help in an emergency. Steve Macko provided the $\delta^{15}N$ values. Financial support was provided by the U.S. Coast Guard, contract 08-4119-79, and by the National Oceanic and Atmospheric Administration, contract NA79 RAC00141.

REFERENCES

1. Parker, P. L. "Ixtoc-I Chemical Characterization and Acute Toxicity Effects," Final report to the U.S. Environmental Protection Agency, contract DOT–CG08–8174 (1979).
2. Calder, J., and P. L. Parker. "Stable Carbon Isotope Ratios as Indices of Petrochemical Pollution of Aquatic Systems," *Environ. Sci. Technol.* 2(7):535–539 (1968).
3. Sofer, Z. "Preparation of Carbon Dioxide for Stable Carbon Isotope Analysis of Petroleum Fractions," *Anal. Chem.* 52:1389–1391 (1980).
4. Holt, B. D., and A. G. Engelkemeir. "Thermal Decomposition of Barium Sulfate to Sulfur Dioxide for Mass Spectrometric Analysis," *Anal. Chem.* 42(12):1451–1453 (1970).
5. Bailey, S. H., and J. W. Smith. "Improved Method for Preparation of Sulfur Dioxide from Barium Sulfate for Isotope Studies," *Anal. Chem.* 44(8):1542–1543 (1972).
6. Macko, S. A. Personal communications and unpublished results (1979).
7. McKinney, C. R., et al. "Improvements in Mass Spectrometers for the Measurement of Small Differences in Isotope Abundance Ratios," *Rev. Sci. Inst.* 21:724–730 (1950).
8. Scalan, R. S., and J. K. Winters. "Quantitation and Organic Geochemical Characterization of Petroleum-Like Material Found on Undisturbed Beaches of Padre Island National Seashore and St. Joseph Island," private communications to Sid Richardson Foundation.

CHAPTER 12

CHARACTERIZATION OF AZAARENES IN IXTOC-I OIL BY GAS CHROMATOGRAPHY AND GAS CHROMATOGRAPHY/MASS SPECTROMETRY

I. R. DeLeon, E. B. Overton, G. C. Umeonyiagu
and J. L. Laseter

Center for Bio-Organic Studies
University of New Orleans
Lakefront
New Orleans, Louisiana 70148

INTRODUCTION

The Ixtoc-I blowout and oil spill in the Gulf of Mexico has provided us with the opportunity to continue our investigations into the chemistry, nature, distribution and fate of petroleum and other fossil fuel hydrocarbons and nonhydrocarbons in the marine environment [1-8]. Our objectives in this effort were the following: (1) to determine "passive tags" which would allow differentiation of Ixtoc-I oil from other petroleum hydrocarbon sources; and (2) to study the effects of weathering on hydrocarbons and nitrogen-, sulfur- and oxygen-containing compounds in Ixtoc-I oil. In this chapter we describe the results of our first objective.

Petroleum is an extremely complex mixture of individual components covering a wide variety of chemical classes. Its diagensis has been influenced by many factors including very complicated biological, chemical and geological processes, and its composition has been derived from living organic matter. Thus, in order for any two oils to be identical, their chemical compositions must have been influenced by exactly the same set of factors [9–10].

The identification and characterization of petroleum and other fossil fuel sources in the marine environment is often a difficult task. These efforts are complicated by the complexity of the samples, the small differences that often exist between samples, the effects of weathering in the environment and the possible contamination of the samples by other organic matter [11].

The process of identifying petroleum sources is called "fingerprinting" or passive tagging of oils. This process is based on the principle that each dissimilar oil differs in its chemical composition, and therefore has a unique fingerprint. Passive tagging of oils can take a variety of forms, depending on the instrumentation and analytical techniques employed. Various oil fingerprinting techniques have been described in the literature [9–15]. The most commonly used methods include separation and spectral techniques, and combinations of both. One of the most powerful of these techniques is the combination of high-resolution gas chromatography (HRGC) and mass spectrometry (MS).

Some of the more routine forms of passive tagging of oils employ the measurement, determination and comparison of atomic ratios such as $^{13}C/$ ^{12}C, $^{34}S/^{32}S$, N/S and Ni/Vd. Other forms are based on measurements by gas chromatography (GC). The earlier GC and GC/MS techniques employed the measurement, determination and comparison of selected hydrocarbon ratios and distributions in oil. Included in these were: (1) pristane/phytane ratios, (2) the combination of pristane/n-C_{17} alkane and phytane/n-C_{18} alkane ratios, (3) stearane and hopane distributions and (4) Rasmussen indices.

The development of high-resolution glass capillary gas chromatographic (GC^2) methods facilitates the use of techniques that rely on aromatic hydrocarbon homolog patterns for the identification of oils. These techniques require silica gel fractionation and the use of HRGC/MS with the display of mass chromatograms for the parent and alkyl aromatic hydrocarbon homologs of such compound families as naphthalene, acenaphthylene, fluorene, phenanthrene and pyrene [8].

More recently, these methods have been extended and applied to study the distribution and patterns of aromatic heterocyclic compounds in oils. These methods work well with fairly nonpolar heterocyclic compounds such as benzothiophenes, dibenzothiophenes and naphthobenzothiophenes, which coelute with the aromatic hydrocarbons. However, these techniques have not been very successful in determining aromatic heterocyclics, which elute in the more polar fraction, because these compounds are usually masked by coelution with the much more abundant fatty acids, other biogenic compounds and phthalate esters.

An approach that we have recently used involves the isolation, determination and characterization of heterocyclic aromatic nitrogen-containing compounds in oil—the azaarenes. The basis for this approach is that once the azaarenes are selectively isolated and analyzed, the differences in their distributions in oils can be used effectively to distinguish one oil from another. In this chapter, we discuss how this approach has been utilized in studies related to the Ixtoc-I oil spill.

EXPERIMENTAL

Reagents

Toluene, benzene, pentane, methylene chloride and chloroform (distilled-in-glass quality, Burdick and Jackson, Muskegon, Michigan) were used as received. All other reagents were ACS reagent grade or equivalent. Glassware was cleaned with detergents and water, followed by dipping in chromic acid cleaning solution and rinsing with deionized water prior to use.

Extraction

A small sample of oil, usually 0.8 g, was dissolved in 20 mL of toluene. The azaarenes (nitrogen-containing aromatics) in the toluene solution were extracted with three 15-mL portions of 4 N HCl. The HCl extracts were combined and washed three times with 15-mL portions of pentane. The pentane-washed HCl extract was slowly neutralized by the direct addition of a 50% NaOH solution. The pH of the mixture was adjusted to very basic (pH 11-12). The basified mixture was extracted three times with 15-mL portions of either chloroform or methylene chloride. The extracts were combined and reduced in volume to approximately 3 mL on a rotary evaporator, followed by further concentration under a gentle stream of dry nitrogen. The halocarbon solvent was displaced by the addition of small volumes of benzene and subsequent concentration under dry nitrogen. The extracts were never taken to complete dryness. The final volume was adjusted to 50 μL. Aliquots of 1-2 μL were analyzed by GC and GC/MS.

Glass Capillary Gas Chromatography

Gas chromatographic analyses were performed on a Hewlett-Packard (HP) 5710A gas chromatograph and a Tracor 560 gas chromatograph. Both gas chromatographs were interfaced to an HP3354B laboratory data system for acquisition, storage and reduction of data.

The HP 5710A gas chromatograph was equipped with a flame ionization detector (FID), and a 0.3-mm × 30-m high-resolution glass capillary column coated with SE-52 methylphenyl silicone. The column temperature was programmed from 50°C to 240°C at 4°/min. The injection port temperature was kept at 250°C, the detector temperature at 300°C. The carrier gas was helium, at a flowrate of 2.0 mL/min at the injection temperature. All injections were splitless and employed a 35-sec delay in venting.

The Tracor 560 gas chromatograph was equipped with both an FID and a model 700A Hall electrolytic conductivity detector (HECD) which was operated in the nitrogen specific detection (NSD) mode. The gas chromatographic separations were performed on a 0.3-mm × 50-m high-resolution glass capillary column coated with SE-52 methylphenyl silicone. The column temperature was programmed from 50°C to 240°C at 8°/min. The injection port temperature was kept at 250°C, the FID temperature at 300°C. The HECD furnace was kept at 800°C. Effluent from the glass capillary column was split between the FID and the HECD. Consequently, simultaneous general and element-specific detection were achieved from the HRGC analysis of these extracts. Other parameters for the simultaneous use of the FID and HECD in the nitrogen-specific mode are described elsewhere [16].

Glass Capillary Gas Chromatography/Mass Spectrometry

Analyses by combined GC/MS were performed on either a Varian MAT 311A GC/MS-Data System and on an HP 5985A GC/MS-Data System. Both GC/MS-DS systems were equipped with 0.3-mm × 50-m high-resolution glass capillary columns coated with SE-52 methylphenyl silicone. The GC column effluent was introduced directly into the mass spectrometer's ion source, which was maintained at 250°C. The ionization potentials were at 70 eV. The Varian mass spectrometer was operated at a resolution of 1000 (M/ΔM at 10% valley), the HP quadrupole mass spectrometer at a unit resolution of approximately 600. All mass spectrometric data were acquired and processed using the respective data systems. Data displays included total ion summation plots, mass chromatograms of selected masses and background-subtracted mass spectra.

RESULTS AND DISCUSSION

The presence of heterocyclic azaarenes (aromatic nitrogen-containing compounds) in petroleum and other fossil fuel sources is well established. In petroleum, this group of compounds represents a minute portion, usually <1%, of the whole oil [17, 18]; other fossil fuels contain relatively higher levels of azaarenes.

As with other compound classes, the azaarene content in petroleum is expected to be unique for each production zone. This uniqueness has been the basis of recent gas chromatographic attempts to fingerprint oils [15, 18, 19]. These passive tagging attempts have been of limited success owing to the general difficulties encountered in the gas chromatography of nitrogen-containing compounds, and the nonutilization of more selective extraction methods.

Analytical Approach

The isolation procedure we applied in this study is very easy to use. It employs the classical technique for extracting nitrogen bases from complex organic mixtures. In this technique, the organic mixture is treated with a mineral acid, which effects the conversion of the nitrogen bases into their corresponding salts. These salts are very soluble in aqueous media and consequently are quantitatively partitioned into the aqueous phase. The aqueous phase is separated, neutralized with base and brought to high pH. This process releases the nitrogen bases as free amines and allows their removal from the aqueous mixture by a simple liquid-liquid extraction with an organic solvent.

Figure 1 shows our analytical approach for the analysis of azaarenes in petroleum. The sample is dissolved in toluene and then partitioned against 4 M HCl. The azaarenes react with the acid and are converted to the water-soluble hydrochloride salts. These salts are subsequently converted to the free amines on neutralization and basification, and are taken up in chloroform or methylene chloride. Analysis of the final extract is by HRGC using the nitrogen-specific detector (NSD) or the flame ionization detector (FID), and by HRGC/MS. Other work-up techniques, such as fractionation of the extracts by high-performance liquid chromatography (HPLC), can and have been employed prior to instrumental analysis [20].

GC Comparison of Oils

Using the acidic extraction procedure, we analyzed and compared the content of azaarene bases in three relevent oils: a Middle Eastern crude, a southern Louisiana crude and Ixtoc-I crude. Data from these analyses are graphically illustrated by the high-resolution FID profiles from the acidic extract of the three oils in Figure 2. The azaarene content of each oil is unique and distinctively characteristic for that oil.

In order to demonstrate that the compounds in these extracts were nitrogen-containing, the samples were analyzed by HRGC using the nitrogen-specific detector and HRGC/MS. Figure 3 is a typical HRGC-NSD profile of the Ixtoc-I azaarene extract. In an attempt to further extend the resolution of

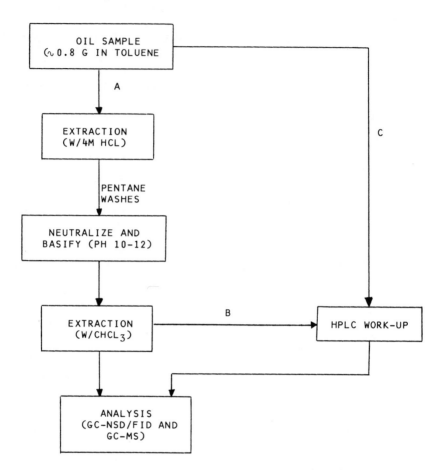

Figure 1.   Schematic diagram for the characterization of azaarenes in oil.

the Ixtoc-I azaarenes, this extract was fractionated by normal-phase HPLC using NH_2-bonded columns into polar and nonpolar fractions. The simultaneous HRGC-FID and HRGC-NSD profiles for these fractions are shown in Figure 4. Note that the volatility range for compounds in both fractions is approximately the same, and that both extracts are equally complex.

Characterization of Individual Azaarenes

Figure 5 gives the ring-structure types of some of the heterocyclic azaarene compounds we identified in Ixtoc-I oil. All of these compound types, except the carbazoles, were isolated and characterized by the method described in this chapter. The carbazoles were isolated and characterized by an alternative method using HPLC techniques that have been described elsewhere [20].

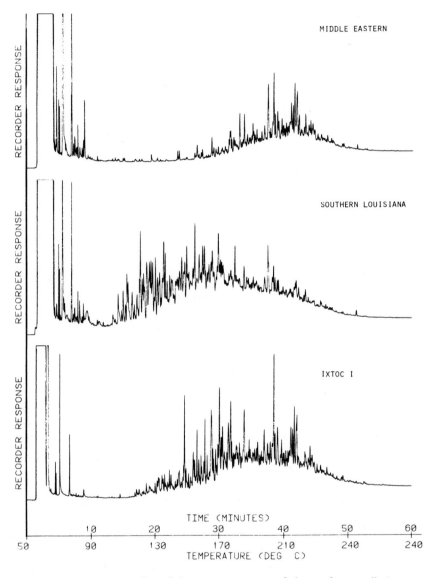

Figure 2. HRGC-FID profiles of the azaarene extract of three reference oils (upper trace = Middle Eastern crude, middle trace = Southern Louisiana crude, lower trace = Ixtoc-I crude).

Table 1 lists the specific component types and the molecular weights for azaarenes characterized in Ixtoc-I oil as part of this study. In most cases, three or more isomers of each component type were observed. Due to the

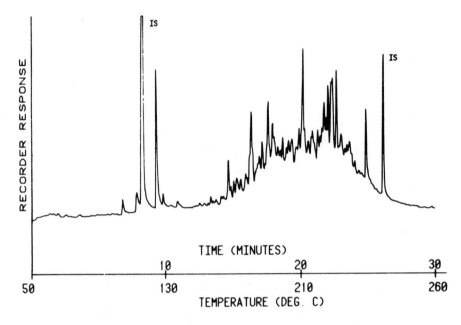

Figure 3. HRGC profile of Ixtoc-I oil using the nitrogen specific detector.

lack of authentic standards for most of the compounds observed, identifications were made on the basis of the mass spectral patterns for compounds in each component type and knowledge of the chemical work-up of the sample. The general classes of azaarenes observed in the Ixtoc-I oil are two- and three-ringed compounds. As with other aromatic compounds in petroleum, the alkyl homologs predominated.

GC/MS Comparison of Oils

Because of our continuing interest in establishing more reliable passive tags with which to identify sources of petroleum hydrocarbons in environmental samples, we sought a method that would provide a greater distinction between the oils. GC/MS methodology involving comparisons of total ion summation plots and of selected mass chromatograms for analysis of the acidic extracts proved to be very effective in achieving this goal. Figures 6–8 illustrate the effectiveness of this approach by comparing the mass chromatograms for selected ions in three homolog families of azaarenes present in both the southern Louisiana crude (SLC) and the Ixtoc-I crude.

For comparative purposes, equal initial quantities of these oils were processed and their extracts were analyzed under identical conditions. Figure 6 compares the mass chromatograms for the molecular ion series charac-

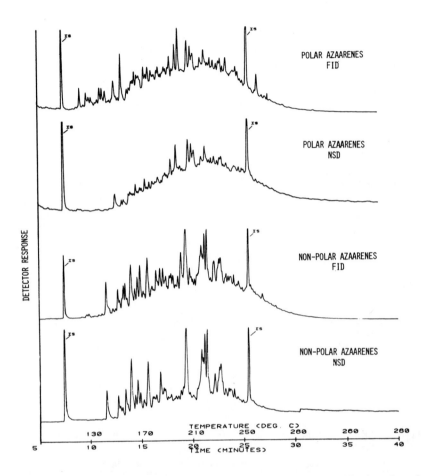

Figure 4. HRGC profiles using simultaneous general and nitrogen-specific detection of the HPLC polar and nonpolar fractions of the azaarene extract of Ixtoc-I oil.

teristic of the azafluorene family (m/z = 167, 181, 195, 209, 223, 237 and 251), Figure 7 compares the mass chromatograms for the molecular ion series corresponding to the quinoline/isoquinoline family (m/z = 129, 143, 157, 171, 185, 199 and 213). Finally, Figure 8 compares the mass chromatograms of the molecular ion series characteristic of the phenanthridine/benzoquinoline family (m/z = 179, 193, 207, 221, 235, 249 and 263). Note that the Ixtoc-I oil is richer than SLC oil in compounds having ions which are characteristic for the homolog series of the quinoline/isoquinoline and phenanthridine/benzoquinoline families. Conversely, the SLC oil is richer than the Ixtoc-I oil in compounds having ions which correspond to the homolog series for the azafluorene family.

QUINOLINE

ISOQUINOLINE

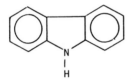

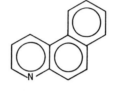

PHENANTHRIDINE/BENZOQUINOLINES

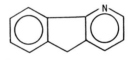

CARBAZOLE

AZAFLUORENE

Figure 5. Ring-structure types for the azaarenes characterized in Ixtoc-I oil by GC/MS methods.

Table 1. Azaarenes Characterized in Ixtoc-I Oil

Compounds	Molecular Weights
Quinoline/isoquinoline	129
C_1 to C_6 alkyl homologs	143, 157, 171, 185, 199, 213
Azafluorene	167
C_1 to C_6 alkyl homologs	181, 195, 209, 223, 237, 251
Carbazole	167
C_1 to C_6 alkyl homologs	181, 195, 209, 223, 237, 251
Phenanthridine/benzoquinoline	179
C_1 to C_6 alkyl homologs	193, 207, 221, 235, 249, 263

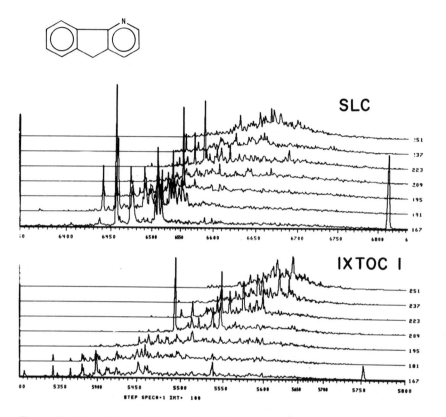

Figure 6. Selected mass chromatograms for the molecular ion series of azafluorene homologs in SLC and Ixtoc-I oils (P = 167, C_1 = 181, C_2 = 195, etc.).

Figure 9 compares the total ion summation plots for the azaarene extract of the SLC and the Ixtoc-I crude. The distributions of azaarenes in these oils, as shown by this figure, are very similar to the distribution for the respective oils in Figure 2. Both figures show that the azaarenes in SLC are composed of compounds having a lower molecular weight range and a higher volatility range than compounds found in the Ixtoc-I crude. Thus, by this method, we demonstrated that significant differences exist in the azaarene distributions in both oils, and that one oil can be easily distinguished from the other. It is interesting to note that the distribution of sulfur-containing aromatic compounds in these two oils are quite similar and cannot be used to effectively distinguish between them in environmental samples.

In order to determine if these new passive tags were affected by weathering processes, we applied this method to the analysis of a heavily weathered (moussified) Ixtoc-I oil sample and compared the results to those from the

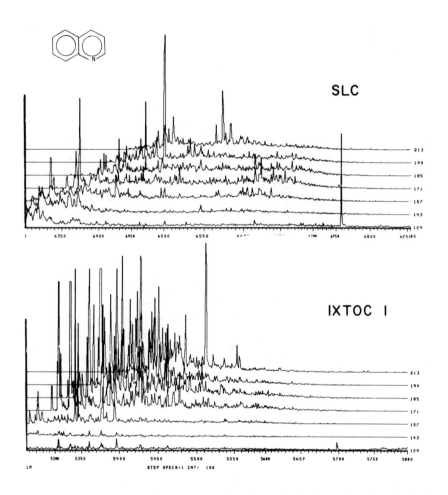

Figure 7. Selected mass chromatograms for the molecular ion series of quinoline/ isoquinoline homologs in SLC and Ixtoc-I oils (P = 129, C_1 = 143, C_2 = 157, etc.).

analysis of a fresh unweathered Ixtoc-I oil sample. Qualitatively, the azaarene distributions in both samples were almost identical. Figure 10 compares mass chromatograms for the C_3- and C_4- phenanthridine/benzoquinoline homolog families. Note the qualitative similarities in both sets of traces. One can readily see from these data that weathering does not adversely affect the distribution of higher-molecular-weight azaarenes in these samples.

We believe this analytical approach can easily distinguish significant differences and identify similarities which exist in the azaarene distributions in different oils. These azaarene distributions appear to be useful tools for the identification or passive tagging of oils in environmental samples.

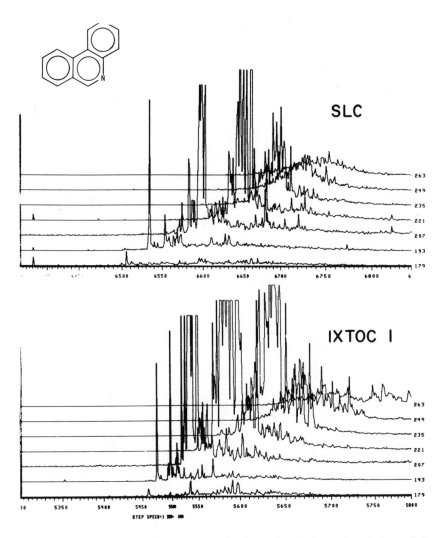

Figure 8. Selected mass chromatograms for the molecular ion series of phenanthridine/benzoquinoline homologs in SLC and Ixtoc-I oils (P = 179, C_1 = 193, C_2 = 207, etc.).

SUMMARY

The data which have been presented in this report lead us to make the following observations: (1) azaarenes in oils can be effectively isolated for characterization by several techniques, (2) a wide variety of extractable azaarenes are present in low concentrations in oils, (3) absolute identification

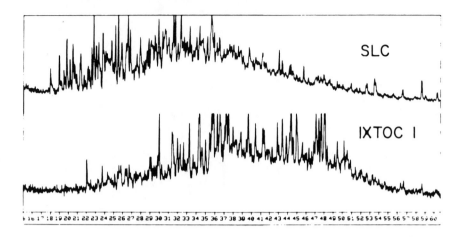

Figure 9. Total summation ion plots for SLC and Ixtoc-I oils.

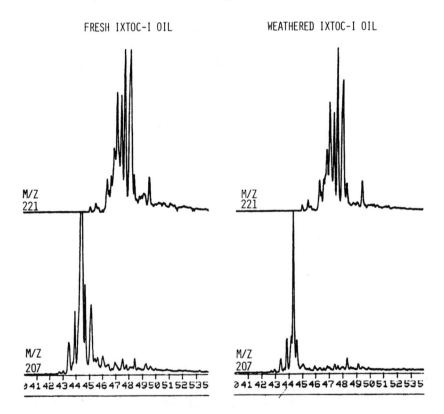

Figure 10. Selected mass chromatograms for the C_2 and C_3 homologs of phenanthridine/benzoquinoline in both fresh and weathered Ixtoc-I oil (C_2 = 207, C_2 = 221).

of azaarenes is difficult to achieve because authentic standards generally are not available and (4) consistent with other aromatic compounds in oil, the aklyl isomers predominate. All data developed to date in this study suggest that azaarenes, isolated and characterized as described in this report, are useful passive tags for tracking oils which are released into the environment.

ACKNOWLEDGMENT

This work was partially supported by the National Oceanic and Atmospheric Administration through contract #NA79RAC00145. We gratefully acknowledge the help of K. Trembley in preparation of this manuscript.

REFERENCES

1. Laseter, J. L. and M. C. Legendre. *Proceedings of the Oceans '76 Conference,* 23C–1 (1976).
2. Overton, E. B., J. L. Bracken and J. L. Laseter. *J. Chromatog. Sci.* 15: 169 (1977).
3. Lawler, G. C., W. A. Loong and J. L. Laseter. *Environ. Sci. Technol.* 12:47 (1978).
4. Lawler, G. C., W. A. Loong and J. L. Laseter. *Environ. Sci. Technol.* 12:51 (1978).
5. Calder, J., J. L. Lake and J. L. Laseter. "NOAA-EPA AMOCO CADIZ Oil Spill Report," NOAA-EPA Special Report (April 1978), pp. 21–84.
6. Lawler, G. C., et al. *Proceedings of the AIBS Conference on the Assessment of Ecological Impacts of Oil Spills* (1979), p. 583.
7. Patel, J. R., et al. *Proceedings of the Annual Conference on Mass Spectrometry and Allied Topics* (1979).
8. Overton, E. B., and J. L. Laseter. in *Petroleum in the Marine Environment,* L. Petrakis and F. T. Weiss, Eds. ACS Series No. 185 (Washington DC: American Chemical Society, 1980).
9. Sleeter, T. *Harvard Environ. Law Rev.* 2:514 (1978).
10. Zafiriou, O., M. Blumer and J. Meyers. "Correlation of Oils and Oil Products by Gas Chromatography," Woods Hole Oceanographic Institution Technical Report, WHOI-72-55 (1972).
11. Rasmussen, D. V., *Anal. Chem.* 48:1562 (1976).
12. Bentz, A. P. *Anal. Chem.* 48:454A (1976).
13. Jones, K. *Int. Environ. Safety* (December 1978), p. 27.
14. Blumer, M., and J. Saas. *Science* 176:1120 (1972).
15. Frame, II, G. M., G. A. Flanigan and D. C. Carmody. *J Chromatog.* 168:365 (1979).
16. McCarthy, L. V., E. B. Overton, M. A. Maberry, S. A. Antoine and J. L. Laseter, *J. High Res. Chromatog. Chem. Commun.* 4:164 (1981).
17. Wakeham, S. G., *Environ. Sci. Technol.* 13:1118 (1979).
18. Lee, M. L., K. D. Bartle and M. V. Novotny. *Anal. Chem.* 47:540 (1975).

19. Frame, G. M., D. C. Carmody and G. A. Flanigan. "An Atlas of Gas Chromatograms of Oils Using Dual Flame Ionization and Nitrogen-Phosphorous Detectors." Coast Guard Report NTIS Association, No. ADA 054 966 (February 1978).
20. Overton, E. B., et al. *Proceedings of the Researcher/Pierce IXTOC-I Symposium* (June 1980), pp. 439-495.

CHAPTER 13

COMPARATIVE EMBRYOTOXICITY AND TERATOGENICITY OF PETROLEUM AND PETROLEUM SUBSTITUTES TO INSECTS DEVELOPING IN CONTAMINATED SUBSTRATES

B. T. Walton

Environmental Sciences Division
Oak Ridge National Laboratory
Oak Ridge, Tennessee 37830

M. V. Buchanan and C.-H. Ho

Analytical Chemistry Division
Oak Ridge National Laboratory
Oak Ridge, Tennessee 37830

INTRODUCTION

One of the major environmental hazards associated with the transportation of petroleum has been the risk of oceanic spills producing adverse effects on marine organisms. But long-term effects of oil spills on terrestrial biota have not been well studied, despite the fact that terrestrial organisms can be exposed either as oil is swept on shore or when oil-contaminated debris is stored or disposed of on land following cleanup activities [1, 2]. There is a need to assess terrestrial impacts of fossil fuel spills because large-scale production of synthetic fuels could become a reality within the next two decades [3, 4], and preliminary assessment has indicated that intracontinental transport of these fuels will result in substantial accidental releases on land [5].

The question of whether the biological effects of synthetic fuel spills will be greater than those of petroleum spills is important. To address this, we compared the toxicities of several synthetic fuels derived from coal and oil shale with toxicities of petroleum fuels on the eggs of a terrestrial insect. Selected shale-, petroleum- and coal-derived oils were class-fractionated, and their constituents were assayed on eggs of the cricket *Acheta domesticus* for embryotoxicity and teratogenicity. Our findings, generalized to other terrestrial invertebrates, suggest that differences in the chemical composition and toxicity of coal liquids and petroleum are significant enough that synthetic fuels spills will lead to greater ecological impact.

MATERIALS AND METHODS

Insects

Acheta domesticus (L.) was selected for experimentation as a representative terrestrial insect that could be exposed naturally to pollutants from oil spills by a number of routes during its life cycle. This cricket is an omnivorous insect found in grass, fields and leaf litter during the nymph and adult stages; its eggs develop in soil. Conditions for laboratory rearing of *A. domesticus* were the same as those previously reported [6, 7].

Bioassay

Embryotoxicity of fossil fuels was determined by the slight modification of a bioassay described for evaluating the toxicity of a water-soluble chemical to cricket eggs [7]. Treatments consisted of adding stock solutions of oils or oil fractions directly to preweighed, autoclaved sand (28 g) in 100-mL beakers and mixing thoroughly. Dichloromethane was used as a solvent to prepare stock solutions. After moistening the sand with 5.0 mL of distilled water, a wire screen was placed on the sand and a gravid female cricket was added to each beaker and permitted to oviposit for 24 hr. Controls were prepared in the same manner with dichloromethane and water.

Oviposited cricket eggs developed for five days in the treated sand at $31 \pm 1°C$ and then were floated on $0.5\ M$ aqueous sucrose to effect separation from the sand. Following this they were rinsed with distilled water and transferred to moist filter paper. The eggs were incubated in petri plates until emergence and during this time were observed for gross morphological abnormalities as an index of teratogenicity; percent hatch also was determined.

Lethal concentrations of whole oils to 50% of the treated eggs (LC_{50}) were interpolated from the toxicity data by computerized probit analysis [8]. The LC_{50} for each oil was based on a minimum of nine toxicity determinations using approximately 300 eggs per treatment.

Whole Crude Oils

Petroleum, shale and coal oils assayed for teratogenic and embryotoxic activity to cricket eggs were obtained from the U.S. EPA/DOE Fossil Fuels Research Materials Facility at Oak Ridge National Laboratory [9]. The samples are identified by repository numbers, are stored under controlled conditions and are available for comparative studies of physical and biological properties. Because some samples were obtained from pilot plants and process development units where the technology for fuel production is under development, they are not necessarily representative of oils that will be produced by commercial-scale facilities. The coal liquid samples include both intermediate and final fuel products.

Diesel fuel No. 2 was obtained from a local distributor. Petroleum-derived light weight mineral oil was assayed to evaluate the effects of sand containing a chemically inert, oily substance on egg development.

Chemical Class Fractionation

Three research fuels—a coal-derived oil (CRM-1), a crude shale oil (CRM-2) and a petroleum crude oil (CRM-3)—were fractionated to obtain classes of structurally similar chemicals for comparative bioassay. The fractionation procedure has been described in detail in the literature [10-12]. Briefly, each sample was evacuated on a vacuum manifold at room temperature and ~ 4.7 kPa (35 torr) overnight and the volatile matter collected in a dry ice/acetone cold trap. The remaining oil was slurried in diethyl ether and contacted overnight with 1 N HCl in a continuous extractor. The aqueous extracts were adjusted to pH 11 and extracted with diethyl ether. The basic constituents were then back extracted into ether to yield ether-soluble bases and insoluble bases. The acid constituents of the ether phase were similarly removed by extraction with 1 N NaOH and adjusted to pH 1 to yield ether-soluble acids and insoluble acids. The neutral constituents remaining in the ether phase were then separated using Sephadex LH-20 into saturates, mono- and diaromatics and polyaromatics by isopropanol. The retained residue was eluted from the column with benzene. Solvents were removed by rotary evaporation.

RESULTS AND DISCUSSION

Petroleum products mixed in sand proved to be of low toxicity to *A. domesticus* embryos. The lack of appreciable egg mortality after treating sand with up to 3500 ppm (μg oil/g dry sand) of mineral oil (Table 1) indicated that embryos were not asphyxiated by oil coating the eggs and preventing oxygen diffusion through the chorion. The other three petroleum products assayed produced $LC_{50} > 180$ ppm (Table 1).

Table 1. Toxicity of Fossil Fuels to *A. domesticus* Eggs

Oil	Repository Number	LC_{50}^a (ppm)	
Petroleum products			
Mineral oil	–	>3500	$(4.7\%)^b$
Diesel fuel No. 2	–	>180	(6.7%)
No. 6 fuel oil	5401	>180	(21%)
Petroleum crude A	CRM–3	>180	(27%)
Petroleum substitutes			
Crude shale oil A	CRM–2	>180	(20%)
Coal liquid A	CRM–1	98	
Coal liquid B^c			
Atmospheric still overheads	1308	72	
Atmospheric still bottoms	1309	75	
Vacuum still overheads	1310	149	
Vacuum still bottoms	1311	>180	(4.1%)
Coal liquid C^d			
Atmospheric still overheads	1312	>180	(32%)
Atmospheric still bottoms	1313	44	
Vacuum still overheads	1314	63	
Vacuum still bottoms	1315	>180	(8.0%)
Coal liquid D^e			
Distillate (raw)	1601	53	
Distillate (low-severity HDT)^f	1602	47	
Distillate (medium-severity HDT)	1603	59	
Distillate (high-severity HDT)	1604	146	
Coal liquid E^g	1701	84	

[a]Concentrations are reported in $\mu g/g$ dry sand. Concentrations were initially calculated as $\mu g/mL$ water but later converted to μg oil/g sand since these units could be more readily related to environmental samples. 180 μg oil/g sand = 1000 μg oil/mL water for above experiments.

[b]Numbers in parentheses are percent mortalities at the concentrations listed.

[c]Experimental run, H-coal Process Development Unit, syncrude mode.

[d]Experimental run, H-coal Process Development Unit, fuel oil mode.

[e]Experimental run, H-coal Process Development Unit, distillate.

[f]HDT = Hydrotreatment.

[g]Experimental run, SRC II Pilot Plant, fuel oil blend.

Crude shale oil (CRM-2) was not very toxic to cricket eggs ($LC_{50} > 180$ ppm) in contrast to the majority of coal-liquefaction products. Of the 14 coal liquids tested, 11 yielded LC_{50} of < 180 ppm (Table 1).

The 1300 series of coal samples were obtained from four different locations within the plant during operation in both syncrude and fuel oil modes. The Vacuum Still Bottoms samples from both operation modes (Repository No. 1311 and 1315) were of very low toxicity as was the sample obtained from the Atmospheric Still Overhead (Repository No. 1312) during operation in the fuel oil mode. All other coal oil samples produced LC_{50} of < 180 ppm. Hydrotreatment of the raw coal distillate (coal liquid D), a process that reduces both heteroatom content and aromaticity of the liquid, lowered embryotoxicity of the fuel after high severity treatment, but low and medium severity hydrotreatment had no effect (Table 1).

Five of the 18 fossil fuels evaluated for embryotoxicity were teratogenic, causing a small percentage of the treated embryos to develop extra compound eyes (Table 2). Supernumerary compound eyes were readily detected as highly pigmented red eyespots typically visible by the eighth day of embryonic development. Although the percentage of treated cricket eggs developing extra compound eyes was low, additional malformations were consistently observed in these treatments. We observed that many substances delayed or arrested embryonic development in this assay; however, those producing multiple eyes also caused formation of twisted, distorted or grossly misshapen cricket embryos. Scoring these abnormals is highly subjective and was not attempted. Instead, the presence of multiple-eyed embryos was used as an objective indicator of a teratogenic oil.

The finding that synthetic fuels were teratogenic to *Acheta domesticus* embryos raised the possibility that other terrestrial invertebrates might be similarly affected by coal-derived fuels. Because studies of the cross-species susceptibility to the teratogen(s) would proceed more rapidly if the structural identity of the active chemical(s) were known, differences in toxicity of constituent chemical classes of selected fuels were compared.

Fractionation of the three research fuels (CRM-1, -2, -3) showed that these samples are quite different in chemical class composition. Saturates comprised 51.14% by weight of petroleum and 44.33% of the shale oil, but only 8.58% of the coal liquid (Table 3). Although the contribution of poly-aromatics to the three fuels was relatively constant (6.56–10.15%), mono- and diaromatics varied from 16.70% by weight of petroleum to 61.40% of the coal liquid. Shale oil contained an intermediate amount of mono- and di-aromatics (35.31%). The remainder of all of the fuels consisted of ether-soluble bases, ether-soluble acids, insoluble bases, insoluble acids, volatiles and residue (Table 3).

The mono- and diaromatic fraction as well as the polyaromatic fraction isolated from the coal liquid were highly toxic (Table 4), but these same fractions from the shale oil were only moderately toxic, and those derived from the petroleum crude were practically nontoxic. The ether-soluble bases derived from both coal and shale fuels produced 100% mortality at 180 ppm. This fraction also was the most toxic of those isolated from petroleum in sufficient quantities to assay (Table 4). Neither the saturates nor the residue contributed substantially to toxicity.

Of the five fractions assayed from each fuel, only the ether-soluble bases from the coal liquid (CRM-1) and shale oil (CRM-2) were teratogenic. The morphologically abnormal crickets observed were identical to those described

Table 2. Teratogenicity of Several Coal Liquefaction Oils to Eggs of the Cricket A. domesticus

Oil	Repository Number	Concentration (ppm)[a]	Number of Eggs	Mortality (%)[b]	Abnormals (%)[c]
Controls	–	–	16,248	–	0
Coal liquid A	CRM-1	18	837	10	0.12
Coal liquid B[d]					
Atmospheric still bottoms	1309	44	863	17	0.12
		89	694	52	0.14
Coal liquid C[e]					
Atmospheric still overheads	1312	180	952	27	0.11
Vacuum still overheads	1314	18	1,050	15	0.10
Coal liquid E[f]	1701	133	1,278	84	0.08

[a]ppm (parts per million) = μg oil/g dry sand.
[b]Percent mortality corrected for control mortality by Abbott's formula [13].
[c]Percent of the treated eggs developing abnormal numbers of compound eyes.
[d]Experimental run, H-coal Process Development Unit, syncrude mode.
[e]Experimental run, H-coal Process Development Unit, fuel oil mode.
[f]Experimental run, SRC II Pilot Plant, fuel oil blend.

Table 3. Percent Contribution of Various Chemical Fractions to Total Weight of Three Fossil Fuels

Fraction	Percentage by Weight[a]		
	Coal Liquid A CRM-1	Shale Oil A CRM-2	Petroleum CRM-3
Volatiles	0.37	trace	trace
Ether soluble bases	5.08	7.16	0.23
Insoluble bases	<0.01	<0.16	3.07
Ether soluble acids	0.66	0.08	0.03
Insoluble acids	0.43	0.34	0.58
Saturates	8.58	44.33	51.14
Mono- and diaromatics	61.40	35.31	16.70
Polyaromatics	10.15	7.32	6.56
Residue	1.88	3.85	6.36
Total	88.55	98.55	84.67

[a]Weight percentage of each fraction is referred to the whole sample (set the whole sample as 100%).

Table 4. Embryotoxicity of Various Chemical Fractions Isolated from Three Fossil Fuels to A. domesticus

Fraction	Percent mortality at 180 ppm[a]		
	Coal Liquid A CRM-1	Shale Oil A CRM-2	Petroleum CRM-3
Ether-soluble bases	100	100	43
Saturates	13	31	11
Mono- and diaromatics	100	24	5.3
Polyaromatics	100	19	5.2
Residue	11	12	19

[a]ppm (parts per million) = $\mu g/g$ dry sand. Percent mortality was corrected for control mortality by Abbott's formula [13].

previously [14] after treatment of A. domesticus eggs with a trace impurity found in commercial samples of acridine. Nymphs with poorly developed second heads emerged from eggs showing four eyespots, and three-eyed nymphs, which emerged from eggs with three eyespots, sometimes had an extra antenna or extra branches of the antennae. As in the previous study, some abnormal insects reached reproductive maturity and most were fertile. Ether-soluble bases (ESB) isolated from coal liquid A (CRM-1) produced three-eyed embryos at a concentration of 11 ppm in sand and from shale oil A (CRM-2) at 35 ppm. At 18 ppm ESB from shale oil A (CRM-2), one single-eyed embryo was observed among 887 treated eggs. However, no abnormals were observed when eggs were treated with the whole shale oil (CRM-2).

CONCLUSIONS

In the present study, coal liquids were found to be more toxic to developing embryos of a terrestrial insect than were shale oil and petroleum. Comparative fractionation and bioassay of a selected coal liquid, a shale oil and petroleum indicated important chemical and toxicological differences among the fuels. Aromatics were both highly toxic and abundant (71.55% by weight) in the coal liquid, less toxic and less plentiful in the shale oil (42.62%), and least toxic and least abundant (23.26%) in the petroleum. Ether-soluble bases, the only fraction positive for teratogenic activity, were present in significant concentrations in coal liquid (5.08%) and shale oil (7.12%) but only in trace amounts in petroleum (0.23%). Furthermore, those from petroleum lacked teratogenic activity.

The possibility for release of mutagens, teratogens or chemosterilants during an oil spill is of special environmental concern because these chemicals could produce undesirable biological effects persisting much longer than the oil itself. Results presented herein suggest that terrestrial invertebrates undergoing embryonic development in soil might be more adversely affected by synthetic oil spills than by petroleum spills. Subfractionation of the ether-soluble bases and characterization of invertebrate teratogens should aid prediction of the chronic ecological effects following synthetic fuel spills on land.

ACKNOWLEDGMENTS

Research was sponsored by the Office of Health and Environmental Research, U.S. Department of Energy, under contract W-7405-eng-26 with Union Carbide Corporation.

We thank E. G. O'Neill, R. B. Quincy, Jr. and M. P. Farrell for technical assistance and the U. S. EPA/DOE Fossil Fuels Research Materials Facility for providing oil samples. Publication No. 1742, Environmental Sciences Division, Oak Ridge National Laboratory.

REFERENCES

1. Farlow, J. S., and C. Swanson, Eds. "Disposal of Oil and Debris Resulting from a Spill Cleanup Operation," Technical Publication 703, American Society for Testing and Materials, Philadelphia, Pennsylvania (1980), 147 pp.
2. McAuliffe, C. D. "The Environmental Impact of an Offshore Oil Spill," in *Background Papers for a Workshop on Inputs, Fates, and Effects of*

Petroleum in the Marine Environment, Vol. I (Washington, DC: National Academy of Sciences, 1973), pp. 224-279.

3. "Achieving a Production Goal of 1 Million B/D of Coal Liquids by 1990," U.S. Department of Energy, DOE/FE/10490-01 (1980).

4. "Synthetic Fuels and the Environment: An Environmental and Regulatory Impacts Analysis," U.S. Department of Energy, DOE/EV-0087 (1980).

5. Leggett, N., et al. "Preliminary Assessment of Spills from the Transportation and Storage of Coal-Derived Synthetic Fuels," ORNL/TM-7606, Oak Ridge National Laboratory, Oak Ridge, TN, in press.

6. Walton, B. T. *Bull. Environ. Contam. Toxicol.* 25:289 (1980).

7. Walton, B. T. *Environ. Entomol.* 9:18 (1980).

8. Finney, D. J. *Probit Analysis* (London: Cambridge University Press, 1971), 333 pp.

9. Griest, W. H., D. L. Coffin and M. R. Guerin. "Fossil Fuels Research Matrix Program," ORNL/TM-7346, Oak Ridge National Laboratory, Oak Ridge, TN (1980), 40 pp.

10. Rubin, I. B., et al. *Environ. Res.* 12:358 (1976).

11. Jones, A. R., M. R. Guerin and B. R. Clark. *Anal. Chem.* 49:1766 (1977).

12. Guerin, M. R., et al. "Distribution of Mutagenic Activity in Petroleum and Petroleum Substitutes," *Fuel*, in press.

13. Abbott, W. S. *J. Econ. Entomol.* 18:265 (1925).

14. Walton, B. T. *Science* 212:51 (1981).

PART 3

IMPACT OF FUGITIVE HYDROCARBON EMISSIONS

CHAPTER 14

FUGITIVE HYDROCARBON EMISSIONS FROM PETROLEUM PRODUCTION OPERATIONS

W. S. Eaton, F. G. Bush III, J. Coster and J. C. Delwiche

Rockwell International
Environmental Monitoring and Services Center
Newbury Park, California 91320

INTRODUCTION

This chapter describes the results of a program funded by the American Petroleum Institute for the development of factors for predicting fugitive hydrocarbon emissions from equipment components in production operations [1]. This study addressed oil and gas production operations, which include well drilling, oil and gas production, and gas plant operations. It involved measurements at twenty-one facilities located in four geographic regions, onshore and offshore. A total of 173,609 components were inventoried and monitored for leaks.

The program involved a statistical selection of facility locations, development of a screening-auxiliary measuring technique and development of a multivariant regression model for prediction of emission factors. Hydrocarbon species C_1-C_5 and C_{6+} were measured and provide a means of predicting emission adjustment at existing and prospective facilities.

A small component maintenance program was implemented to provide additional information about the control of fugitive hydrocarbon emissions and the effectiveness of component repair. Fugitive emissions are defined, for this study, as unintentional emissions without control of rate or direction.

261

PROGRAM DESIGN

The study design and statistical data analysis have been described extensively [1, 2] and consequently are only summarized in this chapter.

Objective

The objective of the study was to develop the capability to predict fugitive hydrocarbon emissions for any site-specific inventory of equipment, operations, products and geographic location.

Site Selection

Site means a geographic location of the field facility. Because it was believed that geographic location may be an important variable, sites from the Gulf Coast, Midcontinent, Rocky Mountain and West Coast were selected. Additionally, onshore and offshore sites were selected to establish any differences in fugitive emissions.

Five major products are handled by oil and gas production facilities: heavy crude ($<20°$ API), light crude ($25° \leq API \leq 40°$), natural gas (gaseous at operating conditions), condensate ($> 40°$ API) and liquefied petroleum gases. These five categories were used for stratification within each geographic region, although not all are indigenous to all regions, and gas plant product composition is, of course, independent of geography.

Statistically, the selected sites should: (1) include the full range of component types and styles and products required, and (2) not introduce biases into the predictions. These conditions were fulfilled by selecting sites randomly from major oil and gas provinces, with division by principal products.

During field measurement, 100% screening and inventory was performed for a site or a representative site section.

Component Types and Styles

Component refers to the potential sources of fugitive hydrocarbon emissions. The components were categorized by "types", i.e., valves, connections, etc. These "component types" were further subdivided into "component styles" such as gate valve or globe valve. A total of 9 "component types" and 40 "component styles" were established.

Data Collection

Reliable inventories or "as-built" drawings of production facilities are not available. Thus, during the visit to the site, a "site section" for monitoring was established first. This site section was inventoried and simultaneously screened for leaks, using a soap solution if possible, and the leaking components were tagged. Components not amenable to soap testing included pits, reciprocating rods and components operating at or near atmospheric pressure. These components were screened using a Century Model 108 Organic Vapor Analyzer (OVA). The "site section" was 100% screened and a semiquantitative score recorded. Soap scores ranged from 0 to 4 as the emissions ranged from zero to more than 1000 cc/min.

A statistically significant sample of the leaking components was selected for quantitative measurement of emission. The quantitative measurements were made by shrouding the component and determining the leakage volume by meter and/or by measuring with a continuous on-line gas chromatographic system. The hydrocarbons were segregated into methane (C_1), ethane (C_2), propane (C_3), butane (C_4), pentane (C_5) and hexane plus heavier hydrocarbons (C_{6+}) fractions.

Data Analysis

Statistical analysis of the data is composed of three parts: (1) correlation of the soap score with the measured leak rates for moderate leaking components, (2) percentage of leaking components and (3) emission rates for leakers.

The soap score was an imprecise quantitative measurement which was correlated with the quantitative measurement. Quantification of a statistically significant sample of known leakers provided a meaningful correlation of the measured emissions and the respective soap scores.

Prediction of Total Hydrocarbon Emissions

Field observations indicated leakage was a function of the product being handled. Several sites were located within the same geographic region and handling the same products. Therefore, it was possible to analyze for differences between sites, to establish if all of the combinations of operation (offshore, onshore, gas plant), product (gas, oil, etc.) and geographic location (midcontinent, Rocky Mountain, etc.) were necessary for accurate prediction of emissions. The analysis indicated the geographic location distinction was not necessary. The two parameters necessary to predict fugitive hydrocarbon emissions from production facilities are operation and product.

Products handled at offshore and onshore operations are similar. A total of 47 different products and/or product combinations were identified in the study. Specifically, these were reclassified as natural gas, water, crude oil, condensate and mixtures of the products. Some onshore sites, primarily in California, handle heavy crude oil. Thus, there were six possible product categories established which can contribute to oil and gas production fugitive hydrocarbon emissions.

Gas plants mainly handle pure paraffinic hydrocarbons and water. Accordingly, the products contributing to fugitive emissions are different at gas plants than at producing sites. Six groupings of these products were established for gas plants.

RESULTS AND DISCUSSION

Fugitive hydrocarbon emission measurements were made at 21 production equipment sites, which included 1 drilling rig, 18 producing and 2 gas plant sites. In addition, one of the onshore sites was selected for a component maintenance study. A total of 173,609 components were screened and inventoried.

A total of 8466 individual field quantitative measurements, emission rate and gas chromatograms, were made. This includes chromatograms for 1914 components representing 22.6% of all leaking components included within this study.

Component Leak Incidence

Table 1 shows components grouped into nine types. The percent of leaking components varied by type, but the overall percent of leaking components was 4.7. This compares favorably with the 4.2% determined by K.V.B., Inc. [3].

Production equipment leakage was observed to be a function of the type of operation involved. A comparison of onshore and offshore leaks for all component types, presented in Table 2, shows a lower incidence of leaks at offshore facilities.

Table 1. Distribution of All Components Tested for Emission Correlation[a]

Component	Total	Leaking	% Leaking
Valve	25,089	2,098	8.4
Connection	138,510	4,668	3.4
Sightglass	676	9	1.3
Hatch	358	22	6.1
Seal Packing	1,246	323	25.9
Diaphragm	1,643	318	19.4
Pit[b]	31	31	–
Meter	92	5	5.4
Sealing Mechanism	5,591	609	10.9
Total	173,236	8,803	4.7

[a]Does not include drilling or maintenance components since these were biased selections.
[b]Inventoried only when unweathered oil was present; not used for percentage calculation.

Table 2. Number of Each Component Type Tested and Percentage Leaking, by Operation

	Onshore Production			Offshore Production			Gas Plant		
	Number of Components	Number Leaking	Percent Leaking	Number of Components	Number Leaking	Percent Leaking	Number of Components	Number Leaking	Percent Leaking
Valve	9,427	603	6.4	11,289	312	2.8	4,304	1,183	27.5
Connection	54,694	1,545	2.8	65,504	1,763	2.7	18,396	1,360	7.4
Sightglass	158	0	0	242	3	1.2	276	6	2.2
Hatch	170	4	2.4	133	13	9.8	54	5	9.3
Seal Packing	474	145	30.6	486	73	15.0	285	105	36.8
Diaphragm	750	134	17.9	808	155	19.2	80	29	36.2
Pit*	26	26	—	2	2	—	3	3	—
Meter	21	0	0	71	5	7.0	0	0	0
Sealing Mechanism	2,386	259	10.9	2,328	111	4.8	869	239	27.5
Total	68,106	2,716	4.0	80,863	2,437	3.0	24,267	2,930	12.1

*Inventoried only when unweathered oil was present; not used for percentage calculation.

The highest overall incidence of leaks was found in gas plant operations. This is a direct consequence of handling gaseous products, which have a higher propensity to leak than liquid products.

Table 3 shows that 93% of the leaks occur in 64% of the production equipment that handles natural gas. The remaining 7% of the leaks occur in the 36% of the components handling water, crude oil, condensate and mixtures of natural gas and the liquids.

A statistical bias toward gas service components leakage is suggested by the values shown in Table 3. This bias is indicated by the difference in ratio of the leaking components to the inventoried components. A ratio of one indicates an equal distribution of leakers over all products. A comparison of the ratios indicates relatively more components leak in gas service, from 5 to 22 times more.

An analogous examination of gas plant leaks indicates components in methane (Group 1) and ethane through butane (Group 2) service are about 24 times more likely to leak than components handling water.

These analyses clearly indicate the major sources of fugitive hydrocarbon emissions are components in gas service.

Gauge Equations

The number of components screened and inventoried (173,609) represents the most comprehensive investigation of production equipment reported to date. Gauge equations were developed for five of the nine component types by product, size and style. These were valves, connections, seal packings, diaphragms and sealing mechanisms. No gauge equations were developed for sightglasses or meters, since so few of these leaking components were located in field service. Hatches, pits and sumps could not be soap-scored and, accordingly, gauge equations were not applicable.

The gauge equation was developed as a means of reducing the number of field quantitative measurements required to provide a statistically defensible prediction equation of fugitive emissions. The validity of the gauge equations approach to establishing accurate estimates of emissions can be reinforced by an analysis of its contribution to the resulting prediction equations. The gauge equation was used only for estimating emissions for moderate leaking components, soap score 1, 2 and 3. Measurements of leakage rates were made for a statistically selected sample of soap score 1, 2 and 3 for each site tested to permit correlation of the measured emission rate with soap scores (see Table 4). Soap score 1, 2 and 3 components in gas service represent 85.3% of all leaking components; however, the percentage of emissions was only 15.6. Accordingly, even a moderate error in the gauge equation has only a small impact on the prediction equation. An attempt was made to quanti-

Table 3. Fractions of Leaking Components and Inventoried Components, by Product

Producing Operations

Product	Gas	Water	Heavy Crude	Light Crude	Condensate	Mixtures
Fraction[a] of Leaking Components	0.933	0.001	0.001	0.002	0.007	0.057
Fraction[a] of Inventory	0.639	0.015	0.005	0.030	0.026	0.285
Ratio[b]	1.46	0.067	0.200	0.067	0.269	0.200

Gas Plant

Product[c]	Group 1	Group 2	Group 3	Group 4	Group 5	Group 6
Fraction[a] of Leaking Components	0.442	0.338	0.098	0.097	0.024	0.001
Fraction[a] of Inventory	0.336	0.243	0.185	0.075	0.143	0.018
Ratio[b]	1.32	1.39	0.530	1.29	0.168	0.056

[a]Fraction of leaking components divided by fraction of inventory.
[b]Not counting pits, sumps, etc.
[c]Refer to Table 2-2 in Ref. 1.

Table 4. Leaking Component Total Hydrocarbon Emission Summary –
Producing Operations

Gas Service Leaking Components			
Soap Score	Average Emission per Leaking Component (lb/day)	Percentage of All Leaking Components	Percentage of All Emissions
1	0.030896	37.5	0.6
2	0.15647	29.2	2.3
3	1.3735	18.6	12.7
4	24.578	6.3	76.6
5	5.9571	1.2	3.6
Other Service Leaking Components			
Soap Score	Average Emission per Leaking Component (lb/day)	Percentage of All Leaking Components	Percentage of All Emissions
1	0.032138	1.8	0.02
2	0.31020	1.5	0.2
3	0.80765	0.9	0.4
4	13.342	0.5	3.5
NA	0.32710	1.7	0.3
Quantifiable Liquid Leakers	0.015925	0.9	0.007

tatively test all soap score 4 (heavy leakers) and quantifiable liquid leaking components.

Predicted Emissions

Prediction equations were developed on the basis of the six products associated with producing operation and gas plant operations, as previously described. The resulting values were tabulated for user convenience, by component type, component style and product handled [1]. A prediction of the total fugitive emission is obtained by applying the tabulated values to the component inventory developed for a site. It is important to recognize that emissions are possible from several components and/or styles of a single piece of equipment. For example, a threaded plug valve has a contribution from the valve stem, the threaded connections and the bottom sealing flange, if present. Thus, the contribution from all valves, connections, flanges, etc. must be summed over the products being handled for the entire site on the basis of onshore or offshore operation. This technique applies to producing and drilling operations. Similarly, the predicted emissions from a gas plant can be calculated on the basis of the entire stream composition.

The individual contribution to emissions from water, crude oil, condensate and mixtures is small, as shown by Table 5. For example, the emission rate for gas service is about 70 times that for mixture. Contributions from water, light crude and condensate are even less for a gate valve.

Calculation of the fugitive emission is simplified by grouping water, crude oil, condensate and mixtures into an "other" category. Emission factor values for "gas/other" have been calculated on this basis as well. Different emission values result from the use of the six product and the gas/other composition tables [1] shown in Table 5. This is the result of using a statistical analysis modified to include only two different compositions. However the differences are small. Average product composition measured and average product molecular weight are provided with each set of tabulated emission valves. Emission factors are tabulated in both mass and volume. Thus, the emission predictions can be classified by individual hydrocarbon species. This permits determination of nonreactive and reactive C_{3+} hydrocarbons. Also, it permits calculating emissions for a product-specific application, i.e., a propane refrigeration system.

Table 5 shows that offshore emissions are about half of those predicted for onshore operations for gate valves. This same trend is apparent for all offshore components [1]. The methane/ethane composition of gas is ~ 70-80%, therefore the reactive (nonmethane/nonethane) emission is between 20 and 30% of those shown in Table 5 for gas.

Prediction Reliability

The data required for calculation of the variance and standard error of a prediction of the emissions for a site are given in Reference 1. These statistical values provide a measure of the reliability of the prediction. The reliability of the prediction for a new site is a function of the product and component mix and thus is different for each site.

An evaluation of the calculated emission values versus the predicted emission values for each site tested provides an indication of the reliability of the predictions. The calculated emission values were based on the inventory, products and actual measured values supplemented by the use of gauge equations.

The percent difference for each individual site ranged between -80.8 and +185. Part of the analysis is shown in Table 6. This range indicates substantial error can result in the individual prediction, although this is of the same order of magnitude of other emission factors developed [4]. However, the average percent difference indicates a slight overprediction for sites on the average. A weighted average, "percent error in prediction total" suggests the prediction factors underestimate the emissions from the sites.

Table 5. Gate Valve Predicted THC Emissions

	Operation	Emissions Using 6 Product Breakdown[a] (lb/day)					Emissions Using Gas/Other (lb/day)	
		Gas	Water	Light Crude	Condensate	Mixtures	Gas	Other
Lease Production	Onshore	0.427	0.00000842	0.0000350	0.00584	0.00666	0.415	0.00593
Lease Production,	Offshore	0.217	0.00000412	0.0000166	0.00267	0.00320	0.212	0.00284
Gas Plant	Group 1 1.72	Group 2 0.684	Group 3 0.492	Group 4 0.347	Group 5 0.0282	Group 6 0.00000991		

[a]No heavy crude present.

Table 6. Calculated vs Predicted Emissions[a]
Site Totals (THC lb/day)
Lease Production, Onshore

Site Number	Calculated	Predicted ± Standard Error	% Δ[b]
2	962	467 ± 207	−51.5
3	363	777 ± 783	114
7	688	337 ± 741	−51.0
8	650	1304 ± 744	101
9	748	316 ± 182	−57.8
10	124	342 ± 267	176
11	652	421 ± 360	−35.4
18	1568	301 ± 156	−80.8
21	44.8	99.6 ± 75.0	122
22	342	231 ± 294	−32.5
Totals	6141.8	4595.6 Avg. % Δ[c]	20.4
		Percent Error in Prediction Total[d]	−25.2

[a]Cf. pp. 3–14, Ref. 1.

$$^{b}\% \Delta = \frac{(predicted - calculated)\ 100}{calculated}$$

$$^{c}\% \Delta = \frac{\sum\limits_{i=1}^{n} \% \Delta_i}{n}$$

$$^{d}Percent\ error\ in\ predicted\ total = \frac{(\Sigma\ predicted - \Sigma\ calculated)\ 100}{\Sigma\ calculated}.$$

Component Correlation

Predictions of fugitive hydrocarbon emissions for any site on the basis of the method outlined in this report require an inventory of components. However, field inspection and tabulation of the inventory for existing sites are very time consuming.

Production equipment needs vary with product, pressure, field size, climate and space availability. Thus, the onshore and offshore sites monitored were analyzed for a possible component inventory correlation. None was found for onshore sites, probably due to the method of data collection, i.e., site sections. However, at offshore sites, there was an excellent correlation between wells and number of components, if one atypical site was eliminated. This study included single and multiple platforms at which both producing and processing activities are performed. The correlation derived applies only to this type of offshore facility which represents the vast majority of offshore sites. It has been concluded that a site inventory is essential in making a site emission prediction for both on and offshore locations.

Maintenance

A small maintenance study was initiated to investigate changes that were observed in leak rate as well as the longevity of component repair. This study included 162 components and was divided into two groups, a study and a control group. The study group consisted of leaking components which were repaired. The control group consisted of nonleaking and unrepaired leaking components located during screening.

Each group was rescreened over a 10-month period using both soap solution and an OVA. In summary, leaking components may or may not spontaneously stop leaking. Repaired components generally remain leak free if no leaks redevelop during the first few weeks after repair. A significantly larger study which would include quantitative measurement and statistical analysis will be required to assess the effects of an inspection and maintenance program and its contribution, if any, to fugitive hydrocarbon emissions reduction.

CONCLUSIONS

The specific conclusions drawn from this study are:

1. Previously published emission factors for production equipment are not applicable to production operations [3, 5].
2. Emission factors for refinery equipment are not applicable to production operations [4].
3. Emission rates determined in this program are an order of magnitude lower than those calculated using previous petroleum production or refinery studies [3, 4, 6].
4. Less than 5% of the more than 173,000 components screened at field locations leaked.
5. Nonreactive hydrocarbons (methane and ethane [7]) are a significant fraction of the fugitive hydrocarbon emissions.
6. Components in gas service leak about 10 times more frequently than similar components in liquid service.
7. Emission rates from components at offshore producing facilities are significantly less than from similar components in onshore producing facilities.
8. Large differences in leakage rate are observed for component types of different styles in the same product service.
9. Component repair should be monitored immediately after completion to establish the effectiveness of the remedial action.
10. In most cases, no leaking of repaired components was observed for the duration of the 10-month test if the repaired component remained leak-free for several weeks after repair.
11. A correlation was observed between the number of components and the number of wells at offshore sites.

ACKNOWLEDGMENTS

The authors express sincere appreciation to the American Petroleum Institute and the Task Force on Hydrocarbon Emissions from Production Operations for their cooperation, effort and support in the successful performance of this program. The American Petroleum Institute Task Force and Rockwell International's Environmental Monitoring and Services Center acknowledge Dr. H. O. Hartley for his development of the statistical design and analysis. Deep appreciation is expressed to the personnel at all sites visited for their cooperation, help and guidance in the study.

REFERENCES

1. Eaton, W. S., et al. "Fugitive Hydrocarbon Emissions from Petroleum Production Operations," API Publication No. 4322 (March 1980).
2. Eaton, W. S., et al. "Experimental Approach and Techniques for the Measurement of Fugitive Hydrocarbon Emissions from Production Operations," presented at the 71st Annual AICHE Meeting, November 12–16, 1978.
3. K.V.B., Inc. "Control of Hydrocarbon Emissions from Stationary Sources in the California South Coast Air Basin," prepared for the California Air Resources Board, K.V.B., Inc., Tustin, CA (June 1978).
4. Radian Corporation. "Emission Factors and Frequency of Leak Occurrence for Fittings in Refinery Process Units," U.S. EPA Report–600/2-79-044 (February 1979).
5. Meteorology Research, Inc. "Total Hydrocarbon Emission Measurements of Valves and Compressors at ARCO's Elwood Facility," Report No. 911–115–1661 (1976).
6. "Joint District, Federal and State Project, for the Evaluation of Refinery Emissions," Los Angeles County Air Pollution Control District, nine reports 1957–1958.
7. U.S. Environmental Protection Agency, *Federal Register,* 42FR 35314 to 35316 (July 8, 1977).

CHAPTER 15

REDUCTION OF FUGITIVE VOLATILE ORGANIC COMPOUND EMISSIONS BY ON-LINE MAINTENANCE

Robert C. Weber*

U.S. Environmental Protection Agency
Office of Research and Development
Industrial Environmental Research Laboratory
Cincinnati, Ohio 45268

G. J. Langley and R. G. Wetherold

Radian Corporation
Austin, Texas 78766

INTRODUCTION

The U.S. Environmental Protection Agency (EPA) Office of Air Quality Planning and Standards (OAQPS) is currently in the process of formulating regulations for the control of fugitive emissions of volatile organic compounds. This study was undertaken by the Office of Research and Development to assist OAQPS. The project was intended to develop data to determine the effectiveness of routine (on-line) maintenance in the reduction of fugitive VOC emissions from in-line valves. The overall effectiveness of an inspection/maintenance program was examined by studying:

- immediate emission reduction due to maintenance,
- propagation of the leaks after maintenance, and
- rate at which new leaks occur, for both pumps and valves.

---

*Present address: Environmental Research and Technology, Inc., Pittsburgh, PA.

This study was conducted by the Radian Corporation (Austin, Texas) under contract to EPA (Contract No. 68-03-2776-04). The project began in 1979 when the scope and technical approach were developed through several meetings with the Chemical Manufacturers Association and the Texas Chemical Council, as well as with individual chemical companies. The field work was completed in 1980. The detailed results and experimental methods will be available later in 1981 as a project report available from the National Technical Information Service (NTIS). The purpose of this chapter is to summarize the significant results of this program.

EXPERIMENTAL TECHNIQUES

The experimental design called for the study of three types of organic chemical production units at each of two locations. The processes chosen were: (1) ethylene production, (2) cumene production from benzene and propylene, and (3) vinyl acetate production. The processes chosen represent a wide range of conditions found in organic chemical manufacturing plants. Ethylene was chosen because typically these units are large and widespread, operate with a wide range of process conditions, and handle very volatile materials. Cumene was of interest because this type of unit (one using the reaction of benzene and propylene) handles a hazardous air pollutant, benzene. Production of vinyl acetate from the reaction of ethylene and acetic acid was chosen because some of the process streams are corrosive.

The sampling and analytical procedures have been summarized in a previous paper [1]. The details of the methods and the statistical techniques for data analysis also can be found in the contractor's final report [2]. The sampling techniques included both "screening" and the actual measurement of hydrocarbon emission rates.

Screening was done with a Century Systems Corporation OVA-108 and a Bacharach TLV Sniffer. The valves were screened by traversing 360 degrees around the stem seal and the seam where the packing gland merges with the valve bonnet. The point of maximum hydrocarbon concentration was identified. The OVA-108 was used before the TLV Sniffer to quickly identify the area of maximum concentration because of its faster response time. The sample probes were placed as close to the maximum leak as possible. The recorded screening value was the highest reading obtained twice during an interval of approximately one minute. Pumps were screened at the outer shaft seals by completely traversing 360 degrees with the OVA to locate the point of maximum concentration. Occasionally a 12-inch Teflon® extension was added to the OVA probe in order to extend past safety screens on vertical pumps.

Selected valves and pumps were sampled to determine the mass emission rate using the flow-through method which has been reported previously [3]. The general sampling procedure was: (1) the source was screened with the OVA-108 and TLV Sniffer and the values and time of day were recorded; (2) the source was tented with Mylar® and duct tape and sampled; (3) the tent was removed and the source was rescreened as in the first step, above. The sequence of screening and sampling is illustrated in Figure 1.

Valve maintenance consisted of tightening packing glands while monitoring the leak. The term "directed maintenance" refers to this procedure when a hydrocarbon detector is used during the maintenance activity. The leak is monitored with the instrument until no further reduction of screening values is observed or until the valve stem rotation starts to be restricted. The type of maintenance personnel performing the repairs depended on the type of valves that were to be maintained. Control valves required instrument personnel experienced with the process unit and with the precautions necessary to safely maintain operations while repairing control valves. Block valves were maintained by regular maintenance personnel such as pipefitters.

The maintenance procedures generally consisted of first screening the valve with the OVA-108 and recording the value. The packing gland nuts were then tightened a little at a time while monitoring the leak with the OVA. Tightening was continued to the point of either minimizing the leak, causing the stem movement to tighten or grab, or reaching the bottom of the packing bolts. The valve was then operated, if the process permitted it, and rescreened. If the leak remained or worsened, the packing was further tightened until the limits described above were reached. In no case was the packing tightened such that the operation of the valve was impaired.

Certain valves could not be maintained due to their locations in the process stream. These were in critical service where sticking, jamming or breaking of the valve might precipitate a unit shutdown.

RESULTS AND CONCLUSIONS

Three aspects of the effect of valve maintenance on fugitive emissions were studied:

1. the immediate effect based on measured leak rates,
2. the long term effect based on measured leak rates, and
3. the immediate effect based on screening values.

Analysis of the immediate effect of maintenance using measured leak rates produced an overall estimate of 71.3% reduction in fugitive emissions (95%

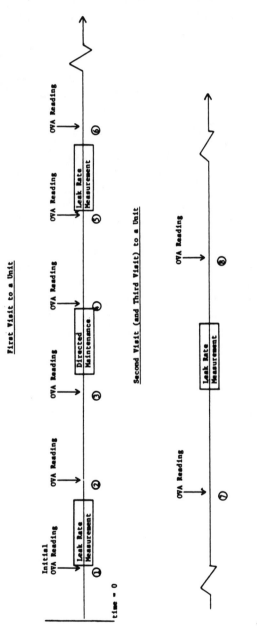

Figure 1. Sequence of emissions measurements and screening values.

Legend

1. initial before maintenance OVA reading;
2. after tenting, before maintenance OVA reading;
3. before maintenance OVA reading—the screening value obtained immediately before maintenance;
4. after maintenance, first OVA reading—the screening value obtained immediately after maintenance;
5. before tenting, after maintenance OVA reading;
6. after tenting, after maintenance OVA reading;
7. before tenting OVA reading;
8. after tenting OVA reading.

confidence limits of 54–88%) immediately after maintenance. This estimate is the weighted percent reduction (WPR), calculated by:

$$WPR = \frac{\left(\begin{array}{c} \overset{m}{\Sigma} \text{ mass emissions} \\ \text{before maintenance} \end{array} - \begin{array}{c} \overset{m}{\Sigma} \text{ mass emissions} \\ \text{after maintenance} \end{array} \right)}{\overset{m}{\Sigma} \text{ mass emissions before maintenance}} \times 100$$

where m = number of valves maintained

Paired observations of measured leak rates were available for 155 valves. Weighted percent reductions were calculated for various groupings of the 155 valves, and are given in Table 1. A graphical presentation is given in Figure 2. Since none of the WPR estimates are statistically different for any of the groupings, the overall estimate is the most appropriate estimate for application in other organic chemical units.

It was also of interest to investigate the change in the WPR estimates for varying screening action levels. When only valves with the immediately-before-maintenance screening values $\geqslant 10,000$ ppmv were considered, the WPR estimate decreased slightly to 70.1% with a 95% confidence interval of 46–95%. This estimate is almost identical to the overall estimate of 71.3%. The WPR estimate for those sources where the before-maintenance OVA reading was $< 10,000$ was 82.4% with an approximate 95% confidence interval of 57–100%. These two WPR estimates are not statistically different.

Later sampling of the maintained sources to study the long-term effect of maintenance indicated that the reduction estimates obtained for immediate effects of maintenance held for the length of the study (up to six months). To put the long-term effect of maintenance into perspective, it is helpful to compare the emissions from the maintained valves to those from a control group of unmaintained valves over a period of time. Figure 3 is a graphic display of this comparison. The major conclusion that can be drawn from Figure 3 is that the immediate effect of maintenance discussed previously was

Table 1. Immediate Effect of Valve Maintenance

	Number of Valves Maintained	Weighted Percent Reduction (WPR)	95% Confidence Limits for WPR
All Valves	155	71.3	(54, 88)
Cumene Units	54	81.6	(67, 96)
Ethylene Units	69	56.6	(22, 91)
Vinyl Acetate Units	32	72.9	(34, 100)
Gas Service Valves	71	84.5	(74, 95)
Liquid Service Valves	84	42.0	(0.4, 84)

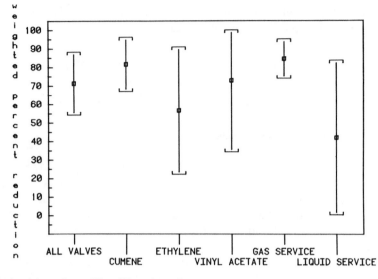

Bracketed intervals are 95% confidence intervals.

Figure 2. Immediate effect of valve maintenance.

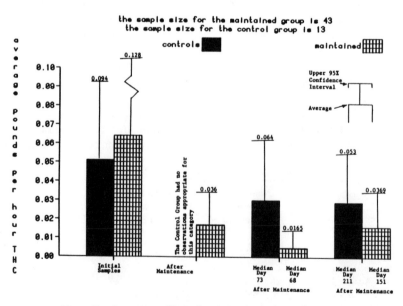

Figure 3. Long-term effect of maintenance vs control group.

sustained for the duration of the project. The minor changes in the control group and the maintenance group after the initial sampling visit (during which the maintenance occurred) are not statistically significant.

The immediate effect of maintenance also can be viewed in terms of screening values only. This can be used to evaluate a leak detection and repair strategy, where a leak is defined by a screening value greater than some specified value. For example, given a definition of a leak as a screening value $\geqslant 10,000$ ppmv, the effectiveness of maintenance was evaluated. Analysis of the immediate effect of maintenance based on screening data and using $\geqslant 10,000$ ppmv as the definition of a leak produced an estimate of $\sim 30\%$ reduction in the number of leaking sources as a result of the maintenance. This indicates that $\sim 70\%$ of the leaking sources could not be repaired (where repaired is defined as screening $< 10,000$ ppmv after maintenance). However, it should be pointed out that even though the screening values were reduced to $< 10,000$ ppmv for only 30% of the valves, this corresponded to a 70% reduction in mass emissions.

To study the recurrence of leaks after maintenance, data from the 155 maintained valves were examined. For this analysis, only those valves which screened $\geqslant 10,000$ ppmv immediately before maintenance and screened $\leqslant 10,000$ ppmv immediately after maintenance were considered as having a potential to recur. This eliminated all but 28 valves from the analysis. Of these 28 valves, eight were seen to recur (i.e., screen $\geqslant 10,000$ ppmv at some time following the after-maintenance screening). Of the eight valves whose leaks recurred, four recurred within a few days after maintenance. The other four recurrences were spread over the study period (up to 7 months). Because of the two distinct groupings of recurrences over time, a mixed model was used in estimating the recurrence rate. This mixed model consists of a uniform distribution for recurrence times within five days after repair and an exponential distribution for recurrence times greater than five days after repair.

A graphical presentation of the modeled percentages for recurrence along with an approximate 95% confidence region is given in Figure 4. The empirical distribution function (actual data) is indicated by the dotted line.

In Table 2, 30-, 90- and 180-day recurrence rate estimates are given along with their approximate 95% confidence limits. In comparison to occurrence rates which are discussed later, recurrence rates are much higher. However, the sample population is very small.

The rate of occurrence of leaks was studied using pumps and valves that initially screened at $\leqslant 10,000$ ppmv and were not maintained during the project. An exponential model was used to approximate the actual distribution of the time to first occurrence of a leak (screening value $\geqslant 10,000$ ppmv). This model is widely used to summarize data similar in nature to leak occurrences if the assumption can be made that the occurrence rate remains

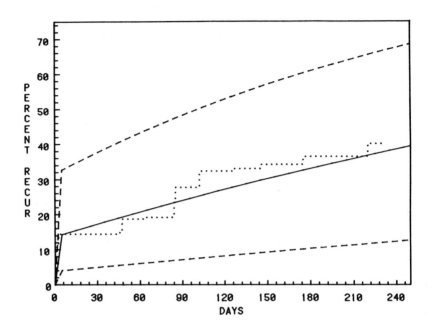

Figure 4. Recurrence rate estimate vs empirical distribution function. (Dotted line is actual data; dashed lines indicate a 95% confidence region.)

Table 2. Valve Leak Recurrence Rate Estimates

	Recurrence Rate Estimate	95% Confidence Limits on the Recurrence Rate Estimates
30–day	17.2%	(5, 37)
90–day	23.9%	(7, 48)
180–day	32.9%	(10, 61)

constant: A major advantage of the exponential model over other statistical distribution models is that a single parameter fully and completely describes a given exponential distribution.

Tables 3 through 5 show the estimated 30-, 90- and 180-day occurrence rates by process for valves, pump seals, valves in gas service, valves in liquid service, block valves and control valves. For example, from Table 3 it is estimated that 29.0% of the pump seals which initially had screening values <10,000 ppmv will screen ≥10,000 ppmv after 180 days. The confidence

Table 3. Occurrence[a] Rate Estimates for Valves and Pumps by Process

	30-Day Estimate	95% Confidence Interval	90-Day Estimate	95% Confidence Interval	180-Day Estimate	95% Confidence Interval
Valves						
Cumene units	1.9	(0.2, 5.9)	5.6	(0.6, 17)	10.8	(1.3, 30)
Ethylene units	2.0	(0.9, 3.6)	6.0	(2.7, 10)	11.6	(5.3, 20)
Vinyl acetate units	0.3	(0.0, 0.6)	0.8	(0.1, 1.9)	1.5	(0.3, 3.8)
All units	1.3	(0.7, 2.1)	3.8	(2.0, 6.0)	7.4	(4.0, 12)
Pumps						
Cumene units	5.8	(0.7, 20)	16.3	(2.1, 49)	30.0	(4.2, 74)
Ethylene units	18.4	(2.8, 42)	45.7	(8.2, 8)	70.5	(16, 96)
Vinyl acetate units	2.8	(0.8, 6.2)	8.1	(2.2, 17)	15.6	(4.4, 32)
All units	5.5	(2.2, 10)	15.7	(6.6, 27)	29.0	(12, 47)

[a] A leak from a source was defined as having occurred if it initially screened <10,000 ppmv and at some later date screened ⩾10,000 ppmv.

Table 4. Occurrence[a] Rate Estimates for Valves in Gas and Liquid Service by Process

	30-Day Estimate	95% Confidence Interval	90-Day Estimate	95% Confidence Interval	180-Day Estimate	95% Confidence Interval
Gas Service						
Cumene units	2.8	(0.1, 9.5)	8.3	(0.4, 29)	15.9	(0.9, 45)
Ethylene units	0.9	(0.3, 2.6)	2.6	(0.9, 7.5)	5.1	(1.9, 15)
Vinyl acetate units	0.7	(0.1, 1.9)	2.0	(0.2, 5.6)	4.0	(0.5, 11)
All units	1.0	(0.4, 2.0)	3.1	(1.2, 6.0)	6.1	(2.4, 12)
Liquid Service						
Cumene units	0.6	(0.1, 10)	1.7	(0.2, 28)	3.4	(0.4, 48)
Ethylene units	4.1	(1.5, 7.8)	11.9	(4.5, 22)	22.4	(8.8, 39)
Vinyl acetate units	0.2	(0 , 1.0)	0.6	(0 , 3.0)	1.3	(0 , 5.9)
All units	2.4	(1.2, 4.0)	7.0	(3.6, 11)	13.6	(7.1, 22)

[a] A leak from a source was defined as having occurred if it initially screened <10,000 ppmv and at some later date screened ≥10,000 ppmv.

Table 5. Occurrence[a] Rate Estimates for Block and Control Valves by Process

	30-Day Estimate	95% Confidence Interval	90-Day Estimate	95% Confidence Interval	180-Day Estimate	95% Confidence Interval
Block Valves						
Cumene units	2.0	(0.3, 6.2)	6.0	(0.8, 18)	11.7	(1.5, 32)
Ethylene units	2.3	(0.8, 4.8)	6.9	(2.3, 14)	13.3	(4.6, 26)
Vinyl acetate units	0	(0 , 0.7)	0	(0 , 2.1)	0	(0 , 4.1)
All units	1.2	(0.5, 2.2)	3.6	(1.6, 6.5)	7.0	(3.1, 13)
Control Valves						
Cumene units	0	(0 , 2.6)	0	(0 , 7.5)	0	(0 , 15)
Ethylene units	1.7	(0.4, 3.9)	5.0	(1.2, 11)	9.7	(2.5, 21)
Vinyl acetate units	1.3	(0.2, 3.2)	3.7	(0.7, 9.3)	7.3	(1.3, 18)
All units	1.6	(0.5, 3.2)	4.6	(1.6, 9.3)	9.0	(3.1, 18)

[a] A leak from a source was defined as having occurred if it initially screened <10,000 ppmv and at some later date screened ≥10,000 ppmv.

interval for this estimate is 12–47%. Apart from the estimates themselves, the main result is that pump seals have a statistically significant higher rate of occurrence than do valves.

Plots of the cumulative distribution functions are shown in Figures 5 through 10. The predicted occurrence rate for periods up to eight months can be obtained directly from these curves. The fact that the curves are not straight lines is a consequence of the effectively decreasing population size, since sources which begin leaking are no longer included in the population. It should be kept in mind that the underlying occurrence rates are always assumed to be constant, however. As a check on the model, the observed and predicted occurrence rates were compared. This is shown in Figure 11.

A statistical test was used to evaluate pairwise differences in occurrence rate estimates. Table 6 presents the results of this analysis. The statistical tests showed that pump seals have a higher occurrence rate than valves.

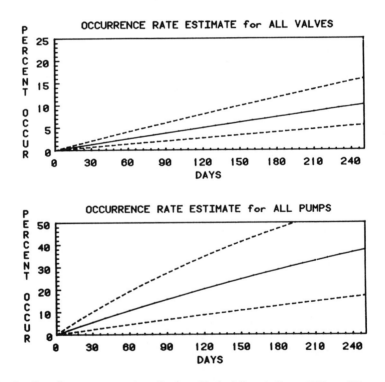

Figure 5. Overall occurrence rate estimates. (Dashed lines indicate 95% confidence region.)

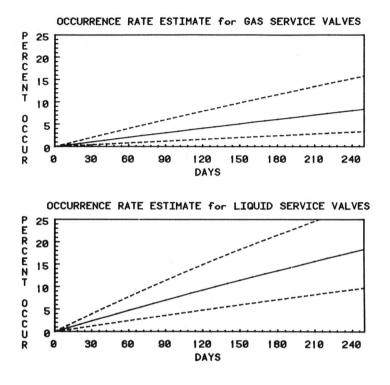

Figure 6. Occurrence rate estimates by service. (Dashed lines indicate 95% confidence region.)

Similarly, valves in liquid service have a higher occurrence rate than those in gas service. Valves in the vinyl acetate process units have a lower occurrence rate than both those in ethylene process units and those in cumene process units. The pump seals in vinyl acetate units have a lower occurrence rate than the pump seals in the ethylene units.

Finally, to aid in assessing the costs of valve maintenance, the total time (in minutes) associated with maintenance was recorded each time that a series of valves was maintained. The maintenance time ranged from 3.7 to 28.7 min/valve with an average of 9.6 min/valve (95% confidence interval for average: 8.6–10.6 min/valve). These data indicate that 10 min/valve would be a reasonable maintenance time to use in assessing costs of valve maintenance.

The activities included in maintenance for this study were restricted to tightening packing gland bolts to compress the packing material around the valve stem and seat while the valve was in service. This operation is a simple on-line maintenance procedure.

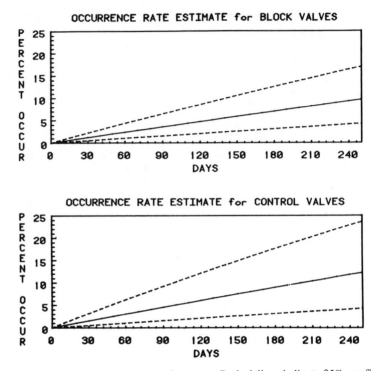

Figure 7. Occurrence rate estimate by valve type. (Dashed lines indicate 95% confidence region.)

Other on-line maintenance procedures could have been used and are currently practiced in industrial plants. Although some of these other methods may be more time consuming, they have been demonstrated as effective. Some valves have lubricated packing and are equipped with fittings to inject lubricant into the packing gland while the valve is in service. There are also valves equipped with backseating capabilities which allow replacement of the packing without dismantling the valve. Also available are commercial leak sealing services which can inject sealant into a valve to seal the leak while the valve is in service. Finally, some process units have piping configurations which allow bypass or isolation of a valve for repacking while it is in place, even though it is not on-line.

Some conditions encountered in the field during this study prevented attempts at maintenance. Of the 172 valves selected for maintenance, 17 were not maintained due to the physical condition or current service of the valve.

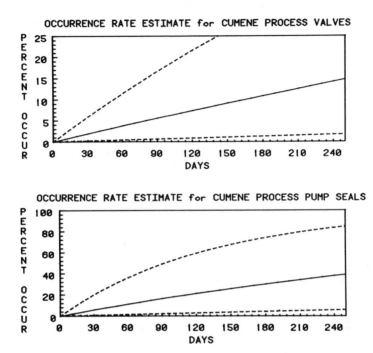

Figure 8. Occurrence rate estimate for cumene process units. (Dashed lines indicate 95% confidence region.)

Table 6. Summary of Results on Pairwise Comparison of Occurrence Rate Estimate

Test Grouping	Categories Being Tested for Statistically Different Occurrence Rates	Result of Test at the 0.05 Significance Level[a] for Each Test Grouping
1	Valves vs Pump Seals	Significant
2	Block vs Control Valves	Not Significant
3	Valves in Gas vs Liquid Service	Significant
4	Valves: Ethylene vs Cumene	Not Significant
	Valves: Ethylene vs Vinyl Acetate	Significant
	Valves: Cumene vs Vinyl Acetate	Significant
5	Pumps: Ethylene vs Cumene	Not Significant
	Pumps: Ethylene vs Vinyl Acetate	Significant
	Pumps: Cumene vs Vinyl Acetate	Not Significant

[a]Means the probability of incorrectly classifying a difference within a category is significant.

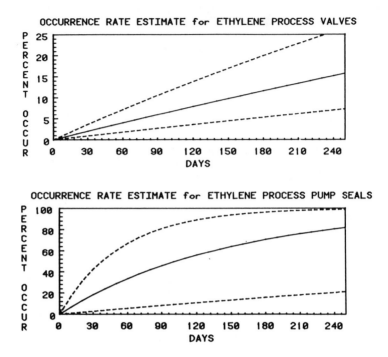

Figure 9. Occurrence rate estimate for ethylene process units. (Dashed lines indicate 95% confidence region.)

However, because time was spent examining or attempting maintenance, they were included in the time estimates. Some of the conditions encountered were rusty bolts on the packing glands, cracked packing glands, and missing gland nuts. Maintenance of valves in these conditions was prevented because of the fear of causing catastrophic failure by disturbing them. Another problem encountered was the inability to tighten packing gland bolts due to the fact that epoxy had been painted over them. Finally, there were some valves in very critical service. These valves were not disturbed because of the fear that tampering with them might cause disruption of the process.

Normally, maintenance on control valves was performed by an instrument mechanic, and block valves were maintained by pipefitters. There did not appear to be significant differences in the time required for maintenance between the two types of valves.

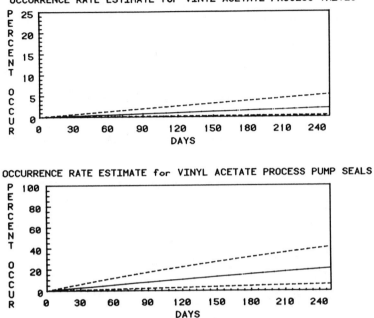

Figure 10. Occurrence rate estimate for vinyl acetate process units. (Dashed lines indicate 95% confidence region.)

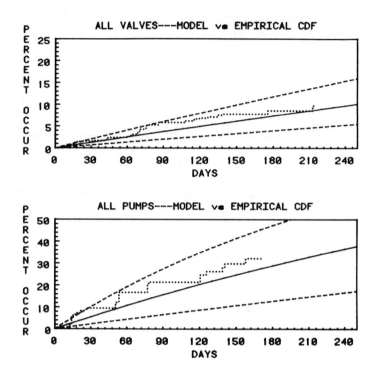

Figure 11. Occurrence rate estimate vs empirical CDF. (Dotted line is plot of actual data.)

REFERENCES

1. Weber, R. C., and C. D. Smith. "Control of Fugitive Volatile Organic Compound Emissions," Paper No. 80-10-3, presented at the 73rd Annual Meeting of the Air Pollution Control Association, Montreal, Quebec, June 1980.
2. Langley, G. J., and R. G. Wetherold. "Evaluation of Maintenance for Fugitive VOC Emissions Control," U.S. EPA 600/52-81-080 (1980).
3. Wetherold, R. G., and D. D. Rosebrook. "Assessment of Atmospheric Emissions from Petroleum Refining," U.S. EPA No. 600/2-80-075a, NTIS PB 80-225-253 (1980).

CHAPTER 16

RESPONSE FACTORS FOR VOC ANALYZERS
USED IN FUGITIVE EMISSION MONITORING

G. E. Harris

Radian Corporation
Austin, Texas 78759

Bruce A. Tichenor

Industrial Environmental Research Laboratory
Office of Environmental Engineering and
  Technology
U.S. Environmental Protection Agency
Research Triangle Park, North Carolina 27711

INTRODUCTION

The study of fugitive volatile organic chemical (VOC) emissions has rapidly gained emphasis over the past several years. Several research efforts designed to measure fugitive emission levels and to assess the effectiveness of proposed control schemes are in various stages of completion. This chapter presents the results of a laboratory study of the sensitivity of two types of portable hydrocarbon detectors to a variety of organic chemicals.

The development of regulations to control fugitive emissions has begun at both the state and federal levels. As a result of this work, a qualitative method for leak detection has been developed, which is commonly called "screening." The screening process is simply the measurement of the VOC concentration at a point close to the potential leak site (valve stem, pump seal, etc.). The measurement is made with a portable VOC detector. While

293

many instruments could potentially be used for this service, the bulk of the existing data has been taken using either the J. W. Bacharach TLV Sniffer, the Century Systems OVA-108, or the OVA-128.

These instruments will respond to almost any combustible material, but they exhibit different sensitivities to various types of hydrocarbons and substituted organic chemicals. The "response factor" is a correction factor that quantifies these differences in sensitivity.

The primary objective of this work was to determine experimentally the response factors for a large number of commonly encountered chemical species. Such response factors are presented here for both the OVA and the TLV. Chemicals were tested at several concentrations to assess variations in the response factor over the range of interest.

The experimental determination of a response factor simply requires reading the observed concentration on the instrument when testing a gas sample of known concentration. The response factor is then calculated by:

$$\text{Response Factor (RF)} = \frac{\text{actual concentration}}{\text{observed concentration}}$$

DESCRIPTION OF THE INSTRUMENTS

The Century Systems OVA-108 and the J. W. Bacharach TLV Sniffer were chosen for the laboratory study because they had both been successfully used in previous fugitive emission studies. Two instruments of each type were tested to check for normal variance between individual instruments and to check for any operating problems. One old and one new instrument of each type were selected to determine if there were any variances in response with detector age.

OVA-108 is a portable flame ionization detector (FID) with a gas chromatograph option. In the total hydrocarbon mode used for fugitive emission screening, the chromatograph column is bypassed and the sample gas is introduced directly to the detector. The unit is battery-powered and has its own hydrogen cylinder with enough fuel for 8–10 hr of continuous operation. The unit is certified to be intrinsically safe for operation in explosive atmospheres.

The readout on the OVA-108 is logarithmic and has a range from 1 to 10,000 ppmv. This range can be extended by use of a dilution probe, but this feature was not used in the response factor study. Since the sample is diluted before it reaches the detector, no new response data would be gained by sampling higher concentrations with the dilution probe.

There are some important differences between the construction of the OVA and that of a "standard" laboratory FID. In the laboratory version, the hydrogen is mixed with the sample gas before combustion, and the flame is supported by an independent air supply. The OVA draws its combustion air through the sample probe, which is thus mixed with the sample gas. The OVA flame will, therefore, be extinguished if the VOC concentration gets too high or if a standard in N_2 is used instead of one in air. More importantly, the OVA is calibrated and reads out in parts per million by volume rather than by weight.

The TLV Sniffer is a combustible gas detector. The sample gas is passed across a wire coated with oxidation-promoting catalyst. Any combustible material is oxidized and the heat released causes the resistance of the coated wire to change. Half of the sample gas is routed through a comparison cell with an identical wire that has no catalyst coating. Both wires are set in a bridge network and the comparative resistance change is read out on a galvanometer. The face of the meter readout is marked directly in ppmv as factory calibrated to hexane. The response of the TLV thus depends on the heat of combustion and heat capacity of the sample gas.

Both instruments were calibrated with methane for this test work. The OVA features a single-point calibration system and an electronic span adjustment. An 8000-ppmv methane-in-air standard was used for calibration, and a 500-ppmv methane standard was used to check the linearity of the calibration. The TLV can be calibrated independently in each of its three multiplier ranges. The 8000 ppmv standard was used to calibrate the high range, the 500 ppmv standard was used in the midrange, and no readings were taken on the low range.

PREPARATION OF STANDARD GASES

The standard gas samples were prepared by one of two methods depending on the volatility of the subject chemical. Liquids with low to medium volatility were injected into a Tedlar bag containing a measured volume of hydrocarbon-free air. The maximum concentration tested was limited to 90% of the saturated concentration at laboratory conditions to prevent condensation or incomplete vaporization. Each chemical was injected with a microliter syringe, which was weighed before and after injection on an electronic balance to get a precise measure of the amount of each chemical introduced into the bag.

Standard gas mixtures for gases and highly volatile liquids were prepared volumetrically. A cylinder of standard gas was purchased at, or above, the highest concentration to be tested. Dilutions of this purchased gas with zero air were made on a manifold with the flows controlled by needle valves and the flowrate of each gas was measured using bubble meters.

RESULTS

Response data were collected for 168 organic chemicals at a minimum of three concentrations each. The response factor of most of these chemicals varied significantly with concentration, as can be seen in Figures 1 through 3. This made it difficult to characterize the response of a chemical with a single number. Accordingly, a correlation was developed between response and concentration, and this correlation was used to estimate the response factor at an actual concentration of 10,000 ppmv, as shown in Table 1.

CORRELATION OF RESPONSE FACTOR TO CONCENTRATION

The user can generate estimates of the response factor at any concentration by the following equation:

$$\ln(IR_c) = A + B \cdot \ln C + Se^2/2$$

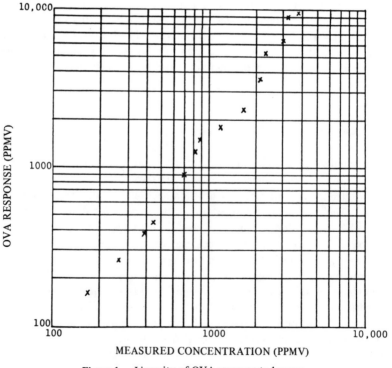

Figure 1. Linearity of OVA response to hexane.

where IR_c is the instrument response to a concentration of "C" ppmv, and A, B, and Se are constants given for each chemical in Tables 2 and 3 for the OVA and TLV, respectively. The 95% confidence intervals bounding this response factor estimate can be calculated by:

$$\ln (C_u) = \ln (IR_c) + t \cdot S_{IR_c}$$

and

$$\ln (C_L) = \ln (IR_c) - t \cdot S_{IR_c}$$

where C_u = the upper confidence interval

C_L = the lower confidence interval

t = the tabulated student's t with N–2 degrees of freedom for 95% confidence, and

$$S_{IR_c} = \sqrt{S_e^2 \cdot [(1/n) + (\ln C - \overline{X})^2 / SS_X]}$$

where S_e, n, $\overline{X}$, and SS_X are constants for each chemical given in Tables 2 and 3.

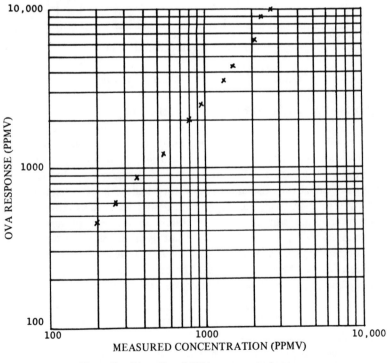

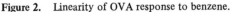

Figure 2. Linearity of OVA response to benzene.

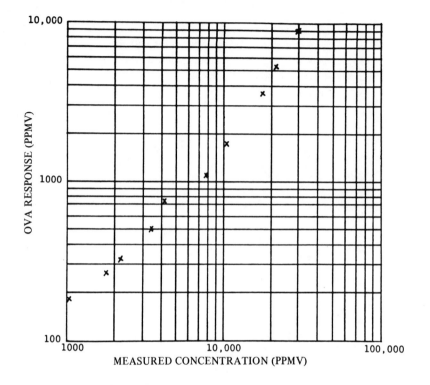

Figure 3.   Linearity of OVA response to methanol.

RESPONSE FACTORS FOR CHEMICAL MIXTURES

A limited amount of testing was performed to determine the response characteristics of chemical mixtures. Benzene, methyl ethyl ketone and methanol were chosen for the mixtures work because they differed widely in their individual response and because they had been well characterized in the early testing. A triangular experiment design was used as shown in Figure 4. This resulted in testing 21 binary mixtures and 4 ternary mixtures.

Several mathematical models, which are typically used to represent the blended properties of a mixture, have been reviewed to try to predict the response of these chemical mixtures. The best fit to date has come from the "weighted average model" based on total concentration:

$$IR = p_1 A_1 C^{B_1} + p_2 A_2 C^{B_2}$$

where IR = the OVA response to the mixture
   C = the total concentration of the mixture ($C = C_1 + C_2$)
   p_i = the proportion of chemical ($p_i = C_i/C$)
A_i and B_i = the constants from the exponential equation given in Tables 2 and 3 for chemical i.

Table 1. Fitted Response Factors with 95% Confidence Intervals (estimated at 10,000 ppmv)

OCPDB[a] ID Number	Compound Name	Volatility Class[b]	OVA		TLV	
			Response Factor	Confidence Intervals	Response Factor	Confidence Intervals
70	Acetic acid	LL	1.83	(1.17–2.87)	5.70	(4.16–7.81)
80	Acetic anhydride	LL	1.36	(1.12–1.65)	2.89	(1.62–5.14)
90	Acetone	LL	0.79	(0.53–1.19)	1.22	(0.81–1.83)
100	Acetone cyanohydrin	HL	3.42	(0.70–16.83)	7.84	(1.40–43.78)
110	Acetonitrile	LL	0.94	(0.85–1.05)	1.17	(0.96–1.42)
120	Acetophenone	HL	10.98	(5.56–21.68)	54.86	(38.35–78.48)
125	Acetyl chloride	LL	1.99	(1.76–2.26)	2.59	(1.81–3.71)
–	3-Acetyl-1-propanol	LL	10.87	(6.58–17.95)	25.81	(13.70–48.63)
130	Acetylene	G	0.37	(0.32–0.43)	11.95	(10.41–13.73)
160	Acrylic acid	LL	4.65	(3.85–5.61)	36.95	(5.34–255.69)
170	Acrylonitrile	LL	0.96	(0.79–1.17)	2.70	(0.44–16.36)
–	Allene	G	0.55	(0.49–0.61)	5.78	(5.36–6.24)
200	Allyl alcohol	LL	0.94	(0.69–1.30)		
250	n-Amyl alcohol	HL	0.69	(0.47–1.00)	1.78	(1.19–2.66)
2855	Amylene	LL	0.31	(0.19–0.51)	1.03	(0.46–2.32)
330	Anisole	LL	0.92	(0.64–1.32)	2.69	(1.69–4.27)
360	Benzaldehyde	HL	2.36	(1.52–3.68)	6.30	(5.77–6.88)
380	Benzene	LL	0.21	(0.19–0.24)	1.07	(0.97–1.19)
450	Benzonitrile	HL	2.24	(1.18–4.24)	9.13	(2.82–29.62)
490	Benzoyl chloride	HL	6.40	(3.20–12.80)	6.60	(3.41–12.79)
530	Benzyl chloride	HL	4.23	(2.74–6.52)	4.87	(2.50–9.48)
570	Bromobenzene	LL	0.36	(0.28–0.46)	1.16	(0.69–1.94)

Table 1, continued

OCPDB[a] ID Number	Compound Name	Volatility Class[b]	OVA		TLV	
			Response Factor	Confidence Intervals	Response Factor	Confidence Intervals
590	1,3-Butadiene	G	0.37	(0.34–0.42)	6.00	(5.64–6.39)
—	n-Butane	G	0.38	(0.31–0.46)	0.68	(0.61–0.75)
640	n-Butanol	LL	1.43	(0.66–3.13)	2.80	(1.49–5.27)
650	sec-Butanol	LL	0.70	(0.63–0.78)	1.26	(1.00–1.57)
660	tert-Butanol	S	0.44	(0.26–0.74)	2.19	(1.54–3.11)
592	1-Butene	G	0.51	(0.45–0.58)	2.97	(2.73–3.24)
600	n-Butyl acetate	LL	0.60	(0.46–0.78)	1.30	(1.18–1.42)
630	n-Butyl acrylate	LL	0.64	(0.56–0.74)	1.98	(0.89–4.43)
—	n-Butyl ether	LL	2.70	(1.95–3.73)	2.66	(1.42–4.95)
—	sec-Butyl ether	LL	0.26	(0.11–0.60)	1.13	(0.78–1.64)
670	n-Butylamine	LL	0.63	(0.43–0.92)	1.91	(1.38–2.65)
680	sec-Butylamine	LL	0.67	(0.53–0.84)	1.50	(1.26–1.79)
690	tert-Butylamine	LL	0.58	(0.52–0.65)	1.80	(1.53–2.11)
—	tert-Butylbenzene	HL	1.27	(0.94–1.73)	6.42	(1.82–22.67)
750	n-Butyraldehyde	LL	1.39	(1.12–1.71)	1.89	(1.60–2.23)
760	Butyric acid	HL	0.74	(0.22–2.52)	4.58	(2.39–8.78)
780	Butyronitrile	LL	0.46	(0.31–0.70)	1.33	(0.44–3.99)
790	Carbon disulfide	LL	571.92	(279.65–1169.7)	2.96	(2.62–3.34)
810	Carbon tetrachloride	LL	21.28	(10.95–41.36)	30.52	(8.96–103.88)
830	Chloroacetaldehyde	LL	13.40	(9.85–18.24)	5.07	(3.65–7.06)
890	Chlorobenzene	LL	0.36	(0.28–0.48)	0.88	(0.77–1.00)
1740	Chloroethane	G	0.67	(0.02–22.54)	2.16	(0.31–14.95)
930	Chloroform	LL	4.48	(3.60–5.56)	8.77	(5.65–13.64)
960	o-Chlorophenol	HL	3.33	(1.88–5.92)	5.87	(2.45–14.06)
—	1-Chloropropene	LL	0.59	(0.53–0.66)	0.86	(0.64–1.15)

210	3-Chloropropene	LL	0.75	(0.65–0.87)	1.24	(1.09–1.41)
970	m-Chlorotoluene	LL	0.43	(0.39–0.47)	0.92	(0.70–1.21)
980	o-Chlorotoluene	LL	0.45	(0.38–0.54)	1.05	(0.70–1.58)
990	p-Chlorotoluene	S	0.52	(0.47–0.58)	1.15	(0.62–2.12)
1010	o-Cresol		0.95	(0.59–1.52)	3.98	(0.22–72.91)
1040	Crotonaldehyde	LL	1.32	(0.78–2.23)	8.54	(2.33–31.36)
1060	Cumene	LL	1.92	(1.13–3.28)	12.49	(5.62–27.74)
1120	Cyclohexane	LL	0.36	(0.26–0.50)	0.72	(0.63–0.82)
1130	Cyclohexanol	HL	0.82	(0.58–1.17)	4.92	(1.18–20.48)
1140	Cyclohexanone	LL	1.50	(1.01–2.22)	3.99	(1.79–8.91)
1150	Cyclohexene	LL	0.40	(0.32–0.51)	1.84	(1.65–2.05)
1160	Cyclohexylamine	LL	0.47	(0.28–0.79)	1.38	(1.31–1.45)
—	Decane	HL	0.00	(0.00–3.31)	0.20	(0.00–31.01)
1190	Diacetone alcohol	HL	1.53	(1.01–2.31)	0.98	(0.29–3.36)
—	Diacetyl	LL	1.61	(1.31–1.99)	2.81	(2.15–3.67)
—	2,3-Dichloro-1-propanol	LL	61.51	(30.13–125.55)	34.34	(23.80–49.55)
1270	2,3-Dichloro-1-propene	LL	0.70	(0.47–1.05)	1.62	(1.22–2.14)
—	1,3-Dichloro-2-propanol	LL	29.34	(17.42–49.40)	18.63	(12.95–26.82)
1215	m-Dichlorobenzene	HL	0.66	(0.56–0.78)	1.89	(1.37–2.60)
1216	o-Dichlorobenzene	HL	0.70	(0.48–1.03)	1.22	(0.80–1.86)
—	1,1-Dichloroethane	LL	0.77	(0.60–0.98)	1.80	(1.58–2.06)
1244	1,2-Dichloroethane	LL	0.95	(0.77–1.18)	2.08	(1.73–2.51)
1235	cis-1,2-Dichloroethylene	LL	1.31	(1.08–1.58)	1.93	(1.02–3.66)
1236	trans-1,2-Dichloroethylene	LL	1.13	(0.99–1.30)	1.86	(0.55–6.29)
2620	Dichloromethane	LL	2.26	(1.94–2.64)	3.63	(2.72–4.84)
3110	1,2-Dichloropropane	LL	1.03	(0.82–1.29)	1.80	(1.12–2.89)
1440	Diisobutylene	LL	0.24	(0.17–0.35)	1.39	(0.97–2.01)
—	1,3-Diisopropyl benzene	LL	9.43	(3.68–24.18)	24.96	(6.07–102.67)
1870	1,2-Dimethoxy ethane	LL	1.28	(0.57–2.90)	1.43	(1.16–1.75)
1490	N,N-Dimethylformamide	LL	3.89	(3.11–4.85)	2.95	(2.86–3.03)
1495	1,1-Dimethylhydrazine	LL	1.04	(0.71–1.52)	2.74	(1.61–4.67)

Table 1, continued

OCPDB[a] ID Number	Compound Name	Volatility Class[b]	OVA		TLV	
			Response Factor	Confidence Intervals	Response Factor	Confidence Intervals
—	2,4-Dimethylstyrene	LL	37.09	(25.75–53.42)	78.06	(38.51–158.23)
1520	Dimethylsulfoxide	HL	0.00	(0.00–4300.8)	4.88	(1.41–16.92)
1560	Dioxane	LL	1.58	(1.10–2.28)	1.23	(1.07–1.41)
1650	Epichlorohydrin	LL	1.72	(1.61–1.83)	2.02	(1.84–2.22)
	Ethane	G	0.57	(0.26–1.26)	0.73	(0.21–2.52)
1660	Ethanol	LL	2.04	(1.83–2.28)		
1910	2-Ethoxyethanol	LL	1.68	(1.38–2.06)	1.61	(1.42–1.84)
1670	Ethyl acetate	LL	0.84	(0.74–0.94)	1.37	(1.08–1.74)
1680	Ethyl acetoacetate	HL	3.02	(1.98–4.62)	3.13	(2.71–3.63)
1690	Ethyl acrylate	LL	0.72	(0.55–0.93)		
1750	Ethyl chloroacetate	LL	1.97	(1.76–2.21)	1.47	(0.97–2.23)
1990	Ethyl ether	LL	0.97	(0.73–1.28)	1.11	(1.01–1.23)
1710	Ethylbenzene	LL	0.70	(0.45–1.09)	3.14	(1.43–6.92)
1770	Ethylene	G	0.52	(0.39–0.68)	1.49	(1.27–1.74)
1980	Ethylene oxide	G	2.72	(2.31–3.20)	2.43	(1.92–3.07)
1800	Ethylenediamine	LL	1.78	(1.39–2.28)	2.46	(1.83–3.32)
2060	Formic acid	LL	34.87	(29.00–41.94)	33.21	(16.78–65.73)
1221	Freon 12	G	9.65	(6.99–13.33)	11.83	(3.63–38.59)
2073	Furfural	HL	7.96	(3.69–17.15)	10.01	(9.72–10.31)
2105	Glycidol	LL	8.42	(5.14–13.81)	5.23	(2.68–10.22)
	n-Heptane	LL	0.30	(0.12–0.75)	0.75	(0.56–1.00)
	n-Hexane	LL	0.31	(0.26–0.36)	0.72	(0.65–0.79)
	1-Hexene	LL	0.39	(0.26–0.58)	2.92	(0.86–9.89)
	Hydroxyacetone	LL	8.70	(6.51–11.64)	9.34	(6.21–14.06)
	Isobutane	G	0.30	(0.09–0.95)	0.61	(0.44–0.86)

No.	Compound	Type	RF	Range	RF	Range
2200	Isobutylene	G	2.42	(1.01–5.84)	6.33	(0.09–460.41)
2350	Isoprene	LL	0.48	(0.32–0.72)	1.35	(0.98–1.86)
2360	Isopropanol	LL	0.90	(0.68–1.18)	1.25	(1.11–1.40)
2370	Isopropyl acetate	LL	0.68	(0.58–0.80)	0.99	(0.81–1.19)
2390	Isopropyl chloride	LL	0.62	(0.52–0.73)	2.04	(1.19–3.51)
–	Isovaleraldehyde	LL	0.55	(0.45–0.66)	3.12	(1.68–5.79)
2450	Mesityl oxide	LL	1.12	(0.94–1.33)	3.10	(1.65–5.83)
–	Methacrolein	LL	1.27	(0.91–1.77)	6.61	(0.37–116.80)
2460	Methacrylic acid	HL	0.71	(0.06–7.84)	1.88	(1.62–2.18)
2500	Methanol	LL	5.69	(4.96–6.52)	2.19	(1.85–2.60)
1930	2-Methoxyethanol	LL	2.70	(1.99–3.67)	1.76	(1.47–2.12)
2510	Methyl acetate	LL	1.80	(1.55–2.08)	3.92	(3.59–4.29)
–	Methyl acetylene	G	0.53	(0.50–0.57)	2.45	(1.62–3.73)
2560	Methyl chloride	G	1.75	(1.44–2.14)	1.12	(0.93–1.35)
2640	Methyl ethyl ketone (MEK)	LL	0.57	(0.40–0.81)		
2645	Methyl formate	LL	3.47	(2.87–4.19)	1.93	(1.76–2.11)
2665	Methyl methacrylate	LL	0.99	(0.89–1.10)	2.36	(1.54–3.62)
2650	4-Methyl-2-pentanol	LL	1.70	(1.33–2.17)	1.94	(1.46–2.58)
2660	4-Methyl-2-pentanone	LL	0.49	(0.38–0.64)	1.54	(1.22–1.94)
–	2-Methyl-2,4-pentanediol	LL	96.34	(72.63–127.80)	67.07	(36.31–123.87)
2550	2-Methyl-3-butyn-2-ol	LL	0.51	(0.32–0.79)		
–	Methylal	LL	1.46	(1.09–1.96)	1.41	(1.23–1.62)
2540	N-Methylaniline	HL	4.13	(3.75–4.54)	5.25	(0.99–27.84)
2570	Methylcyclohexane	LL	0.38	(0.12–1.16)	0.85	(0.67–1.07)
–	1-Methylcyclohexene	LL	0.33	(0.24–0.45)	2.22	(1.70–2.91)
2670	Methylpentynol	LL	1.17	(0.71–1.92)	2.82	(1.92–4.14)
2690	α-Methylstyrene	LL	10.24	(8.30–12.63)	31.46	(16.08–61.53)
1660	Monoethanolamine	LL	28.04	(6.39–123.04)	25.83	(11.82–56.44)
2700	Morpholine	LL	0.92	(0.64–1.32)	1.93	(0.36–10.40)
2770	Nitrobenzene	HL	29.77	(3.92–226.03)	40.61	(4.68–352.68)

Table 1, continued

OCPDB[a] ID Number	Compound Name	Volatility Class[b]	OVA		TLV	
			Response Factor	Confidence Intervals	Response Factor	Confidence Intervals
2790	Nitroethane	LL	1.40	(1.23–1.59)	2.54	(1.60–4.04)
2791	Nitromethane	LL	3.32	(3.03–3.64)	5.25	(2.32–11.86)
2795	Nitropropane	LL	1.06	(0.79–1.42)	1.77	(1.22–2.57)
–	n-Nonane	LL	1.62	(0.93–2.81)	5.54	(2.82–10.89)
–	n-Octane	LL	1.04	(0.88–1.22)	2.08	(1.76–2.46)
2851	n-Pentane	LL	0.42	(0.30–0.58)	0.62	(0.55–0.69)
2910	Phenol	S	11.75	(7.55–18.29)	12.01	(3.57–40.43)
–	2-Phenyl-2-propanol	LL	89.56	(69.84–114.83)	76.57	(46.38–126.42)
2973	2-Picoline	LL	0.34	(0.27–0.41)	1.17	(1.10–1.25)
–	n-Propane	G	0.88	(0.10–7.54)	0.63	(0.24–1.64)
3063	Propionaldehyde	LL	1.19	(1.02–1.39)	1.65	(1.16–2.36)
3066	Propionic acid	LL	1.34	(1.06–1.69)	3.51	(0.90–13.62)
3070	Propyl alcohol	LL	0.91	(0.72–1.15)	1.55	(1.41–1.71)
–	n-Propylbenzene	LL	0.44	(0.37–0.53)	5.97	(0.33–108.71)
3090	Propylene	G	0.79	(0.39–1.57)	2.80	(0.96–8.20)
3120	Propylene oxide	LL	0.80	(0.69–0.93)	1.15	(0.68–1.93)
3130	Pryidine	LL	0.41	(0.33–0.51)	1.17	(1.03–1.32)
3230	Styrene	LL	4.16	(3.68–4.71)	36.83	(7.32–185.28)
3290	1,1,1,2-Tetrachloroethane	LL	3.00	(1.27–7.07)	6.52	(3.82–11.11)
3291	1,1,2,2-Tetrachloroethane	LL	6.06	(4.78–7.68)	14.14	(8.51–23.51)
2860	Tetrachloroethylene	LL	3.16	(1.92–5.21)	11.46	(8.64–15.20)
3349	Toluene	LL	0.33	(0.29–0.38)	2.32	(0.79–6.82)
3393	1,2,4-Trichlorobenzene	HL	1.35	(0.13–14.00)	0.39	(0.10–1.57)
3395	1,1,1-Trichloroethane	LL	0.79	(0.70–0.89)	2.41	(1.96–2.96)
3400	1,1,2-Trichloroethane	LL	1.26	(1.08–1.47)	3.68	(3.05–4.43)

3410	Trichloroethylene	LL	0.94	(0.81–1.08)	3.35	(2.68–4.19)
3420	1,2,3-Trichloropropane	LL	0.95	(0.54–1.69)	2.23	(1.47–3.40)
3450	Triethylamine	LL	0.46	(0.32–0.67)	1.41	(1.03–1.92)
3510	Vinyl acetate	LL	1.31	(0.97–1.76)	3.99	(1.49–10.69)
3520	Vinyl chloride	G	0.65	(0.50–0.84)	1.10	(0.79–1.55)
—	Vinyl propionate	LL	0.94	(0.40–2.23)	0.70	(0.02–22.87)
3530	Vinylidene chloride	LL	1.15	(0.86–1.54)	2.38	(1.95–2.90)
3570	p-Xylene	LL	2.27	(1.87–2.76)	5.35	(3.34–8.58)
3550	m-Xylene	LL	0.30	(0.25–0.37)	3.56	(0.93–13.64)
3560	o-Xylene	LL	0.36	(0.16–0.77)	1.40	(0.73–2.70)

[a]Organic chemical producers data base, Radian Corp., Austin, TX.
[b]G = gas, LL = light liquid, HL = heavy liquid, S = solid.

Table 2. Statistics for Computing Estimates and Confidence Intervals—OVA

OCPDB[a] ID Number	Compound Name	Volatility Class[b]	A	B	n	S_e	SS_x	$\bar{X}$
70	Acetic acid	LL	-2.8754	1.22149	23	0.677209	94.0588	6.71835
80	Acetic anhydride	LL	0.375288	0.925664	6	0.0924492	8.75857	6.97689
90	Acetone	LL	-0.42601	1.06086	14	0.44568	56.4432	6.73273
100	Acetone cyanohydrin	HL	-1.0724	0.98053	6	0.206338	1.07377	5.97147
110	Acetonitrile	LL	-0.80987	1.09408	6	0.0616236	12.5475	7.23366
120	Acetophenone	HL	-1.013	0.817811	14	0.768287	62.071	6.71836
125	Acetyl chloride	LL	-0.36506	0.964417	6	0.0778431	14.6944	7.21322
–	3-Acetyl-1-propanol	LL	1.07586	0.611782	12	0.476851	50.5077	6.46174
130	Acetylene	G	0.485665	1.05433	6	0.0377039	1.44015	7.44562
160	Acrylic acid	LL	-1.627	1.0091	6	0.112184	11.9679	7.31447
170	Acrylonitrile	LL	-0.57262	1.06539	6	0.118698	14.1474	7.2202
–	Allene	G	-2.8183	1.37155	6	0.0573086	4.4727	7.88619
200	Allyl alcohol	LL	-3.6	1.39534	6	0.192964	13.7776	7.22368
250	n-Amyl alcohol	HL	-2.3527	1.29529	6	0.131896	5.39958	6.68114
2855	Amylene	LL	-2.5916	1.40493	6	0.231679	8.23563	7.0079
330	Anisole	LL	0.083984	0.998552	6	0.166868	8.64685	6.92252
360	Benzaldehyde	HL	-0.44202	0.953919	6	0.114793	3.55556	6.34483
380	Benzene	LL	-0.87171	1.26135	31	0.148218	51.4347	6.63953
450	Benzonitrile	HL	1.60572	0.735434	12	0.226988	6.18285	6.08178
490	Benzoyl chloride	HL	1.74643	0.599652	6	0.410048	13.2587	7.18737
530	Benzyl chloride	HL	2.76717	0.528787	12	0.511603	21.483	8.00839
570	Bromobenzene	LL	-0.04983	1.11651	6	0.11169	7.72926	6.8984
590	1,3-Butadiene	G	-5.7128	1.72671	6	0.0471682	3.14273	7.7553
–	n-Butane	G	-2.9002	1.42036	10	0.130245	3.61904	8.06308
640	n-Butanol	LL	-0.96371	1.06262	3	0.233069	6.0393	7.04506
650	sec-Butanol	LL	-2.4932	1.30867	6	0.0571548	10.6652	7.1325
660	tert-Butanol	S	-1.7618	1.27647	6	0.248874	8.58158	7.03513

592	1-Butene	G	-0.75886	1.15507	6	0.0686523	4.98197	7.7713
600	n-Butyl acetate	LL	-1.5906	1.22728	6	0.149389	11.2932	7.15691
630	n-Butyl acrylate	LL	-1.6738	1.22939	6	0.0703242	9.90051	7.02981
—	n-Butyl ether	LL	-1.3459	1.03767	3	0.114048	6.2084	7.50326
—	sec-Butyl ether	LL	-1.2453	1.27583	4	0.286528	4.87663	7.17385
670	n-Butylamine	LL	-2.0713	1.27286	6	0.212276	11.2392	7.13838
680	sec-Butylamine	LL	-0.90586	1.1415	6	0.13293	12.27	7.15
690	tert-Butylamine	LL	-1.2282	1.19206	6	0.0627487	11.0874	7.17992
—	tert-Butylbenzene	HL	0.985678	0.864887	6	0.186581	13.8471	7.21205
750	n-Butyraldehyde	LL	-2.8809	1.27649	6	0.126596	12.3356	7.29208
760	Butyric acid	HL	-2.9839	1.35542	6	0.143141	0.946825	5.84132
780	Butyronitrile	LL	-0.92245	1.18116	6	0.203173	9.07656	7.0786
790	Carbon Disulfide	LL	-2.1633	0.535771	6	0.424013	12.8302	7.22315
810	Carbon tetrachloride	LL	2.13296	0.427673	6	0.400795	13.7915	7.20605
830	Chloroacetaldehyde	LL	-4.2311	1.17545	8	0.198182	12.6252	7.1765
890	Chlorobenzene	LL	0.590851	1.04424	10	0.15665	11.8805	6.68244
1740	Chloroethane	G	1.32501	0.5541	8	2.52458	40.5528	6.09409
930	Chloroform	LL	1.50605	0.67282	6	0.132055	13.3425	7.26212
960	o-Chlorophenol	HL	0.679952	0.793318	6	0.202122	5.50648	6.66217
—	1-Chloropropene	LL	-2.2549	1.30157	7	0.0754264	12.5689	7.30057
210	3-Chloropropene	LL	-2.3441	1.2843	12	0.136945	52.1424	6.45557
970	m-Chlorotoluene	LL	-0.46281	1.14167	6	0.039029	7.74444	6.9301
980	o-Chlorotoluene	LL	0.09082	1.07555	6	0.0746	7.74409	6.88833
990	p-Chlorotoluene	LL	-0.48446	1.12273	6	0.0479559	7.70788	6.93231
1010	o-Cresol	S	-4.3483	1.4774	6	0.111396	2.9799	6.30801
1040	Crotonaldehyde	LL	-2.5728	1.24273	8	0.351319	12.9835	7.24563
1060	Cumene	LL	-1.2383	1.04932	12	0.509365	51.1558	6.46426
1120	Cyclohexane	LL	-2.0872	1.33603	6	0.146801	7.77672	6.96164
1130	Cyclohexanol	HL	-2.0239	1.24055	6	0.0899117	3.57046	6.28751
1140	Cyclohexanone	LL	-0.38416	0.995445	6	0.205081	10.3796	7.06142
1150	Cyclohexene	LL	-1.4625	1.25627	7	0.125436	10.0889	7.13077

Table 2, continued

OCPDB[a] ID Number	Compound Name	Volatility Class[b]	A	B	n	S_e	SS_x	$\bar{X}$
1160	Cyclohexylamine	LL	-2.4639	1.34576	6	0.267438	9.7518	7.07654
–	Decane	HL	-14.781	3.45409	6	0.906647	0.701072	5.67254
1190	Diacetone alcohol	HL	-1.7754	1.14617	6	0.109833	3.44138	6.4446
–	Diacetyl	LL	-1.5731	1.11652	12	0.204878	49.1017	6.60012
–	2,3-Dichloro-1-propanol	LL	-3.0521	0.873976	6	0.432654	13.6007	7.23629
1270	2,3-Dichloro-1-propene	LL	-1.9415	1.24672	6	0.217292	11.2697	7.10441
–	1,3-Dichloro-2-propanol	LL	-0.79172	0.70996	8	0.410248	15.4488	7.55474
1215	m-Dichlorobenzene	HL	0.986623	0.937663	6	0.0547674	5.4863	6.59837
1216	o-Dichlorobenzene	HL	1.10876	0.917648	6	0.0947496	3.19816	6.33962
–	1,1-Dichloroethane	LL	-0.00758	1.02819	6	0.130716	10.5488	7.11116
1244	1,2-Dichloroethane	LL	-0.30622	1.03745	6	0.131104	14.1446	7.2661
1235	cis-1,2-Dichloroethylene	LL	-1.4618	1.12855	8	0.15056	15.325	7.54257
1236	trans-Dichloroethylene	LL	-2.0185	1.20515	8	0.0862401	13.1057	7.02025
2620	Dichloromethane	LL	1.0934	0.792123	6	0.0915693	13.1072	7.23499
3110	1,2-Dichloropropane	LL	-0.88505	1.09215	6	0.123334	10.8605	7.1272
1440	Diisobutylene	LL	-2.0148	1.37176	6	0.165233	7.56692	6.9367
–	1,3-Diisopropyl benzene	LL	1.76459	0.547649	6	0.561958	28.5679	6.27642
1870	1,2-Dimethoxy ethane	LL	-2.536	1.23573	6	0.483998	13.2176	7.19222
1490	N,N-Dimethylformamide	LL	-0.87643	0.947127	6	0.109855	9.19596	7.02873
1495	1,1-Dimethylhydrazine	LL	-3.7531	1.40074	6	0.230674	13.1967	7.26392
–	2,4-Dimethylstyrene	LL	-1.5367	0.771785	6	0.224922	13.6858	7.28067
1520	Dimethylsulfoxide	HL	-16.573	3.43902	6	1.0935	0.409881	5.66978
1560	Dioxane	LL	-2.0521	1.17043	6	0.215831	13.0262	7.20306
1650	Epichlorohydrin	LL	-0.75538	1.02334	6	0.0376406	13.356	7.22066
–	Ethane	G	-2.0456	1.27648	6	0.325301	2.31961	7.84885
1660	Ethanol	LL	-2.9244	1.23973	6	0.068237	15.4173	7.19679

ID	Compound	Code						
1910	2-Ethoxyethanol	LL	-2.3344	1.19637	6	0.101094	9.75276	7.00472
1670	Ethyl acetate	LL	-1.2288	1.15152	18	0.15158	66.2747	6.71408
1680	Ethyl acetoacetate	HL	0.494514	0.825583	6	0.107831	3.5444	6.29103
1690	Ethyl acrylate	LL	-2.1188	1.26487	6	0.144331	11.6226	7.14221
1750	Ethyl chloroacetate	LL	-0.60685	0.99184	6	0.0551139	8.85726	6.99629
1990	Ethyl ether	LL	-1.8661	1.20484	6	0.155052	11.6344	7.12129
1710	Ethylbenzene	LL	-1.0274	1.13086	20	0.597093	66.947	6.80288
1770	Ethylene	G	-8.0566	1.94541	6	0.157469	3.84623	8.01344
1980	Ethylene oxide	G	-1.9953	1.10758	6	0.102798	4.79615	8.11482
1800	Ethylenediamine	LL	-1.0209	1.04732	6	0.146117	12.4844	7.23272
2060	Formic acid	LL	-6.6959	1.33858	6	0.107919	12.0961	7.24228
1221	Freon 12	G	1.31433	0.609471	6	0.174541	7.09205	7.51936
2073	Furfural	HL	1.04653	0.650043	6	0.452342	12.6998	7.21469
2105	Glycidol	LL	-3.1147	1.10534	6	0.163724	4.78354	6.66621
—	n-Heptane	LL	-1.8365	1.32734	3	0.223944	3.74967	6.98098
—	n-Hexane	LL	-1.7626	1.31685	37	0.199415	40.4766	6.87982
—	1-Hexene	LL	-2.1275	1.33237	6	0.179963	7.54485	6.94449
—	Hydroxyacetone	LL	-3.2967	1.12007	8	0.232517	15.5756	7.59178
—	Isobutane	G	-2.2393	1.35284	8	0.557399	2.43992	7.90812
2200	Isobutylene	G	1.16949	0.775741	6	0.147294	2.43034	5.46695
2350	Isoprene	LL	-2.8398	1.38474	6	0.215396	10.3612	7.11844
2360	Isopropanol	LL	-1.9189	1.21872	6	0.166719	13.4625	7.20498
2370	Isopropyl acetate	LL	-0.80703	1.12898	6	0.0896555	12.0222	7.11716
2390	Isopropyl chloride	LL	-1.6028	1.22527	8	0.127767	14.9689	7.52885
—	Isovaleraldehyde	LL	-2.7791	1.36676	6	0.102188	11.1909	7.14904
2450	Mesityl oxide	LL	-2.5944	1.26898	6	0.10352	12.9133	7.1985
—	Methacrolein	LL	-2.8695	1.28287	6	0.233299	12.6663	7.72282
2460	Methacrylic acid	HL	-6.6118	1.74781	6	0.366906	1.37953	6.10761
2500	Methanol	LL	-2.3981	1.06519	48	0.345716	185.666	7.35637
1930	2-Methoxy-ethanol	LL	-3.127	1.22972	6	0.181911	13.4296	7.2025
2510	Methyl acetate	LL	-1.1087	1.0562	6	0.0881283	12.9204	7.26467

Table 2, continued

OCPDB[a] ID Number	Compound Name	Volatility Class[b]	A	B	n	S_e	SS_X	$\overline{X}$
—	Methyl acetylene	G	-1.8489	1.26873	6	0.0333733	3.13703	7.8821
2560	Methyl chloride	G	-5.5539	1.54125	6	0.124856	3.89091	8.21681
2640	Methyl ethyl ketone (MEK)	LL	-1.8876	1.26262	9	0.243063	25.2199	6.54915
2645	Methyl formate	LL	-2.1378	1.09632	6	0.123123	16.9883	7.23917
2665	Methyl methacrylate	LL	-1.8063	1.19693	6	0.0609566	12.9869	7.19995
2650	4-Methyl-2-pentanol	LL	-0.916	1.03982	8	0.196403	15.6961	7.57994
2660	4-Methyl-2-pentanone	LL	-1.3252	1.22039	6	0.135332	10.2416	7.05588
—	2-Methyl-2,4-pentanediol	LL	-0.20173	0.524322	6	0.17221	13.1088	7.288
2550	2-Methyl-3-butyn-2-ol	LL	-1.8974	1.27675	6	0.232082	10.1905	7.05895
—	Methylal	LL	-2.2406	1.19782	12	0.281975	52.6838	6.41101
2540	N-Methylaniline	HL	-0.71359	0.923574	6	0.0227449	3.11941	6.24884
2570	Methylcyclohexane	LL	-2.3663	1.34952	6	0.494885	7.87591	6.87425
—	1-Methylcyclohexene	LL	-2.1159	1.34919	6	0.153802	8.95789	6.98327
2670	Methylpentynol	LL	-0.06858	0.985905	6	0.289271	12.5573	7.19601
2690	α-Methylstyrene	LL	-1.2874	0.884034	14	0.240239	66.6129	6.69408
1660	Monoethanolamine	LL	-2.4791	0.855164	10	0.979098	22.2589	6.384433
2700	Morpholine	LL	-0.9405	1.10892	6	0.215556	13.3896	7.18013
2770	Nitrobenzene	HL	3.37879	0.255857	5	0.403458	3.7675	5.51561
2790	Nitroethane	LL	-0.38829	1.00525	6	0.0769128	14.102	7.17721
2791	Nitromethane	LL	-0.77951	0.954076	6	0.0563271	13.5221	7.27112
2795	Nitropropane	LL	-1.043	1.10518	6	0.179594	13.5435	7.26298
—	n-Nonane	LL	-1.5826	1.10603	14	0.501526	411.1309	6.3957
—	n-Octane	LL	-1.9407	1.20639	6	0.0908505	11.4258	7.14769
2851	n-Pentane	LL	-2.059	1.31598	6	0.165283	9.68585	7.05094
2910	Phenol	S	0.710164	0.654066	10	0.155352	7.92326	5.72457
—	2-Phenyl-2-propanol	LL	1.00078	0.402044	6	0.151805	13.1932	7.2875
2973	2-Picoline	LL	-1.7278	1.30562	6	0.100047	8.91213	7.02075

ID	Name	Code						
—	n-Propane	G	0.76145	0.882834	10	0.949772	3.20321	7.48077
3063	Propionaldehyde	LL	-2.8873	1.29443	6	0.0929661	12.9242	7.24108
3066	Propionic acid	LL	-1.4915	1.12963	6	0.0994827	7.43093	6.89446
3070	Propyl alcohol	LL	-2.4137	1.27078	6	0.142748	14.0456	7.22723
—	n-Propylbenzene	LL	-1.2593	1.22527	6	0.0785943	7.78882	6.93443
3090	Propylene	G	0.864818	0.926563	6	0.321062	4.94104	7.47693
3120	Propylene oxide	LL	-1.5249	1.1886	7	0.103767	13.6953	7.43403
3130	Pyridine	LL	-0.59723	1.16018	6	0.0990446	8.47388	6.97017
3230	Styrene	LL	-1.3382	0.990213	6	0.0751161	13.777	7.23935
3290	1,1,1,2-Tetrachloroethane	LL	1.50932	0.697932	10	0.589679	17.5221	6.82611
3291	1,1,2,2-Tetrachloroethane	LL	-0.63198	0.872029	8	0.137087	12.074	6.91454
3349	Tetrachloroethylene	LL	-2.0129	1.05606	26	0.831676	113.842	6.77884
3349	Toluene	LL	-0.55953	1.18126	6	0.0633951	8.59474	6.89291
3393	1,2,4-Trichlorobenzene	HL	-0.17734	0.982448	3	0.265473	1.23926	6.19842
3395	1,1,1-Trichloroethane	LL	-0.58382	1.08916	6	0.0701193	12.7084	7.19098
3400	1,1,2-Trichloroethane	LL	-0.78558	1.05963	6	0.0916246	12.8405	7.20735
3410	Trichloroethylene	LL	-1.6403	1.18484	6	0.0879318	13.453	7.21675
3420	1,2,3-Trichloropropane	LL	-2.9814	1.32332	6	0.31741	7.54119	7.53407
3450	Triethylamine	LL	-0.68947	1.15664	6	0.17419	9.19563	6.95765
3510	Vinyl acetate	LL	-1.4414	1.12549	6	0.178741	13.2953	7.23837
3520	Vinyl chloride	G	-12.813	2.43718	3	0.0838916	2.44541	8.00167
—	Vinyl propionate	LL	-0.9365	1.10343	3	0.286336	6.59355	7.29263
3530	Vinylidene chloride	LL	-2.3696	1.24041	6	0.172576	11.489	7.34056
3570	p-Xylene	LL	-1.6901	1.09082	18	0.264759	76.5381	6.73317
3550	m-Xylene	LL	-1.758	1.32044	6	0.084293	7.86889	6.8857
3560	o-Xylene	LL	-1.2606	1.24282	6	0.340035	7.74549	6.8848

[a] Organic chemical producers data base, Radian Corp., Austin, TX.
[b] G = gas, LL = light liquid, HL = heavy liquid, S = solid.

Table 3. Statistics for Computing Estimates and Confidence Intervals –TLV

OCPDB[a] ID Number	Compound Name	Volatility Class[b]	A	B	n	S_e	SS_x	$\bar{X}$
70	Acetic acid	LL	1.53532	0.633431	20	0.446849	81.6063	6.75306
80	Acetic anhydride	LL	2.64396	0.596488	3	0.149324	4.37929	6.97689
90	Acetone	LL	-0.22007	0.991616	14	0.452099	56.4432	6.73273
100	Acetone cyanohydrin	HL	0.900022	0.67604	6	0.222799	1.07377	5.97147
110	Acetonitrile	LL	0.206418	0.959987	6	0.113292	12.5475	7.23366
120	Acetophenone	HL	1.57858	0.384909	14	0.40411	62.071	6.71836
125	Acetyl chloride	LL	-0.53649	0.95225	6	0.221368	14.6944	7.21322
–	3-Acetyl-1-propanol	LL	1.85273	0.426187	12	0.60199	50.5077	6.46174
130	Acetylene	G	2.77721	0.428709	5	0.0808672	4.9029	8.1201
160	Acrylic acid	LL	3.0261	0.207029	6	1.15552	11.9679	7.31447
170	Acrylonitrile	LL	1.08345	0.70882	6	1.10159	14.1474	7.2202
–	Allene	G	1.48948	0.647644	4	0.0420336	4.48003	8.31859
250	n-Amyl alcohol	HL	1.69081	0.753762	3	0.0770913	2.69979	6.68114
2855	Amylene	LL	-2.7517	1.27023	9	0.679982	15.1755	7.62642
330	Anisole	LL	1.5387	0.72482	3	0.1171	4.32343	6.92252
360	Benzaldehyde	HL	2.73967	0.502696	3	0.0123884	1.77778	6.34483
380	Benzene	LL	0.28303	0.960989	28	0.128504	55.3541	6.61165
450	Benzonitrile	HL	3.30419	0.397316	6	0.263045	3.09142	6.08178
490	Benzoyl chloride	HL	2.87184	0.48077	3	0.213308	6.62934	7.18737
530	Benzyl chloride	HL	3.46699	0.438342	6	0.495549	10.7415	8.00839
570	Bromobenzene	LL	1.40637	0.830522	3	0.12348	3.86463	6.8984
590	1,3-Butadiene	G	0.49448	0.751221	4	0.0321868	4.16748	8.22042
–	n-Butane	G	1.74095	0.852814	7	0.0692898	3.32684	8.337
640	n-Butanol	LL	1.36408	0.738053	3	0.188523	6.0393	7.04506
650	sec-Butanol	LL	-0.32886	1.00927	8	0.180589	15.8209	7.59599
660	tert-Butanol	S	-0.90524	1.01084	5	0.202464	8.24512	7.75919
592	1-Butene	G	2.4338	0.617307	5	0.0537661	4.64345	8.27484

600	n-Butyl acetate	LL	1.47233	0.811784	4	0.0440696	8.27958	7.62533
630	n-Butyl acrylate	LL	1.69695	0.738919	3	0.222439	4.95025	7.02981
—	n-Butyl ether	LL	0.973474	0.78565	3	0.218728	6.2084	7.50326
—	sec-Butyl ether	LL	0.854386	0.890522	6	0.242698	8.93521	7.78741
670	n-Butylamine	LL	0.058361	0.921838	4	0.155349	8.07529	7.59075
680	sec-Butylamine	LL	0.364395	0.916162	3	0.543876	6.13498	7.15
690	tert-Butylamine	LL	0.514191	0.880031	4	0.076921	8.00765	7.63305
—	tert-Butylbenzene	HL	2.31474	0.515112	6	0.763323	13.8471	7.21205
750	n-Butyraldehyde	LL	1.53293	0.76441	3	0.0547613	6.1678	7.29208
760	n-Butyric acid	HL	1.90498	0.627895	3	0.0415519	0.473413	5.84132
780	Butyronitrile	LL	0.400785	0.919176	3	0.345687	5.88803	7.22597
790	Carbon disulfide	LL	0.809447	0.794289	3	0.039928	6.41512	7.22315
810	Carbon tetrachloride	LL	4.65946	0.093298	6	0.738964	13.7915	7.20605
830	Chloroacetaldehyde	LL	-1.6493	1.00029	8	0.21254	12.6252	7.1765
890	Chlorobenzene	LL	0.337457	0.976794	11	0.116228	20.2555	7.35836
1740	Chloroethane	G	2.43057	0.548012	8	1.38856	40.5528	6.09409
930	Chloroform	LL	1.1337	0.637191	6	0.267978	13.3425	7.26212
960	o-Chlorophenol	HL	1.99922	0.589287	3	0.167483	2.75324	6.66217
—	1-Chloropropene	LL	-1.2944	1.15423	8	0.239914	16.3457	7.58835
210	3-Chloropropene	LL	-0.29344	1.00783	12	0.123718	52.1424	6.45557
970	m-Chlorotoluene	LL	0.974584	0.902659	3	0.0661162	3.87222	6.9301
980	o-Chlorotoluene	LL	1.12209	0.871937	3	0.0965194	3.87205	6.88833
990	p-Chlorotoluene	LL	0.643704	0.914041	3	0.149187	3.85394	6.93231
1010	o-Cresol	S	2.24855	0.598375	3	0.373867	1.48995	6.30801
1040	Crotonaldehyde	LL	2.10615	0.49768	8	0.866604	12.9835	7.24563
1060	Cumene	LL	0.372112	0.653917	12	0.762507	51.1558	6.46426
1120	Cyclohexane	LL	1.02342	0.924071	10	0.119838	15.7682	7.68236
1130	Cyclohexanol	HL	3.00453	0.49349	6	0.363684	3.57046	6.28751
1140	Cyclohexanone	LL	1.20283	0.709531	6	0.419218	10.3796	7.06142
1150	Cyclohexene	LL	1.36081	0.785839	6	0.0734484	9.15524	7.81843
1160	Cyclohexylamine	LL	-0.39214	1.00749	5	0.0315026	8.99739	7.81516

Table 3, continued

OCPDB[a] ID Number	Compound Name	Volatility Class[b]	A	B	n	S_e	SS_x	$\bar{X}$
—	Decane	HL	-3.6251	1.55525	6	0.484709	0.701072	5.67254
1190	Diacetone alcohol	HL	-2.3153	1.24765	6	0.324757	3.44138	6.4446
—	Diacetyl	LL	0.13275	0.869795	12	0.259957	49.1017	6.60012
—	2,3-Dichloro-1-propanol	LL	2.70208	0.319956	6	0.222292	13.6007	7.23629
1270	2,3-Dichloro-1-propene	LL	0.779298	0.861583	6	0.169488	9.26572	7.79473
—	1,3-Dichloro-2-propanol	LL	0.358175	0.639059	8	0.286746	15.4488	7.55474
1215	m-Dichlorobenzene	HL	1.75138	0.740694	3	0.0598515	2.74315	6.59837
1216	o-Dichlorobenzene	HL	1.02702	0.866847	3	0.0570792	1.59908	6.33962
—	1,1-Dichloroethane	LL	-0.15658	0.952456	8	0.105846	15.7135	7.57506
1244	1,2-Dichloroethane	LL	-0.34562	0.957124	6	0.116586	14.1446	7.2661
1235	cis-1,2-Dichloroethylene	LL	-3.9143	1.33988	8	0.500384	15.325	7.54257
1236	trans-1,2-Dichloroethylene	LL	-2.9985	1.22745	8	0.752501	13.1057	7.02025
2620	Dichloromethane	LL	-0.89162	0.955286	6	0.173045	13.1072	7.23499
3110	1,2-Dichloropropane	LL	-2.2472	1.17229	8	0.380407	15.8547	7.58337
1440	Diisobutylene	LL	-0.09895	0.969107	10	0.318329	15.3066	7.62355
—	1,3-Diisopropyl benzene	LL	2.06897	0.387379	6	0.843952	28.5679	6.27642
1870	1,2-Dimethoxy ethane	LL	1.0292	0.849491	3	0.0667093	6.60878	7.19222
1490	N,N-Dimethylformamide	LL	2.15791	0.648372	3	0.0077422	4.59798	7.02873
1495	1,1-Dimethylhydrazine	LL	-1.1679	1.01568	3	0.17593	6.59834	7.26392
—	2,4-Dimethylstyrene	LL	1.84209	0.316572	6	0.435465	13.6858	7.28067
1520	Dimethylsulfoxide	HL	1.19045	0.698461	3	0.049848	0.204941	5.66978
1560	Dioxane	LL	1.90645	0.770596	3	0.0435575	6.5131	7.20306
1650	Epichlorohydrin	LL	-0.63762	0.992733	6	0.0562992	13.356	7.22066
—	Ethane	G	1.67058	0.849183	3	0.28197	1.15981	7.84885
1910	2-Ethoxyethanol	LL	1.38151	0.798001	3	0.0354359	4.87638	7.00472
1670	Ethyl acetate	LL	0.671462	0.887545	18	0.309859	66.2747	6.71408
1680	Ethyl acetoacetate	HL	1.96467	0.66266	3	0.0202922	1.7722	6.29103

1750	Ethyl chloroacetate	LL	1.10037	0.837749	3	0.108622	4.42863	6.99629
1990	Ethyl ether	LL	1.60083	0.814314	4	0.0494937	8.56318	7.59965
1710	Ethylbenzene	LL	0.665875	0.735578	19	1.11767	68.9211	6.97741
1770	Ethylene	G	0.594303	0.891134	6	0.145849	6.26961	8.8141
1980	Ethylene oxide	G	-1.0089	1.01284	3	0.0811012	2.39808	8.11482
1800	Ethylenediamine	LL	1.26645	0.764087	3	0.0956445	6.24219	7.23272
2060	Formic acid	LL	-0.27968	0.641373	6	0.39944	12.0961	7.24228
1221	Freon 12	G	0.067945	0.702149	6	0.639168	7.09205	7.51936
2073	Furfural	HL	2.48411	0.480107	3	0.0094205	6.34988	7.21469
2105	Glycidol	LL	-1.2608	0.954506	6	0.221402	4.78354	6.66621
—	n-Heptane	LL	0.990665	0.92378	3	0.0713575	3.74967	6.98098
—	n-Hexane	LL	1.36541	0.886527	26	0.126527	35.4753	7.23407
—	1-Hexene	LL	1.17156	0.6936	10	1.07612	15.3057	7.65326
—	Hydroxyacetone	LL	-0.64006	0.821067	8	0.326534	15.5756	7.59178
—	Isobutane	G	2.21202	0.809978	7	0.236544	4.02085	8.25287
2200	Isobutylene	G	2.47459	0.502984	6	0.71852	2.43034	5.46695
2360	Isopropanol	LL	0.450358	0.916636	6	0.192117	13.4625	7.20498
2370	Isopropyl acetate	LL	1.49877	0.813083	4	0.0570508	8.54523	7.5767
2390	Isopropyl chloride	LL	-0.24062	1.02655	8	0.148871	14.9689	7.52885
—	Isovaleraldehyde	LL	0.012977	0.910504	8	0.438335	16.1261	7.60251
2450	Mesityl oxide	LL	-1.1452	0.993585	8	0.363946	12.9133	7.1985
—	Methacrolein	LL	-0.47037	0.905636	6	0.643145	18.2865	8.20673
2460	Methacrylic acid	HL	-0.69002	0.85944	8	0.438606	1.37953	6.10761
2500	Methanol	LL	0.181625	0.906937	32	0.296364	119.802	7.24719
1930	2-Methoxy-ethanol	LL	2.07507	0.689261	3	0.0552239	6.7148	7.2025
2510	Methyl acetate	LL	0.165303	0.919727	6	0.109923	12.9204	7.26467
—	Methyl acetylene	G	1.26462	0.714083	5	0.0537374	3.42491	8.3449
2560	Methyl chloride	G	-5.2929	1.47615	3	0.143325	1.94546	8.21681
2640	Methyl ethyl ketone (MEK)	LL	-0.26525	1.01477	11	0.178905	35.5776	7.00659
2645	Methyl formate	LL	-0.54896	0.988267	6	0.0582411	16.9883	7.23917
2665	Methyl methacrylate	LL	-0.61443	0.970127	6	0.252803	12.9869	7.19995

Table 3, continued

OCPDB[a] ID Number	Compound Name	Volatility Class[b]	A	B	n	S_e	SS_x	$\bar{X}$
2650	4-Methyl-2-pentanol	LL	-0.32448	0.960325	8	0.227184	15.6961	7.57994
2660	4-Methyl-2-pentanone	LL	0.64249	0.880799	10	0.222826	18.0426	7.77479
–	2-Methyl-2,4-pentanediol	LL	1.40603	0.383098	6	0.373948	13.1088	7.288
–	Methylal	LL	0.426532	0.915145	12	0.131833	52.6838	6.41101
2540	N-Methylaniline	HL	1.45363	0.659612	3	0.214953	1.55971	6.24884
2570	Methylcyclohexane	LL	0.270354	0.986468	10	0.209975	16.6993	7.63536
–	1-Methylcyclohexene	LL	1.20211	0.779284	10	0.253704	17.4672	7.72897
2670	Methylpentynol	LL	0.378822	0.843633	6	0.224115	12.5573	7.19601
2690	α-Methylstyrene	LL	0.5176	0.428728	14	0.768371	66.6129	6.69408
1660	Monoethanolamine	LL	3.77579	0.231518	5	0.317533	11.1294	6.38433
2700	Morpholine	LL	0.417395	0.867233	3	0.543624	6.69482	7.18013
2770	Nitrobenzene	HL	5.86537	-0.04625	4	0.364955	2.79181	5.73648
2790	Nitroethane	LL	1.30098	0.753164	6	0.278354	14.102	7.17721
2791	Nitromethane	LL	-0.09862	0.817192	6	0.498902	13.5221	7.27112
2795	Nitropropane	LL	1.14653	0.810834	6	0.22734	13.5435	7.26298
–	n-Nonane	LL	0.745275	0.712676	14	0.614029	41.1309	6.3957
–	n-Octane	LL	-0.54983	0.979727	6	0.0939152	11.4258	7.14769
2851	n-Pentane	LL	0.069793	1.04455	8	0.0904502	15.1917	7.52991
2910	Phenol	S	2.83764	0.418267	5	0.261277	3.96163	5.72457
–	2-Phenyl-2-propanol	LL	2.79853	0.22002	6	0.306155	13.1932	7.2875
2973	2-Picoline	LL	0.044869	0.977725	5	0.0380285	8.34831	7.73468
–	n-Propane	G	1.40707	0.875637	7	0.632665	5.85865	7.95726
3063	Propionaldehyde	LL	0.034692	0.938077	7	0.260854	15.4002	7.48396
3066	Propionic acid	LL	0.705684	0.772068	5	0.526247	7.1852	6.80396
3070	Propyl alcohol	LL	1.48397	0.791088	3	0.0321125	7.0228	7.22723
–	n-Propylbenzene	LL	2.15539	0.480363	6	1.29885	7.78882	6.93443

		[b]						
3090	Propylene	G	3.54911	0.466641	8	0.815257	8.85238	7.88063
3120	Propylene oxide	LL	-0.10224	0.989086	7	0.366882	14.1261	7.45571
3130	Pyridine	LL	-0.24347	1.00886	12	0.130339	18.0116	7.85092
3230	Styrene	LL	1.92813	0.346454	6	0.984542	13.777	7.23935
3290	1,1,1,2-Tetrachloroethane	LL	-1.6641	0.969855	10	0.367616	17.5221	6.82611
3291	1,1,2,2-Tetrachloroethane	LL	-1.0205	0.81849	8	0.293543	12.074	6.91454
2860	Tetrachloroethylene	LL	0.31901	0.688569	26	0.469724	113.842	6.77884
3349	Toluene	LL	0.120702	0.851475	9	0.901616	16.2675	7.5501
3393	1,2,4-Trichlorobenzene	HL	-1.3223	1.24323	3	0.1571	1.23926	6.19842
3395	1,1,1-Trichloroethane	LL	-0.92319	1.00391	6	0.120871	12.7084	7.19098
3400	1,1,2-Trichloroethane	LL	-1.2924	0.99825	6	0.1101	12.8405	7.20735
3410	Trichloroethylene	LL	-0.14259	0.883197	6	0.134591	13.453	7.21675
3420	1,2,3-Trichloropropane	LL	-2.3703	1.16714	6	0.233901	7.54119	7.53407
3450	Triethylamine	LL	0.797863	0.874591	5	0.180692	8.96445	7.71796
3510	Vinyl acetate	LL	0.47409	0.779072	6	0.593492	13.2953	7.23837
3520	Vinyl chloride	G	-7.0574	1.75493	3	0.110247	2.44541	8.00167
—	Vinyl propionate	LL	-11.761	2.24221	3	1.16066	6.59355	7.29263
3530	Vinylidene chloride	LL	-0.75599	0.987246	6	0.117839	11.489	7.34056
3570	p-Xylene	LL	-0.13472	0.813054	17	0.597877	70.8015	6.59625
3550	m-Xylene	LL	0.644307	0.716846	10	1.17881	15.9403	7.60576
3560	o-Xylene	LL	-0.54235	1.01815	4	0.266544	6.46294	7.3494

[a] Organic chemical producers data base, Radian Corp., Austin, TX.
[b] G = gas, LL = light liquid, HL = heavy liquid, S = solid.

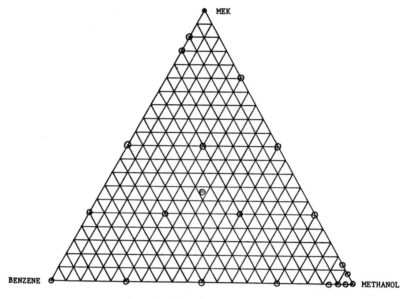

Figure 4. Experimental design.

This model appears to be able to predict mixture response within ∼20% with a slight low bias. Further laboratory work is planned to verify the validity of this model with a broader variety of chemicals and to attempt to "fine tune" it to remove the bias.

CONCLUSIONS

The response factors of portable VOC analyzers to various organic chemicals were found to vary significantly. The response factors for individual chemicals were likewise found to vary considerably with concentration. Response factor estimates were made for an actual concentration of 10,000 ppmv, and a correlation of response factor to concentration was developed. An analysis of the data was also performed to develop response factor estimates for a meter reading of 10,000 ppmv, the results of which were submitted in a separate EPA report. Confidence intervals were computed to provide information on the experimental error associated with the estimated response factors. Preliminary evaluation of chemical mixtures indicated that a weighted average model based on total concentration could be used to estimate response factors to within ∼20% of the experimentally determined values.

ACKNOWLEDGMENTS

This work was funded by the U.S. EPA as part of contract 68-02-3171, Work Assignment No. 1, and is described fully in U.S. EPA Report 600/2-81-002 (NTIS #PB81-136194).

CHAPTER 17

AIRBORNE HYDROCARBON EMISSIONS FROM LANDFARMING OF REFINERY WASTES— A LABORATORY STUDY

J. L. Randall, B. F. Jones, D. D. Rosebrook,
R. G. Wetherold, R. A. Minear and L. P. Provost

Radian Corporation
Austin, Texas 78766

E. W. Cunningham

Standard Oil Company (Ohio)
Cleveland, Ohio 44115

INTRODUCTION

Disposal of refinery oily wastes by landfarming has been practiced for several years. With the passage of the Resource Conservation and Recovery Act of 1976, this and many other sources of fugitive hydrocarbon emissions from refineries came under scrutiny by the U.S. Environmental Protection Agency (EPA).

In 1980, the American Petroleum Institute (API) and the EPA jointly funded the Radian Corporation to study atmospheric emissions from landfarming of refinery oily wastes. The objectives of the study were:

- to determine the extent of hydrocarbon emissions from landfarming oily wastes,
- to identify the sludge characteristics and/or landfarm parameters having the greatest influence on the magnitude of hydrocarbon emissions,
- to develop a method of characterizing oily waste sludges, and
- to develop a method for estimating emissions from landfarming operations based on waste characteristics and operating parameters.

321

Phase I has recently been completed. It consisted of a literature search, the design and construction of laboratory-size simulation devices, a series of simulation experiments providing a systematic study of seven variables, chemical and physical characterization of collected waste and soil materials, and data evaluation/correlation. This chapter presents some of the more important results of that study.

Standard computer search routines were used to obtain literature on the topics involved. Many articles were found dealing with the subjects of microbial degradation of the oil, land reclamation and other qualitative aspects of landfarming. Two articles dealt with hydrocarbon emission from land disposal but neither provided any assistance in simulation of field conditions.

Four refineries were visited during the project and four different methods of waste spreading were observed. Of the various spreading mechanisms, direct application from a vacuum truck and subsurface injection seem to be the most widely used.

During these visits, two API separator sludges (S3 and T), one tank bottom sludge (S1), one straight run gasoline and two landfarm soil samples (S and T) were obtained.

EXPERIMENTAL

Sludge samples were characterized both chemically and physically, with the following tests:

- solids content,
- weight loss on heating,
- centrifugation,
- oil and grease,
- gas chromatography/mass spectrometry (GC/MS),
- stripping of volatiles by nitrogen purge, and
- landfarm simulation under varying conditions.

The total solids were determined by heating samples to constant weight at 180°C. Total volatile solids were determined by heating the residue at 550°C. The initial heating removes all highly volatile organics as well as water.

Suspended solids were isolated through filtration. This was followed by heating the solids at 550°C to determine volatile suspended solids.

The weight of sample lost after heating at various discrete temperatures from 50 to 170°C was used to determine weight loss on heating. Centrifugation at 2500 rpm for 15 minutes was performed on each sludge.

Oil and grease were extracted from the samples with Freon. The extracts were analyzed by two different methods. Using gravimetric procedures, the Freon was evaporated at 180°C and the residue was considered oil and grease. Using an infrared (IR) method, the C-H stretch absorbance of the Freon extract was measured. Quantitation was effected by comparison to a reference standard composed of isooctane, hexadecane and benzene. This method has the advantage of retaining volatile organics with boiling points below 180°C.

Standard EPA priority pollutant routines (method 624 and 625) were used for the GC/MS analyses. Two samples were run by direct injections, and another sample was mixed with 5 mL of organic-free water and then purged onto a Tenax trap.

In an effort to obtain relative volatility measurements, each sludge sample was subjected to a stripping test. A Bellar-Lichtenberg tube and a Byron Model 301C Total Hydrocarbon Analyzer (THC) were used for these experiments. A schematic diagram of the apparatus is given in Figure 1. For each test, the Bellar-Lichtenberg tube was filled with 25 mL of water. Nitrogen was purged at a constant measured rate through the water and into the THC analyzer. When the background hydrocarbon level reached zero, a small volume of sludge was introduced into the Bellar-Lichtenberg tube. Analyzer sampling took place every 4 minutes and was continued for at least 80 minutes after sludge introduction.

The landfarm simulation device developed in this study consisted of a detachable soil chamber, a mixing/sampling chamber and an exit chamber. A diagram of the device is shown in Figure 2.

Two different soil chambers were used which provided either a 4-in. or a 14-in. soil depth. Each chamber had 1-ft^2 soil surface and 4-in. head space above the soil. Soil temperature was maintained by a heater buried one inch below the surface. The heater employed a feedback controller. Soil moisture was controlled manually.

Incoming air was preheated to the desired temperature by a 1650-W space heater in front of the air inlet. Air movement across the soil was provided by a Dayton 2C939A blower with a 2000 cfm capacity. The blower was attached to the air exit chamber. Air velocity was controlled by a series of coverable holes in the exit chamber.

The mixing/sampling chamber was designed to maximize mixing of the hydrocarbons and air while minimizing turbulence. Located at the midpoint of this chamber were probes for air velocity and hydrocarbon instruments as well as an air temperature and relative humidity monitor. The air velocity at this point was measured by a Kurtz (hot wire) anemometer. Hydrocarbons were monitored by a Bacharach TLV Sniffer connected to a strip chart recorder with a disc integrator.

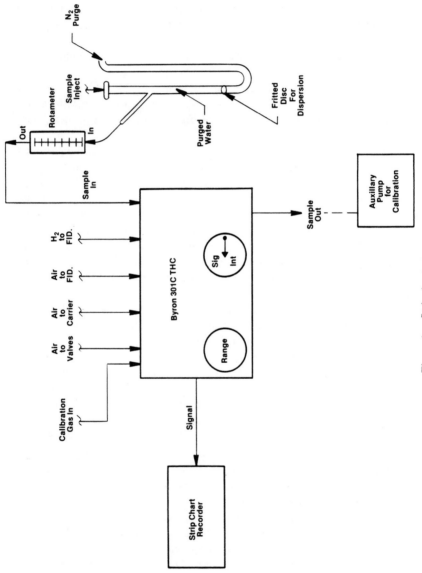

Figure 1. Stripping test apparatus.

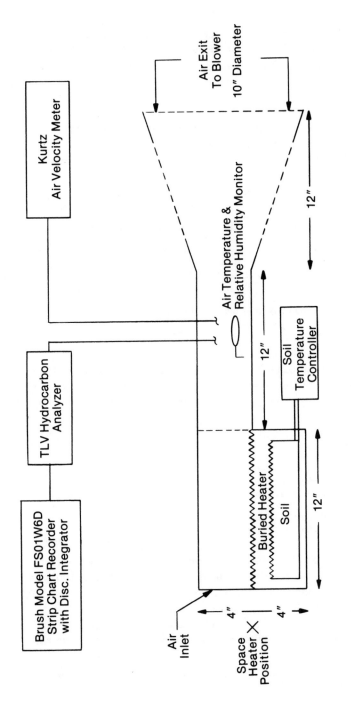

Figure 2.   Landfarm simulation apparatus.

Sludge samples for the simulation experiments were preweighed into a gallon jug. Affixed to the jug was a length of Teflon®* tubing leading to a 12-in. length of 1/2-in. plexiglass tube. The plexiglass tube had a number of 1/8-in. holes along its length and served as a spreader bar for simulating surface application. Subsurface injection was simulated by repeated insertions of the wand from a hand-pressurized garden sprayer.

For the subsurface experiments, tilling was performed manually with a small, short-handled three-tined garden hoe.

A typical simulation experiment was performed as follows:

1. The air and soil heaters were turned on and set for the desired temperatures.
2. The exhaust blower was turned on and the air velocity over the soil adjusted to the desired value.
3. The TLV Sniffer was turned on and zeroed on hydrocarbon-free ambient air. Instrument response was checked with a 533-ppm methane standard.
4. The TLV Sniffer probe was inserted in the mixing chamber. When all controlled parameters were steady at the desired level, the sludge was spread or injected as rapidly as possible. The application usually took less than 60 seconds.
5. The hydrocarbon concentration in the mixing chamber was monitored continuously until the level dropped to or near zero or for at least 30 minutes from spreading.

Twenty-nine simulation experiments were run in this study. A matrix showing the conditions for each run is given in Table 1. Twenty-four runs consisted of sludge application to soil. For the three sludges and two soils, this represented a series to systematically study the following variables:

- soil type,
- air/soil temperature,
- soil moisture,
- relative humidity,
- sludge volatility,
- waste application method,
- wind speed,
- sludge loading level, and
- sludge temperature at spreading

The conditions for experiments 6 and 7 were considered typical or base case conditions. After varying each parameter independently from the base case conditions, the variables identified as having the largest effect on hydrocarbon emissions were studied in more detail.

Since the TLV Sniffer/disc integrator results were in arbitrary units, calibration of the simulation apparatus was necessary to obtain quantitative

---

*Registered trademark of E. I. du Pont de Nemours & Company, Inc., Wilmington, Delaware.

Table 1. Matrix of Experimental Conditions for Landfarm Simulation Runs

Run No.	Waste Material[a]	Waste Loading (lb/ft²)[b]	Soil Type[c]	Soil Moisture (wt %)	Soil/Air Temperature (°F/°F)	Air Velocity (mph)	Relative Humidity (%)[d]	Spreading Mechanism	Remarks
1	Hexane	1.0	None	NA	NA/100	3	Amb	Pour	Pure hexane—no soil
2	Hexane	1.0	S	10.7	120/100	3	Amb	Surface	Pure hexane with soil
3	Gasoline	1.0	None	NA	NA/100	3	Amb	Pour	Pure gasoline—no soil
4	Gasoline	1.0	S	10.7	120/100	3	Amb	Surface	Pure gasoline with soil
5	S1	2.5	None	NA	NA/100	3	Amb	Pour	Sludge S1 with no soil
6	S1	2.5	S	10.7	120/100	3	Amb	Surface	Base case
7	S1	2.5	S	10.7	120/100	3	Amb	Surface	Repeat of base case
8	S1	2.5	S	10.7	120/100	1	Amb	Surface	Reduced air velocity
9	S1	2.5	S	10.7	100/80	3	40	Surface	Reduced soil/air temperature
10	S1	1.0	S	10.7	120/100	3	Amb	Surface	Reduced soil loading
11	S1	1.0	S	10.7	120/100	3	Amb	Surface	Repeat of Run 10
12	S1	2.5	T	4.7	120/100	3	Amb	Surface	Used soil T
13	S1	2.5	S	20.7	120/100	3	Amb	Surface	Increased soil moisture
14	S1	2.5	S	10.7	120/100	3	75	Surface	Increased relative humidity
15	T	2.5	S	10.7	120/100	3	Amb	Surface	Used sludge T
16	S1	2.5	S	10.7	120/100	3	Amb	Surface	Base case with deeper (14 in.) soil box
17	S1	4.0	S	10.7	120/100	3	Amb	Surface	Increased soil loading
18	S1	0.5	S	10.7	120/100	3	Amb	Surface	Lowest soil loading
19	S1	2.5	S	10.7	120/100	3	Amb	Surface	Heat sludge to 140°F before application
20	S3	1.0	S	10.7	120/100	3	Amb	Surface	Sludge S3 at low loading
21	S3	2.5	S	10.7	120/100	3	Amb	Surface	Sludge S3 at base case loading
22	S3	4.0	S	10.7	120/100	3	Amb	Surface	Sludge S3 at increased loading
23	S3	1.0	T	4.7	120/100	3	Amb	Surface	Sludge S3 with soil T-low loading
24	S3	4.0	T	4.7	120/100	3	Amb	Surface	Sludge S3 with soil T-high loading

Table 1, continued

Run No.	Waste Material[a]	Waste Loading (lb/ft^2)[b]	Soil Type[c]	Soil Moisture (wt %)	Soil/Air Temperature ($^\circ$F/$^\circ$F)	Air Velocity (mph)	Relative Humidity (%)[d]	Spreading Mechanism	Remarks
25	S3	2.5+	S	10.7	120/100	3	Amb	Surface	Spike sludge with 0.5 lb of gasoline
26	S3	2.5+	S	10.7	120/100	3	Amb	Surface	Spike sludge with 1.5 lb of gasoline
27	S3	2.5+	S	10.7	120/100	3	Amb	Surface	Spike sludge with 1.5 lb of H_2O
28	S1	2.5	S	10.7	120/100	3	Amb	Subsurface	Till soil after spreading
29	S1	2.5	S	10.7	120/100	3	Amb	Subsurface	Use deeper injection

[a]S1 = tank bottoms sludge, S3 = API separator sludge, T = API separator sludge.
[b]2.5 lb/ft^2 $\cong$ 300 bbl/acre loading in terms of sludge.
[c]S = soil taken from Landfarm #1, T = soil taken from Landfarm #2.
[d]Amb = ambient humidity was typically 50% at room temperature but measured $\ll$30% at 100°F.

results. To accomplish this, the 4-in. soil chamber was fitted with a false bottom at the normal soil level. This modified soil chamber was positioned on a Mettler P4400 balance. The space between the soil chamber and mixing chamber was sealed with lightweight, flexible plastic to eliminate air leaks.

When the atmospheric conditions reached base case conditions, S1 sludge was poured into the false bottom. As the sludge volatilized, weight loss and integrated areas were recorded as a function of time. From these values, a response factor for weight loss per unit area was calculated.

RESULTS

The three sludges were quite different in their physical characteristics. The tank bottoms sludge, S1, contained approximately 92 wt % oil. The API separator sludges contained about 20 wt % oil for S3 and 10 wt % oil for T.

The apparent volatility of the sludges also was markedly different. Results of the stripping tests are shown graphically in Figure 3. Note the logarithmic scale for concentrations. The ranking of volatility is S1 > S3 > T, which is in the same order as the oil content.

The weight loss on heating and the solids measurement provided little or no useful information. Results of the GC/MS scans of the sludges are given in Table 2. These indicate that the major hydrocarbon constituents are substituted aromatics, particularly C_3 benzenes and xylenes. The total hydrocarbon levels follow the rankings above. However, a large difference between the direct injections (S1 and S3) and the water purged (T) sample is obvious.

Table 2. Chemical Composition of Sludge and Soil Samples

	Concentration (ppm)		
Compound (mol wt)	API Separator S3	Tank Bottoms S1	API Separator Sludge T
Benzene (78)	1,400	1,800	N.D.
Toluene (92)	5,600	9,300	0.02
Ethylbenzene (106)	4,300	5,200	0.46
m,p-Xylene (106)	15,000	23,000	1.2
o-Xylene (106)	32,000	42,000	0.55
Saturated Hydrocarbon (85)	24,000	4,000	5.0
Indane (118)	28,000	34,000	9.2
C_3-Benzene (120)	160,000	220,000	7.9
Dimethylpentene (98)	1,900	2,200	5.0
Saturated Hydrocarbon (57)	5,200	6,200	6.6
Total	277,400	347,700	35.9

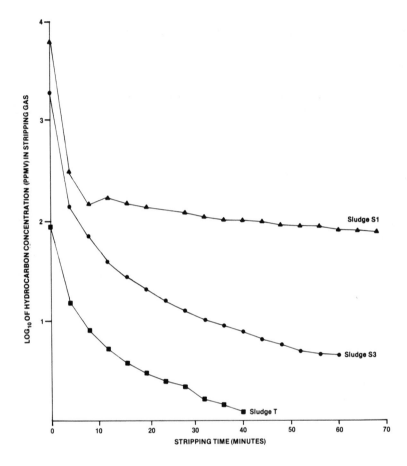

Figure 3. Concentration of hydrocarbons in purge gas during stripping tests.

A typical TLV Sniffer hydrocarbon analyzer strip chart recording of the landfarm simulation experiments is shown in Figure 4. In almost every run, a very rapid rise in hydrocarbon level was obtained. A maximum usually was attained within two minutes of the start of spreading. This was followed by a rapid decline, then a slower decline, and an eventual approach toward zero or ambient air. The hydrocarbon level typically declined to < 5 ppm within 30–60 min from spreading.

Data collected for the simulation runs were the maximum hydrocarbon level, the integrated area after 6 minutes and the integrated area after 30 minutes. Data for the 29 experiments are shown in Table 3. The areas are expressed in arbitrary but consistent units.

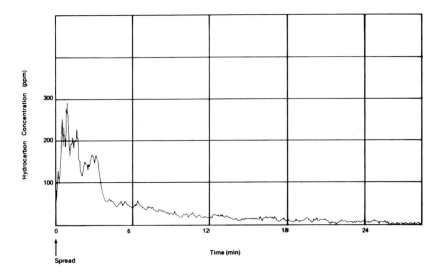

Figure 4. Typical chart recording for a simulation experiment.

To convert the area units to weight of sludge, calibration was effected. Graphical presentation of the calibration runs is given in Figure 5. The slope or calibration factor was determined to be 9.85 area units per gram of sludge.

This factor was used to estimate the weight of sludge volatilized after 30 minutes in each simulation run with the following assumptions:

- The TLV Sniffer response to S1 sludge vapors was linear and equal to the calibration factor,
- The TLV Sniffer response to S3 and T sludge vapors was equal to the response to S1 vapors.

The estimated weight of sludge volatilized is given in Table 4. The data indicate that from 0.5 to 3.2% of the S1 sludge is lost during the first 30 minutes. Since this sludge is 92% oil, the oil percentages volatilized are slightly higher. For sludge S3, only 0.07 to 0.22% of the sludge was volatilized, but this corresponds to values of 0.22 to 1.1% of the oil. Sludge T experienced a 0.09% oil loss or 0.009% sludge loss in the first 30 minutes. Thus the ranking of weight loss in the simulation runs is identical to the other indicators, i.e., S1 > S3 > T.

Further illustration of the volatility effects on fugitive hydrocarbon emissions is given in Figure 6. The maximum concentrations and 6- and

Table 3. Results of Simulation Runs

Run No.	Variable[a]	Maximum Hydrocarbon (ppm)	Integrated Area in 6 min	Integrated Area in 30 min	Remarks
1	Pure hexane (false bottom)	>10,000	2,460	7,980	8580 area at dryness
2	Pure hexane	>10,000	3,900	5,500	
3	Gasoline (false bottom)	> 5,000	690	1,810	2400 area at zero hydrocarbons
4	Gasoline	>10,000	1,750	2,470	
5	Sludge S1 (false bottom)	400	43	342	
6	Base case	215	21	78	
7	Base case (repeat)	200	22	62	
8	Air velocity decreased to 1 mph	310	57	231	
9	Air/soil temperature decreased	260	23	53	
10	Loading decreased	325	49	143	
11	Loading decreased (repeat of 10)	200	31	136	
12	Soil T used	485	86	230	
13	Moisture in soil increased	> 500	85	263	
14	Moisture in air increased	> 500	68	116	
15	Waste T used	35	1	1	
16	Base Case in a deeper box	350	27	71	
17	Loading increased	425	33	81	
18	Loading decreased drastically	250	31	61	
19	Waste heated before spreading	350	45	74	
20	Waste S3 at low loading	23	4	10	
21	Waste S3 used	15	1.5	5	

22	Waste S3 at high loading	25	4	16	
23	Waste S3 at low load on T soil	17	2	5	
24	Waste S3 high load on T soil	34	4	12	
25	Waste S3 with low gasoline spike	7,000	480	760	
26	Waste S3 with high gasoline spike	>10,000	760	1,360	
27	Waste S3 cut with water	25	5	8	
28	Subsurface injection	300	17	114	
28A	Tilling of 28	375	29	68	Tilled 80 min after spreading
29	Subsurface deep injection	0	0	0	
29A		225	15	15	Tilled 4 days after spreading
29B		50	3	3	Tilled 6 days after spreading

aVariable changes from the base case conditions of Runs 6 and 7. All other variables remained as in Runs 6 and 7.

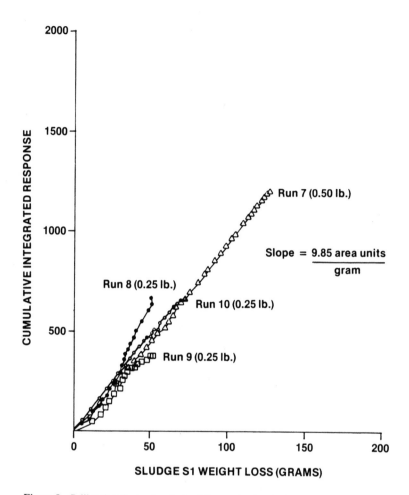

Figure 5. Calibration factor for sludge S1 vaporization in landfarming simulator.

30-min areas are shown. The ranking order of S1 > S3> T is clearly evident for the 6- and 30-min areas.

As a further test of volatility effects, 0.5 and 1.5 lb of gasoline were added to 2.5 lb of S3 sludge. The mixtures were then applied to the soil. The results, given in Figure 6, indicate much higher emissions in all data points for the mixtures as compared with unadulterated S3 sludge.

Additionally, 1.5 lb of water were added to 2.5 lb of S3 sludge. Hydrocarbon emissions from this mixture during simulation showed no apparent effect. This is shown in Figure 6.

Table 4. Estimated Emissions from Sludge Applied to Soil in Landfarming Simulation Tests

Run/Sludge	Area 30	Grams Volatilized[a]	Weight Sludge Applied (g)	% Total Sludge Volatilized	Weight Oil Applied (g)	%Oil Volatilized
6/S1	78	7.9	1135	0.70	1004	0.75
7/S1	62	6.3	1135	0.56	1044	0.60
9/S1	53	5.4	1135	0.48	1044	0.52
10/S1	143	14.5	454	3.2	418	3.5
11/S1	136	13.8	454	3.0	418	3.3
12/S1	230	23.3	1135	2.1	1044	2.2
13/S1	263	26.7	1135	2.4	1044	2.6
14/S1	116	11.8	1135	1.0	1044	1.1
15/T	1	0.10	1135	0.009	114	0.088
16/S1	71	7.2	1135	0.63	1044	0.69
17/S1	81	8.2	1816	0.45	1671	0.49
18/S1	61	6.2	227	2.7	209	3.0
19/S1	74	7.5	1135	0.66	1044	0.72
20/S3	10	1.0	454	0.22	91	1.1
21/S3	5	0.51	1135	0.045	227	0.22
22/S3	16	1.6	1816	0.088	363	0.44
23/S3	5	0.51	454	0.11	91	0.56
24/S3	12	1.2	1816	0.066	363	0.33
27/S3	8	0.81	1135	0.071	227	0.36

[a]Based on response factor given in Figure 5.

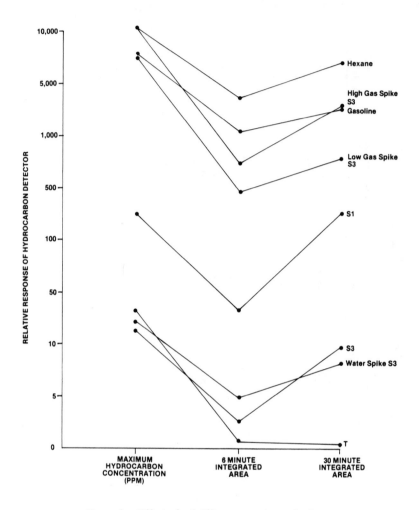

Figure 6. Effect of volatility—comparison of spikes.

The amount of sludge applied to the soil surface (loading) was varied to determine the effect of loading on emission rates. Varied loadings were examined for sludges S1 and S3 only, since the emissions from sludge T were too low to detect any effects.

The effects of loading are given in Table 5. These data indicate lower emissions for higher loading of S1. This same general trend was observed for S3 although it is less obvious, probably due to the higher water content of S3. The practical results of this effect would depend heavily on the mechanical ability to work with sludge laden soil.

Table 5. Effects of Loading on Emissions

Sludge Type	Loading Level lb/ft^2 of soil	% Oil Volatilized
S1	0.5	3.0
S1	1.0	3.5
S1	1.0	3.3
S1	2.5	0.75
S1	2.5	0.69
S1	2.5	0.60
S1	4.0	0.49
S3	1.0	1.1
S3	2.5	0.22
S3	4.0	0.44

Runs 28 and 29 consisted of subsurface injection simulations. In Run 28, injection occurred at a shallow depth of 3-4 in., which allowed some sludge to bubble to the surface. The resultant emissions were relatively high in the application and during the tilling operation performed 80 minutes after application (see Tables 1 and 3).

In Run 29, sludge injection occurred at a depth of ~6 in. below the soil surface. No sludge appeared on the surface and no detectable hydrocarbons were emitted. The soil was tilled 4 days and again 6 days after application. Small amounts of hydrocarbon were emitted during the tilling operations. Figure 7 shows a chart tracings of the three events superimposed on each other. This gives a clear indication that a substantial overall emission reduction can be obtained by subsurface injection followed by a reasonable waiting period before tilling.

Figure 8 shows the effects of various other parameters on the relative magnitude of emissions from landfarming simulation. All the runs shown in the figure were made with sludge S1 and a consistent loading of 2.5 lb/ft^2 In all except one run, the moisture content of the soil was 10.7% by weight. In the experiment with 20.7% soil moisture, there was a large increase in the amount of fugitive hydrocarbon emissions. This effect may be due to the wetter soil being much less permeable to the liquids in the sludge. Pools of liquid were observed on the soil surface. The residence time of these liquid pools was much longer than that observed with the lower moisture soils.

In one run, the sludge was heated to 140°F before applying it to the soil surface. There was no impact of this higher temperature on the 30-min integrated area, probably because of relatively rapid cooling. The maximum concentration and the 6-min integrated area were somewhat higher than the base case.

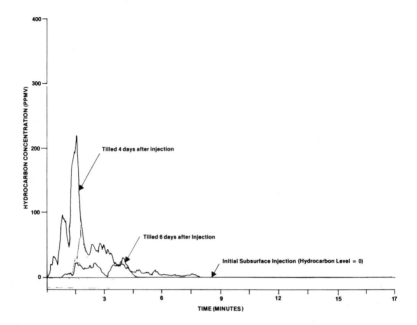

Figure 7. Effect of resting time before tilling on emissions after subsurface injection.

The effect of soil type on the emission rate was not adequately defined in this study. When sludge S3 and Soil T were used in test runs, the emissions were lower than found with soil S. This is consistent with the lower moisture content of soil T. However, emissions were higher with sludge S1 and soil T than with sludge S1 and soil S. At this point, the anomaly cannot be explained.

Figure 8 indicates that emissions increase dramatically when the air velocity is reduced from 3 to 1 mph. However, it should be remembered that the cumulative integrator counts are equivalent to integrated concentrations in the exiting air. Thus, a threefold increase in integrated concentration was detected for a threefold decrease in air flowrate. This is equivalent to the same mass emission for both wind speeds.

All experiments but one were run at ambient humidity ($\sim$ 50%). One run was made at approximately 75% relative humidity. As shown in Figure 8, there appears to be a pronounced effect on the emission rate. The cumulative emissions at 30 minutes from spreading are about twice as great at the elevated humidity level.

One test run was made with the air temperature reduced from 100 to 80°F and the soil temperature reduced from 120 to 100°F. The hydrocarbon

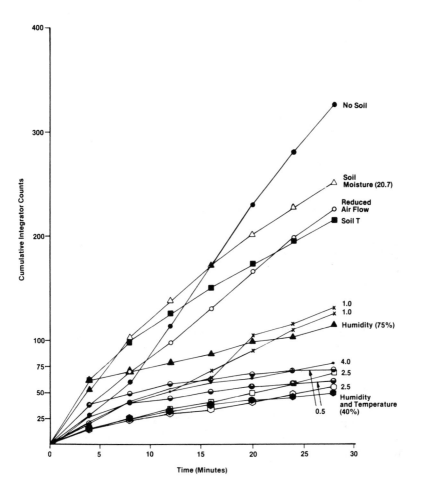

Figure 8. Effect of moisture, loading and air flow on emissions.

emissions for the first 30 minutes following spreading were reduced by about 25% for the test run at lower temperatures.

Much of the variability in the data and results could not be attributed solely to the independent effects of the variables studied. This suggests the possibility of important interactions among these variables. Interactions were not adequately studied in this program because of constraints on the number of tests that could be made.

CONCLUSIONS

The results obtained in this study support the following conclusions:

- The volatility of the sludge is perhaps the most important parameter in estimating the emissions from landfarming of sludge.
- The sludge stripping test developed in this study gives a quantitative measure of the volatility of a sludge. This volatility measure appears to correlate with the landfarming emission rates.
- The laboratory landfarming simulation device developed during this study can provide very reproducible results for sludge emission studies.
- The highest emission rates occur within the first 30 minutes after application of the sludge.
- From 0.01 to 3.2 wt % of the sludges applied to the soil in the landfarming simulations was vaporized. This is equivalent to 0.1 to 3.5% of the oil in the applied sample sludges.
- The emission rates are affected by the sludge loading on the soil. There may exist some low soil loading ranges which are less desirable with respect to atmospheric emissions.
- Within the range studied, air velocity does not affect the quantity of hydrocarbons emitted.
- Subsurface injection of sludges appears to be the preferred spreading technique for minimizing atmospheric emissions from landfarming operations.
- There may be important interactions among some of the variables studied in this project. These could not be investigated within the scope of this study.

ACKNOWLEDGMENTS

The authors would like to thank the American Petroleum Institute and the U.S. Environmental Protection Agency for their support and guidance in this work. We would also like to acknowledge the contribution of the Radian technical staff, notably Mr. K. R. Williams, who conducted most of the simulation and stripping tests.

PART 4

IMPACT OF LURGI COAL GASIFICATION

CHAPTER 18

CHARACTERIZATION OF EMISSIONS FROM A LURGI COAL GASIFICATION SYSTEM AT KOSOVO, YUGOSLAVIA

K. J. Bombaugh and K. W. Lee
  Radian Corporation
  Austin, Texas 78766

T. Kelly Janes
  U.S. Environmental Protection Agency
  North Carolina 27711

INTRODUCTION

There are two aspects to the problem of characterizing atmospheric pollution. One approach deals with what and how much of a substance was discharged, the other with what and how much is present at the sampling site. Both approaches are relevant when an effort is being made to trace pollutant discharges to a particular source, and both approaches were used in the Kosovo study.

An international program was conducted in the Kosovo Region of Yugoslavia to characterize the environmental problems associated with a commercial Lurgi coal gasification system. This program was undertaken because the Lurgi system has significant potential for near-term use. This effort involved characterization of the pollutants in the plant's key discharge streams and determination of the levels of these pollutants in the ambient air in the plant's vicinity [1-5]. Pollutants characterized included sulfur species, aromatic and aliphatic hydrocarbons, nitrogenous species, phenols and particulate matter.

The test facility is an integral part of a large mine-mouth industrial complex, which also includes an ammonia-based fertilizer plant, a Fleissner coal-drying plant, a cryogenic air-separation plant, a steam-generating plant and a large electric-power–generating plant. The gasification plant consists of six pressurized oxygen-blown gas generators with an associated gas-cooling system, a tar and oil separation plant, a Rectisol acid-gas removal plant, a Phenosolvan wastewater extraction plant plus liquid by-products storage facilities. All operating units produce gaseous discharges that are a potential source of atmospheric pollution.

The Kosovo plant is a representative of future gasification facilities, which probably will be constructed and, although its emissions exceed the levels that would be acceptable from a tightly controlled plant, the environmental problems at this facility are representative of problems that must be considered by future operators. In addition, the plant's isolation from other pollution sources provides ideal conditions for a study of atmospheric pollution problems associated with commercial coal gasification technology.

In this, the first of a four-chapter report, emphasis is placed on the plant as the emission source. The three chapters that follow address the ambient air characterization study. Since the ambient air in the plant's vicinity was studied in an effort to define the relationship between the pollutants emitted and the pollutants collected, this chapter provides a base for those which follow.

PLANT DESCRIPTION

The Lurgi gasification facility at Kosovo is comprised of six process units as illustrated in Figure 1. The gasification plant consumes dried lignite and produces two primary products: medium-Btu (23-25 MJ/m^3 @ 25°C) fuel gas and hydrogen for use in ammonia synthesis. It also produces several by-products: tar, medium oil, naphtha and crude phenol. The lignite fed to the Lurgi gas generators is dried by the Fleissner process (high-temperature steam soak) and sized to select particles between 6 and 60 mm in size. Product gas, generated by the reaction of the lignite with oxygen and steam at 2.3 MPa (23 atm) pressure, is cooled and then cleaned to remove acid gases prior to its transportation by pipeline to the utilization site. In the cooling step, tars, oils, naphtha and phenolic water are removed from the gas. In the acid-gas removal step, H_2S and CO_2 are removed by sorption with cold methanol. The rich methanol is regenerated by depressurization and heating. The H_2S-rich waste gas released by the regeneration step is sent to a flare; the CO_2-rich waste gas is vented directly to the atmosphere. Tar and oil are separated from the phenolic water by decantation, after which the water-soluble organics (crude

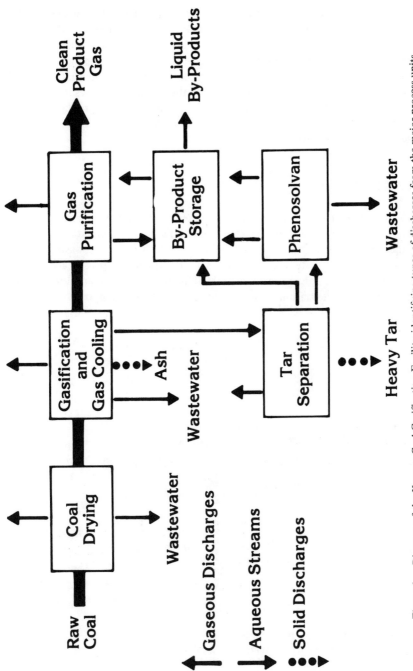

Figure 1. Diagram of the Kosovo Coal Gasification Facility identifying types of discharges from the major process units.

phenols) are removed from the wastewater by extraction with diisopropyl ether. Four liquid by-products (tar, medium oil, naphtha and crude phenol) are collected in storage tanks and used as fuel. Dissolved acid gases and ammonia are removed from the phenolic water by steam stripping and vented to the atmosphere. Each process unit at this plant is a potential source of environmental discharges as indicated by the ascending arrows in Figure 1. Figure 2 shows the mass flows of the plant's major feed and product streams. The flowrates shown are based on design conditions with five of six Lurgi generators in operation.

EXPERIMENTAL

Approximately 35 of the plant's gaseous streams were analyzed preliminarily, then 18 were selected for detailed characterization. Various other streams, such as liquid by-product, aqueous and solid waste, process feed and intermediate streams were included in the study, but are not addressed in this chapter. The characterization consisted of stream flowrate and particulate concentration measurements, plus stream composition analyses

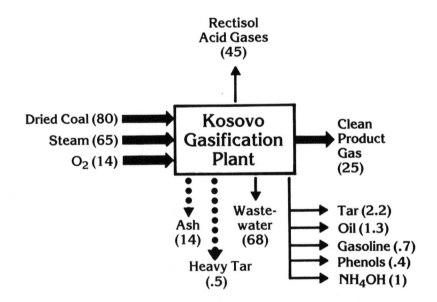

Figure 2. Major stream flowrates in the Kosovo Gasification Plant in metric tons per hour. Approximate heating values (kcal/kg) are: dried coal, 3500; clean product gas, 2200; liquid by-products, 8000.

with emphasis on selected pollutants. Flowrate measurements were made with pitot tubes using adaptations of standard U.S. Environmental Protection Agency methods. Particulate concentrations were determined either by dry filtration or by wet collection in an impinger train illustrated in Figure 3. The choice of collection method depended on stream conditions. Streams containing high concentrations of tar aerosols or moisture were sampled using the wet impinger techniques. Gas composition was determined by gas chromatography (GC) using both universal and species selective detectors. Reactive species, such as HCN, H_2S and ammonia, plus phenol, were sampled with impinger trains and determined by chemical methods. Fixed gases were determined by GC using a thermal conductivity detector (TC/GC) on all grab samples as a quality control measure. Sampling and analysis methods are summarized in Table 1.

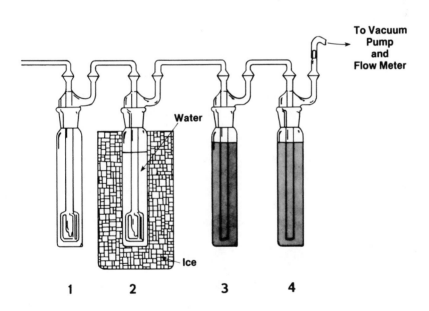

Legend
1. Smith-Greenburg impinger at ambient temperature
2. Smith-Greenburg impinger containing water and refrigerated
3. absorber containing Rohm and Haas XAD-2 resin
4. absorber containing silica gel

Figure 3. Diagram of the impinger train used to collect particulates from gas streams which contained large amounts of moisture and/or suspended tar.

Table 1. Methods Used to Determine the Composition of Gas Streams

Component	Collection Method	Analytical Method
Fixed Gases CO, CO_2, O_2, H_2, CH_4, N_2	Gas conditioning and collection system/ glass bombs	TC-gas chromatography
Hydrocarbons C_1-C_6, C_6, BTX	Same	FID-gas chromatography
Sulfur Species H_2S, COS, RSH, SO_2	Same	FPD-gas chromatography
H_2S	Impinger/CdOAc	Titration with iodine/Hypo
HCN	Impinger/ (CdOAc) (NaOH)	Distillation/titration with $AgNO_3$
NH_3	Impinger/H_2SO_4	Distillation/titration with H_2SO_4
Phenols	Impinger/NaOH	Colorimetric-4 aminoantipyrene
Particulates	Impinger train	Gravimetric

RESULTS AND DISCUSSION

Composition and flowrate data for the major gaseous streams in the Kosovo Coal Gasification Plant are shown in Table 2. Gas compositions are expressed in terms of volume. Data for all of the major gaseous streams, i.e., atmospheric discharges, flare feed, plus crude and finished product streams are included; however, attention will be focused on the atmospheric discharges, since they affect the ambient air composition. Atmospheric discharge streams are identified in Table 2 as vents. The flare feed process streams are identified parenthetically. Data from the flare feed and process streams are included to show the plant's operational status.

Composition of the Atmospheric Discharge System

Several observations can be made about the composition of the atmospheric discharges from the Kosovo plant. For example, the gas discharged from the LP coal lock has essentially the same composition as the generator gas, indicating that generator gas is escaping through the coal lock to the atmosphere. The discharge from the start-up vent has a composition which changes over the start-up period from that of combustion (flue) gas to that of producer gas. When this stream was sampled the generator was still being fired with air (rather than with O_2) as indicated by the 42% N_2 in the vent gas. Several compounds are present in this stream at levels that are greater

Table 2. Kosovo Gaseous Stream Composition Data

Plant Section	Gas Production[a]				Rectisol[a]		
Sample Point	Low Pressure Coal Lock Vent	Gasifier Start-up Vent[b]	High Pressure Coal Lock (Flare Feed Stream)	H_2S-Rich Waste Gas (Flare Feed Stream)	CO_2-Rich Waste Gas Vent	Generator Gas (Process Stream)	Product Gas (Process Stream)
Dry Gas Flowrate (m^3/gasifier-hr @ 25°C)	21	4,000	230	3,600	3,600	18,800	10,900
Temperature (°C)	56	–	54	12	19	22	22
Moisture Content (%)	44	70	11	3.9	5.1	2.5	4.1
Molecular Weight of Dry Gas	23.5	33.1	24.9	43.0	42.2	21.9	10.3
Composition (Dry Basis)							
Fixed Gases (vol %)							
H_2	37	0.09	32	0.70	Tr	38.1	60
O_2	0.27	4.5	0.24	Tr	Tr	0.36	0.44
N_2	0.18	42	0.14	Tr	Tr	0.64	0.38
CH_4	8.6	1.6	10.5	4.3	1.2	11.5	16
CO	14.6	14	12	1.1	Tr	15	22
CO_2	36.5	34	42	88	94	32	0.02
Sulfur Species (ppmv)							
H_2S	12,700	6,300	3,500	45,500	39	6,000	NF
COS	110	110	120	420	62	97	0.17
CH_2SH	420	490	460	2,100	8.5	590	1.1
C_2H_3SH	220	240	210	780	4.4	200	1.0
Hydrocarbons (vol %)							
C_2H_6	0.22	0.15	0.42	0.82	1.6	0.47	0.15
C_2H_4	Tr	0.05	Tr	Tr	Tr	0.04	Tr
C_3	0.14	0.08	0.25	0.63	0.28	0.19	Tr
C_4	0.05	0.03	0.11	0.32	Tr	0.074	Tr

Table 2, continued

Plant Section	Gas Production[a]				Rectisol[a]		
Sample Point	Low Pressure Coal Lock Vent	Gasifier Start-up Vent[b]	High Pressure Coal Lock (Flare Feed Stream)	H$_2$S-Rich Waste Gas (Flare Feed Stream)	CO$_2$-Rich Waste Gas Vent	Generator Gas (Process Stream)	Product Gas (Process Stream)
C$_5$	Tr	0.007	0.01	0.04	Tr	0.044	Tr
C$_6$ +	0.12	0.09	0.08	0.21	NF	0.064	0.03
Aromatic Hydrocarbons (ppmv)							
Benzene	760	90	550	110	1.0	750	–
Toluene	220	10	100	8	Tr	230	–
Xylene and Ethylbenzene	~75	Tr	38	–	1.4	100	–
Phenols	5.7	630	2.5	NF	NF	Tr	NF
Higher Aromatics	–	–	–	–	–	–	–
Nitrogen Species (ppmv)							
NH$_3$	2,400	11,100	NF	2,200	46	3.3	NF
HCN	600	2,900	170	200	13	320	–

Plant Section	Tar Separation[a]				Phenosolvan[a]	By-Product Storage[a]	Flare System[a]
Sample Point	Tar Tank Vent	Medium Oil Tank Vent	Tar Separation Waste Gas (Flare Feed Stream)	Phenolic Water Tank Vent	NH$_3$ Stripper Vent	Naphtha Tank Vent	Combined Gas to Flare
Dry Gas Flowrate (m^3/gasifier-hr @ 25°C)	0.51	1.7	28	5.5	260	4.5	1,330
Temperature (°C)	52	42	40	76	91	32	21
Moisture Content (%)	14	8.4	7.7	42	76	5	2.5
Molecular Weight of Dry Gas	29.1	32.5	39.0	34.4	32.7	33.3	41.7
Composition (Dry Basis)							
Fixed Gases (vol %)							

H$_2$	Tr	Tr	11	Tr	NF	NF	Tr
O$_2$	19	0.45	Tr	13	–	2.6	0.10
N$_2$	77.5	1.1	Tr	39	–	84	0.21
CH$_4$	0.16	7.6	3.5	0.2	Tr	NF	6.2
CO	Tr	5.9	1.1	NF	NF	NF	1.9
CO$_2$	0.86	56	77.5	35	55	0.85	88
Sulfur Species (ppmv)							
H$_2$S	6,900	26,000	9,000	12,600	19,500	NF	10,600
COS	110	96	120	41	NF	NF	250
CH$_3$SH	390	5,200	2,500	2,100	290	2,600	2,500
C$_2$H$_2$SH	240	2,100	1,600	7,200	100	9,700	190
Hydrocarbons (vol %)							
C$_2$H$_6$	Tr	0.34	0.33	0.02	Tr	Tr	0.77
C$_2$H$_4$	–	Tr	Tr			–	Tr
C$_3$	0.01	0.30	0.41	0.02	Tr	0.01	0.65
C$_4$	Tr	0.25	0.41	0.02	Tr	0.07	0.38
C$_5$	Tr	0.09	0.09	0.006	Tr	0.08	0.04
C$_6$+	0.37	2.4	1.3	1.8	NF	5.3	0.06
Aromatic Hydrocarbons (ppmv)							
Benzene	2,000	7.650	9,600	11,000	–	37,600	640
Toluene	960	1,400	1,200	2,300	–	1,900	215
Xylene and Ethylbenzene	220	140	150	280	Tr	60	33
Phenols	22	45	4.2	Tr	6,200	Tr	Tr
Higher Aromatics	2.2	–	4.9	3.1	–	–	–
Nitrogen Species (ppmv)							
NH$_2$	2,600	11.4	19,300	12,000	418,000	NF	4,000
HCN	130	–	65	38	4,000	1,100	100

[a] Tr (Trace) = 0.01 vol. % for fixed gases, 1 ppmv for all others; NF (Not Found) = less than a trace.
[b] Stream vents only during a period of start up.

than would be expected for flue gas, e.g., H_2S at 6300 ppm, mercaptans at 730 ppm, phenols at 630 ppm and NH_3 at 11,100 ppm. This stream, though intermittent, is significant environmentally because a large gasification complex with many generators could be expected to have at least one generator in a start-up mode at all times. In such a complex, the start-up vent could be considered a continuous stream.

The CO_2-rich waste gas that is discharged from the acid-gas removal process (two-stage Rectisol) is the largest atmospheric discharge stream in the plant. The dominant group of contaminants in this stream is the light hydrocarbons. However, several offensive substances such as mercaptans, present at the ppm level, are also environmentally significant.

The discharges from the several tank vents contain high concentrations of light aromatics as well as ammonia and sulfur bearing species. In the phenolic water tank the discharge levels of both benzene and H_2S exceed 1% and in the naphtha tank discharge the benzene level exceeds 3%.

A blanket of nitrogen is maintained over the liquid in the naphtha storage tank. When corrections are applied to these composition data for the 84% N_2-diluent the tank vapor shows the following approximate composition:

Component	Approximate Percentage
Mercaptans	15
Hydrocarbons (C_1–C_6)	19
Benzene	47
Toluene	2
Hydrogen Cyanide	1
Other compounds	16

These results show that the tank's vapors are comprised primarily of light hydrocarbons, of which benzene is the major component, and that sulfur-bearing species are also major components in the vapor.

The ammonia stripper vent discharge contains ~42% ammonia, 55% CO_2 and ~2.0% H_2S. Ammonium carbonate particles were observed in the collection equipment during sampling and they conceivably could be forming in the air. Since the stripping was being done with steam, it was not possible to see if solid particle formation was occurring in the atmosphere.

Stream Flowrates

Gas stream flowrates, shown in Table 2, range from 0.5 m^3/hr for the tar tank vent to 3600 m^3/hr for the CO_2-rich waste gas from the Rectisol acid-gas removal plant. The several streams containing high concentrations of light aromatics had flowrates in the range of 2–5 m^3/hr.

Pollutant Mass

The gas composition values presented in Table 2 are given in volume percent since gas volume is the basic unit of measurement and is a useful form for presenting gas composition. However, mass concentration (μg/m^3) is the preferred unit for use in environmental science and particularly in discharge assessment since mass concentration is directly convertible to mass flow. Therefore, mass concentrations for the major streams that discharged to the atmosphere regularly during normal operation are given in Table 3. The flowrates of each stream are included so that mass flows can be calculated.

Converting stream composition from volume to mass units may alter its apparent composition, particularly when the stream's components cover a wide range of molecular weights. This effect is illustrated by the data in Table 4, which shows a comparison of gas compositions expressed in volume and mass. Those components whose molecular weights are most different from the average molecular weights of the gas stream suffer the greatest change. In the low-pressure (LP) coal lock discharge that is used for the example in Table 4, hydrogen is the major component by volume, but is a minor component by mass. However, of greater significance are the heavy aromatic and sulfur species, whose relative quantities quadruple or triple when they are considered on the basis of their respective masses. Aromatics, COS and ethyl mercaptan all fit into this category.

Attention is called to the CO$_2$-rich waste gas vent from the Rectisol acid-gas removal plant. This is the largest stream that discharges to the atmosphere at the Kosovo plant and, although pollutant concentrations are comparatively low, the mass discharge rates of pollutants such as hydrocarbons and mercaptans are significant.

Pollutant Mass Discharge Rate

A pollutant's mass discharge rate is the product of its concentration (m/v) and its volumetric flowrate. Figure 4 shows a graphic presentation of the mass discharge rates of benzene and mercaptans for the seven major atmospheric discharge streams at the Kosovo gasification plant. These plots show that the stream discharge profiles differ considerably from stream to stream. This variation is to be expected since the plant's unit processes are significantly different from each other, and no single stream can be considered representative of the plant's total discharge. Although a majority of the organic compounds encountered in the plant originate in the gasifier, each

Table 3. Mass Concentration and Dry Gas Flowrate Data for Gaseous Streams Discharging to the Atmosphere at Kosovo

Plant Section	Gas Generation	Rectisol	Tar Separation			Phenosolvan[a]	By-Product Storage[a]
Component ($\mu g/m^3$ @ 25°C) — Sample Point	Low-Pressure Coal Lock Vent	CO_2-Rich Waste Gas Vent	Tar Tank Vent	Medium Oil Tank Vent	Phenolic Water Tank Vent	Ammonia Stripper Vent	Naphtha Tank Vent
Fixed Gases							
H_2	3.05E07	—	2.48E08	5.88E06	8.24E03	NF	NF
O_2	3.53E06	—	8.87E08	1.26E07	1.70E08	—	3.40E07
N_2	2.06E06	—	5.69E05	4.98E07	4.46E08	—	9.61E08
CH_4	5.64E07	8.06E06	—	6.75E07	1.31E06	—	NF
CO	1.67E08	—	—	—	NF	NF	NF
CO_2	6.56E08	1.69E09	1.55E07	1.01E09	6.29E08	9.89E08	1.53E07
Sulfur Species							
H_2S	1.77E07	5.43E04	9.61E06	3.62E07	1.75E07	2.72E07	NF
COS	2.70E05	1.47E05	2.70E05	2.36E05	1.01E05	NF	NF
CH_3SH	8.25E05	1.67E04	7.66E05	1.02E07	4.13E06	5.7 E05	5.11E06
C_2H_3SH	5.58E05	1.12E04	6.09E05	5.33E06	1.83E07	2.54E05	2.46E07
Hydrocarbons							
C_2H_6	2.79E06	1.97E07	—	4.18E06	2.46E05	—	—
C_2H_4	—	—	—	—	—	—	—
C_3	2.52E06	5.04E06	1.80E05	5.40E06	3.60E05	—	1.80E05
C_4	1.19E06	—	—	5.94E06	4.75E05	—	1.66E06
C_5	—	—	—	2.65E06	1.77E05	—	2.36E06
$C_6 +$	4.22E06	NF	1.30E07	8.45E07	6.34E07	NF	1.87E08
Benzene	2.43E06	3.19E03	6.38E06	2.44E07	3.51E07	—	1.20E08
Toluene	8.28E05	3.76E03	3.61E06	5.27E06	8.66E06	—	7.15E06
Xylene and Ethylbenzene	3.25E05	—	9.54E04	6.20E05	1.21E06	—	2.60E05
Phenols	2.19E04	NF	8.46E04	1.73E05	—	2.38E07	—

Nitrogen Species						
NH_3	1.67E06	3.20E04	1.81E06	7.93E03	8.35E06	2.91E08	NF
HCN	6.62E05	1.44E04	1.44E05	—	4.20E04	5.30E06	1.21E06
Dry Gas Flowrate (m^3/gasifier-hr @ $25°C$)	21	3,600	0.51	1.7	5.5	260	4.5

[a]NF = Not Found = less than a trace.

Table 4. A Comparison of Gas Compositions Expressed in Volume Percent and Weight Percent Using the Composition of the LP Coal Lock Vent Gas

	Volume Percent	Weight Percent	W%/V%
H_2	37.0	3.22	0.09
CH_4	8.60	5.96	0.69
CO	14.60	17.66	1.21
CO_2	36.50	69.36	1.90
H_2S	1.27	1.87	1.47
COS	0.01	0.03	3.00
CH_3SH	0.04	0.09	2.25
CH_2CH_3SH	0.02	0.06	3.00
HC (2–6)	0.53	1.12	2.11
Benzene	0.08	0.26	3.25
Toluene	0.02	0.09	4.09
Xylene	0.007	0.03	4.29
NH_3	0.24	0.18	0.75
HCN	0.06	0.07	1.17

Table 5. Ratios of Discharge Rate for Benzene to Xylene from the Kosovo Plant's Major Atmospheric Discharges

Discharge Source	Stream Flowrate	Mass Discharge Ratio
LA Coal Lock	21.0	7.5
Tar Tank Vent	0.5	67
Medium Oil Tank Vent	1.7	39
Phenolic Water Tank Vent	5.5	29
Gasoline Tank Vent	4.5	461
Weighted Average-Mass Discharge Ratio		33

downstream process changes its feed streams to the extent that each discharge stream must be viewed separately when performing an environmental assessment.

Benzene-to-Xylene Ratio

Attention is called to the ratios of benzene to xylene in the plant's atmospheric discharges. Xylene was the lightest aromatic hydrocarbon that was trapped quantitatively during the ambient air sampling program, i.e., toluene and benzene were only partially retained. Therefore, a comparison is made here of the mass discharge ratios of benzene to xylene so that an estimate can be made of the probable level of benzene in the air at the collection point. Table 5 shows the benzene-to-xylene ratios for the major discharge sources

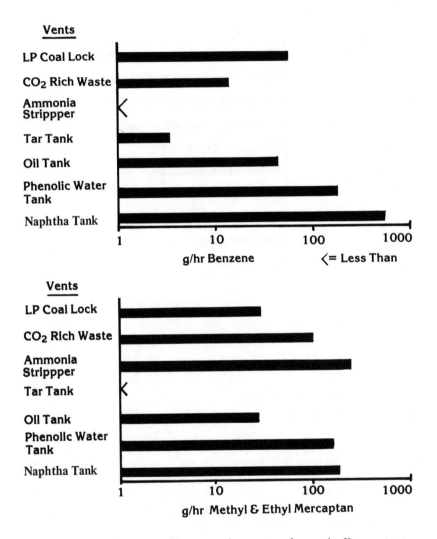

Figure 4.   Mass discharge rate of benzene and mercaptans from major Kosovo streams.

using data from Table 3. Those results indicate that the mass discharge rate of benzene from these six streams is approximately 33 times the xylene discharge rate. It is not unreasonable to assume that the same mass concentration ratio would be found at a point 1–2 km downwind from the source where the ambient air samples were collected.

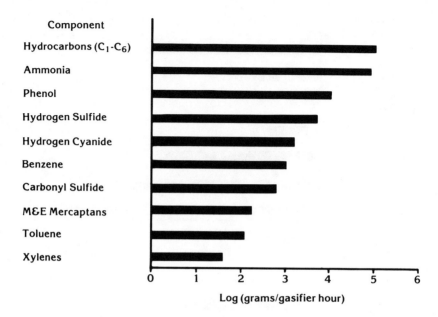

Figure 5. Kosovo mass discharges–plant wide.

Total Discharge to the Atmosphere– Based on Seven Major Atmospheric Discharge Streams

Kosovo plant-wide discharge rates for several components are presented in Figure 5. Hydrocarbons and ammonia are the major atmospheric discharges in terms of mass–being approximately ten times the quantity of any other class of substances. Phenol, discharged at the rate of ~11 kg/hr is released primarily from the Phenosolvan water extraction plant through the stripper vent. It probably is stripped as a vapor and transported through the atmosphere as an aerosol.

Particulate Discharges

Particulate concentration and stream flowrate data for the six discharge streams that were tested during this study are shown in Table 6. The particles, or aerosols, in most streams consisted primarily of suspended droplets of tar/oil which could not be sampled reliably by EPA Method 5 (dry filter) and were, therefore, collected with an impinger-train which yielded three fractions: condensed organics, dissolved solids and filtered solids. The mass

Table 6. Particulate Concentration and Flow Rate Data for Kosovo Gaseous Streams

Stream Type	Atmospheric Discharge Streams			Flare Feed Streams		
Sample Points	Coal Room Vent	Low-Pressure Coal Lock Vent	Gasifier Start-up Vent[a]	High-Pressure Coal Lock	Tar Separation Waste Gas	Combined Gases to Flare
Dry Gas Flowrate (m^3/gasifier-hr @ 25° C)	7200	21	4000	230	28	1330
Total Particulates (mg/m^3 @ 25° C)	48	8100	9450	960	920	380
Condensed Organics (tars and oils)	—	7300	8980	660	660	290
Dissolved Solids	6	650	400	240	230	49
Filtered Solids	42	220	61	61	29	43

[a]Stream discharges to atmosphere during start up until a combustible gas is produced.

Table 7. Hazardous PNA in Kosovo Tar and Oil (μg/g)

	Tar	Oil
7,12-Dimethylbenz(a)anthracene	1090	62
Benz(a)anthracene	490	156
Benzo(b)fluorene	306	115
Benzo(a)pyrene	210	68
Dibenzo(a)anthracene	23	7
3-Methylcholanthrene	26	NF
252 Isomer Group	945[a]	282[a]

[a]GC/MS analyses show that BAP comprises 24% of the molecular weight 252 isomer group in both tar and oil

concentration in each fraction and for the total stream is shown in Table 6 for each stream that was measured.

The particles contained in the Lurgi process and discharge streams are of interest environmentally, because these particles consist primarily of tars and oils that may contain high levels of polynuclear aromatics (PNA). These aerosols, when discharged to the atmosphere, as through the LP coal lock vent, can increase the level of PNA in the ambient air in the plant's vicinity. Analytical results are not yet available for the condensed liquids that were collected from these discharge sources. Therefore, by-product analyses were used to make judgments about the potential levels of PNA in these discharges. The concentrations of several hazardous PNA in Kosovo tar and Kosovo oil are shown in Table 7. These results were obtained by gas chromatography/mass spectrometry (GC/MS) using a liquid crystal column and chemical ionization detection. The by-product PNA results were used to estimate the levels of these same PNA in atmospheric discharges. Results are shown in Table 8, where values indicate that aerosols discharged from the gas generating plant may transport comparatively large amounts of PNA to the environment.

For example, using the tar-based value shown in Table 8, a BAP concentration of 1.53 mg/m^3 in the gas that is discharged from the LP coal lock would produce a BAP mass flow of 32 mg/hr for each generator. During this test three of the six generators were usually in operation. At full production, five generators would be in operation and the sixth would probably be in some phase of startup.

During several hours of the start-up period, BAP could be discharged through the start-up vent at the rate of 7560 mg/hr. These levels of BAP discharge could produce significant concentrations at a collection point 1 km downwind from the source.

Table 8. Projected Concentrations of Several PNA in Kosovo Atmospheric Discharges[a]

| | Estimated Concentration (mg/m^3) | | | |
| | LP Coal Lock Vent | | Start Up Vent | |
Substance	Tar	Oil	Tar	Oil
7,12-Dimethylbenz(a)anthracene	7.96	0.45	9.79	0.55
Benz(a)anthracene	3.58	1.14	4.40	1.40
Benzo(b)fluorene	2.23	0.84	2.75	1.03
Benzo(a)pyrene	1.53	0.50	1.89	0.61
Dibenzo(a)anthracene	0.17	0.05	0.21	0.06
252 Isomer Group[b]	6.90	2.06	8.49	2.53

[a]Values are based on the levels of those substances in by-product tar and by-product oil and, therefore, represent probable maximum and minimum values for each substance.
[b]BAP GC/MS analysis shows that the BAP comprises 24% of the molecular weight 252 isomer group in both tar and oil.

SUMMARY

During this study, characterization effort was concentrated on the major discharge streams, i.e., those showing the greatest pollutant mass flowrate. Of the 35 gaseous discharge streams that were screened preliminarily, only the streams described here were subjected to a detailed characterization. However, these 7 major discharge sources provide a reliable profile of the plant's emissions. The numerous small discharges that did not receive a detailed characterization were judged to be similar in composition to the streams that were characterized. Consequently, the data presented here should provide a reliable indication of the Kosovo coal gasification plant's emission profile and should provide a reliable data base for an ambient air characterization study.

ACKNOWLEDGMENT

This work was sponsored by the Industrial Environmental Research Laboratory of the U.S. EPA. The authors express their thanks to the following organizations and people for their contribution to this work: Radian Corporation (Austin, Texas), W. E. Corbett, R. V. Collins and R. A. Magee; Rudarski Institute (Belgrade, Yugoslavia), Mira Mitrovic and Dragan Petkovic; REMHK Kosovo (Pristine, Yugoslavia), Shani Dyla and Emelia Boti; and the Institute of Forestry and Agriculture (Belgrade, Yugoslavia), S. Kapor.

REFERENCES

1. Salja, B., and M. Mitrovic. "Environmental and Engineering Evaluation of the Kosovo Coal Gasification Plant (Test Plan)," Proceedings of the Environmental Aspects of Fuel Conversion Technology III, U.S. EPA–600/7–78–063 (April 1978).
2. Salja, B., and M. Mitrovic. "Environmental and Engineering Evaluation of the Kosovo Coal Gasification Plant—Yugoslavia (Phase 1)," Symposium Proceedings: Environmental Aspects of Fuel Conversion Technology IV, U.S. EPA–600/7–79–217 (April 1979).
3. Bombaugh, K. J., and W. E. Corbett. "Kosovo Gasification Test Program Results—Part II Data Analysis and Interpretation," Symposium Proceedings: Environmental Aspects of Fuel Technology IV, U.S. EPA–600/7–79–217 (April 1979).
4. Bombaugh, K. J., W. E. Corbett and M. D. Mattson. "Environmental Assessment: Source Test and Evaluation Report—Lurgi (Kosovo) Medium-Btu Gasification, Phase 1," U.S. EPA–600/7–79–190 (August 1979).
5. Bombaugh, K. J., et al. "Aerosol Characterization of Ambient Air Near a Commercial Lurgi Coal Gasification Plant, Kosovo Region, Yugoslavia," Final Report, U.S. EPA–600/7–80–77 (1980), and presented at the Second Chemical Congress of the North American Continent, Las Vegas, Nevada, August 24–29, 1980.

CHAPTER 19

A COMPARISON OF ORGANICS COLLECTED FROM THE AMBIENT AIR WITH THE BY-PRODUCTS OF A LURGI COAL GASIFICATION PLANT

K. W. Lee, K. J. Bombaugh, C. H. Williams, Jr.,
D. S. Lewis and L. D. Ogle

Radian Corporation
Austin, Texas 78766

INTRODUCTION

Coal gasification is becoming increasingly important in the United States and the world as an alternative energy source to petroleum-based fuels and also as a potential source of chemical feedstocks. Accompanying this increasing importance is an increasing concern for the impact that coal gasification plants will have on their environment. Because the Lurgi gasification process is an established technology with a high potential for future use, it was considered important to characterize emissions from Lurgi-type facilities in order to assess their potential impacts on their environment. A part of this emission assessment involved monitoring ambient air to determine the impact of a Lurgi-type gasification facility on surrounding air quality.

An air monitoring program was conducted at a commercial Lurgi-type facility in the Kosovo region of Yugoslavia as a cooperative effort between Yugoslavian and American scientists under the sponsorship of the U.S. Environmental Protection Agency (EPA) [1]. A significant aspect of the study was to determine if the emissions from a Lurgi-type coal gasification plant could be identified in ambient air in the vicinity of the facility and also to determine if these emissions could be distinguished from other emission sources that contributed to the area's pollutants.

A combination of gas chromatography (GC) and GC/mass spectrometry (GC/MS) was used for these studies. Profiles of the organic species in the plant's by-products and in the collected airborne organics were compared to detect similarities in composition. Species-specific detection in the sulfur-specific and nitrogen-specific modes was used to provide profiles that established distinguishable correlations between the plant's by-products and the collected organics.

EXPERIMENTAL

Sampling

Vapor-phase organics were collected from the air in glass cartridges filled with approximately 3 g of 60/80 mesh Tenax resin. Two cartridges were used in series, so that any breakthrough of organics from the first cartridge would be detected in the second. The samples were taken at a flowrate of $\cong$ 4 L/min over a 24-hr period, resulting in a sample volume of $\cong$ 5.8 m^3. Sampling was carried out over a period of 16 days at five stations located on a perimeter 1–2 km from the center of the gasification complex.

Sample Selection

Using meteorological data, samples were rated according to the percentage of time the collection site was downwind. The most upwind and most downwind samples were selected for this study.

Sample Preparation

Extraction of Tenax resin. Resin samples were first homogenized, after which a 1.0- to 1.5-g aliquot of the resin sample was extracted with cyclopentane for 24 hours in a Soxhlet apparatus. The extract was concentrated in a Kuderna-Danish concentrator fitted with a 3-ball Snyder column and then in a graduated receiver fitted with a 2-ball micro-Snyder column to a final volume of 0.5 mL.

Thermal desorption for gas-liquid chromatographic/flame ionization detector (GLC-FID) profile. A 0.05-g aliquot of a Tenax resin sample was placed in a thermal desorption injector at 275°C. The desorbed vapors were collected for 2 minutes on a precolumn in a dry ice/acetone bath. At the completion of desorption, the bath was removed and the GLC oven program was initiated.

Analysis. GC profiles of the principal by-products: naphtha, medium oil and tar, and of the collected organic vapors were prepared using the conditions given in Table 1. Chromatographic profiles of the major organics in the samples were produced with a flame ionization detector. In addition, separate profiles were produced using a Hall electrolytic conductivity detector (HECD) in sulfur- and nitrogen-specific modes. The HECD has a response ratio for the specific heteroatoms relative to hydrocarbons of $\cong 10^5$ for sulfur and $\cong 10^6$ for nitrogen.

Compound Identification

A combination of techniques was used to identify organic species in the samples. First, boiling point versus retention time for normal alkanes was plotted and used to predict the boiling points of eluting compounds. These tentative identities were then substantiated by peak enrichment with known compounds. This technique was successful in the tentative identification of benzene, toluene, xylenes, naphthalene and four of the alkyl-substituted thiophenes and benzothiophenes. The predicted boiling points were used along with a knowledge of expected emissions from the gasification facility

Table 1. Conditions Used with the Three Different Detectors (Modes)

Detector (Mode):	Flame Ionization Detector (FID)	Hall-Detector Sulfur (HECD-S)	Hall-Detector Nitrogen (HECD-N)
Condition			
Carrier Flow	N_2, 30 mL/min	He, 35 mL/min	He, 35 mL/min
Injector Temperature	250°C	250°C	250°C
Detector Temperature (Reactor)	300°C	225°C (850°C)	225°C (850°C)
Oven Program	50°C for 5 min; 20°C/min to 250°C, hold for 60 min	50°C for 5 min; 20° C/min to 250°C, hold for 60 min	50°C for 5 min; 20° C/min to 250°C, hold for 60 min
Column	10% OV101 on 80/100 mesh Supelcoport; 10 ft x 2 mm i.d. glass column	10% OV101 on 80/100 mesh Supelcoport; 10 ft x 2 mm i.d. glass column	10% Apiezon L, 2% KOH on 80/100 mesh Chromosorb, W-AW, 10 ft x 2 mm i.d. glass column

to make tentative identifications. In addition, sample peak relative retention times were compared to those of known compounds. Tentative identities of key compounds were then confirmed by GC/MS including such materials as C_1- through C_3-alkylthiophenes, pyridine, C_1- through C_3-alkylpyridines and C_1- and C_2-alkylquinolines.

RESULTS AND DISCUSSION

Collection Efficiency

The results from the evaluation of collection efficiency are contained in Table 2. Twelve components that were thermally desorbed from both the front and the back cartridge of the collection train are compared in terms of chromatographic peak height which is directly proportional to the amount desorbed. The evaluation indicates that materials with boiling points lower than C_8 were not collected quantitatively by the combined cartridges; C_8 materials broke through the first cartridge but probably were retained by the second. Materials above C_8 probably were retained quantitatively by the first cartridge.

Composition of By-Products

The bulk composition of the major by-products from the Kosovo plant are shown in Figure 1. The by-products contain virtually the same components that are found in the plant's major atmospheric discharge streams. The bulk composition of the downwind collection is shown in Figure 2. These two figures show that the downwind collection contains many components from the same boiling range as the by-product medium oil and that medium oil is a reasonable choice for a detailed characterization with a species specific detector.

A comparison of the HECD-nitrogen profiles of medium oil and of a downwind vapor sample is shown in Figure 3. These chromatographic profiles show a high degree of similarity but are not identical. There are two significant reasons for their differences. The first is that, in this comparison, a vapor is being compared to a liquid. Since the amount of a compound's concentration in the vapor state is directly proportional to its vapor pressure, the higher boiling compounds which elute later in the profiles are at lower concentrations in the ambient air than in the medium oil. The second reason is that significant amounts of the lower boiling compounds may have been lost by breakthrough during sample collection, and by evaporation during extraction of the resin and concentration of the extract. It is assumed that this

Table 2. Sample Vapor Breakthrough: Quantitation by GLC-FID of Corresponding Components on a Downwind Front-Rear Tenax Cartridge Set

Tentative Peak Identification	Retention Time[a] (min)	Peak Height (mm)		Percent of Total Peak on Rear Cartridge
		Front Cartridge	Rear Cartridge	
1. Benzene	3.9	126	117	48%
2. C$_7$-alkane	5.1	40	27	40%
3. Toluene	6.5	1335	1243	48%
4. C$_8$-alkane	7.3	184	61	25%
5. Ethylbenzene	8.1	720	137	16%
6. p- and m-Xylenes	8.3	2420	289	11%
7. o-Xylene	8.7	608	110	15%
8. Naphthalene	12.4	980	19	2%
9. Methylnaphthalene	13.5	512	10	2%
10. Anthracene	18.0	128	6	4%
11. Methylanthracene	18.9	35	2	5%
12. Unidentified	21.6	32	2	6%

[a]Values given are within ±15%.

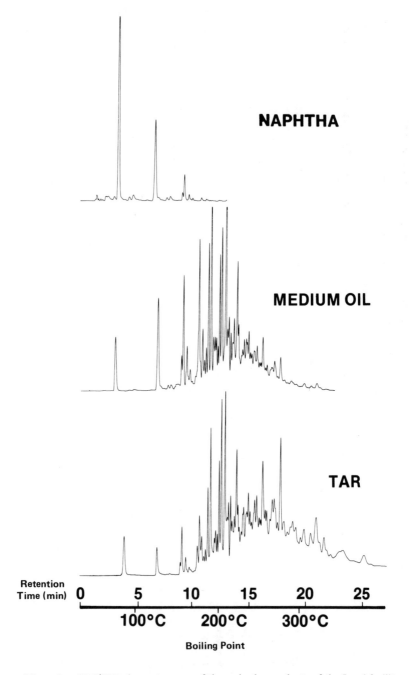

Figure 1. GLC/FID chromatograms of the major by-products of the Lurgi facility.

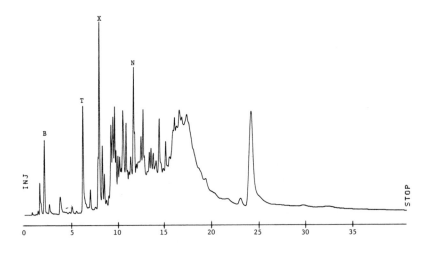

Figure 2. Thermal desorption GLC/FID chromatogram of a Tenax vapor sample taken downwind from the Lurgi facility.

accounts for the absence of a pyridine peak (retention time $\cong 7$ min) in the chromatogram for the vapor sample in Figure 3.

A comparison of the HECD-nitrogen profiles of an upwind vapor sample and a resin blank is shown in Figure 4. The two profiles are very similar and show very few peaks. Only two peaks (eluting at $\cong 15$ and $\cong 19$ min) are common to both the upwind and downwind vapor chromatograms. These peaks may well be of natural origin rather than from the gasifier.

Semiquantitative results are given in Table 3 for nitrogen-containing organics in: three downwind samples, an upwind sample and a resin blank. These data, together with the chromatographic profiles, indicate a high degree of correlation between the nitrogen containing organics in the air downwind from the gasifier but virtually no correlation between the upwind sample and the gasifier.

The results from a parallel experiment using the sulfur specific detector support the results that were obtained with the nitrogen detector. The sulfur profiles in Figure 5 allow a comparison between a downwind vapor sample and medium oil. A high degree of similarity between the two profiles can be observed. The sulfur profiles of an upwind sample and a resin blank are shown in Figure 6. No peaks were detected in the upwind sample that were not detected in the blank. The only major peaks common to both the upwind and downwind samples were also present in the blank. These profiles show that a high degree of similarity exists between sulfur-bearing organics in the medium oil and the sulfur-containing organics in the air downwind from the

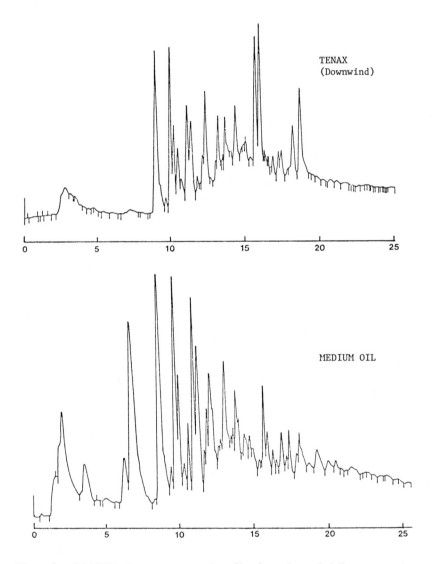

Figure 3. GC-HECD nitrogen compound profiles for a downwind Tenax vapor trap extract and for Kosovo medium oil.

gasifier. Semiquantitative analytical data for the several components in these profiles are given in Table 4. These data also show that a high degree of correlation exists between the concentrations of these components and the percent of the time the site was downwind from the plant and that no correlation exists between the gasifier and the upwind sample.

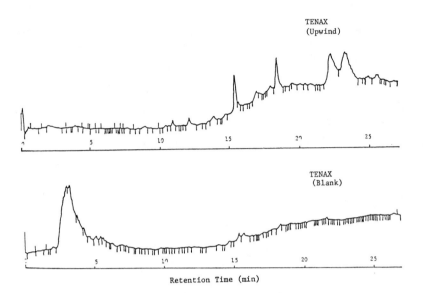

TENAX
(Upwind)

TENAX
(Blank)

Retention Time (min)

Figure 4. GC-HECD nitrogen compound profiles for an upwind and a blank Tenax vapor trap extract.

SUMMARY

In summary, aromatic hydrocarbons, thiophenes, pyridines and quinolines were all detected both in downwind ambient air samples and Kosovo medium oil. The presence of these compounds along with the similarities between the selective detector chromatographic profiles for the samples show that a strong correlation exists between the ambient air downwind of the Lurgi facility and medium oil from that facility. This correlation is upheld by the absence of profile similarities between upwind air samples and medium oil. Such correlation is strong evidence that the collected species originated with the gasifier. The technique of organic speciation with selective detector gas chromatography is a promising method for distinguishing between the organic emissions from adjacent sources.

ACKNOWLEDGMENT

This work was sponsored by the Environmental Science Research Laboratory of the U.S. Environmental Protection Agency. The authors express their thanks to Mr. Ronald K. Patterson for his contribution to this work.

Table 3. Correlation of Concentration[a] with Recent Downwind for Several Tentatively Identified Nitrogen Containing Organic Compounds in Kosovo Air

Compound(s)	Blank[b] (μg/m^3)	0% Upwind (μg/m^3)	18.7% Downwind (μg/m^3)	40% Downwind (μg/m^3)	45.2% Downwind (μg/m^3)	Correlation Coefficient
Pyridine	<DL	<DL	0.005	<DL	<DL[c]	—
2-Methylpyridine	<DL	<DL	0.04	0.36	0.62	0.92
3-Methylpyridine	<DL	<DL	0.04	0.19	0.55	0.83
Dimethyl- or Ethylpyridine(s)	<DL	<DL	0.11	0.21	0.46	0.90
Dimethyl- or Ethylpyridine(s)	<DL	<DL	ND[d]	0.17	0.49	0.83
Alkylpyridine(s)	<DL	<DL	0.06	0.20	0.56	0.84
Alkylpyridine(s)	<DL	<DL	0.02	0.11	0.46	0.77
Alkylpiridine(s)	<DL	<DL	0.11	0.16	0.74	0.76
Quinoline	0.13	<DL	0.21	0.46	0.94	0.91
Alkylquinoline(s)	<DL	<DL	<DL	0.19	0.54	0.82
						0.84
						±0.06

[a]Concentrations of all compounds based on response factor for 3-nitrotoluene.
[b]Concentration based on a hypothetical air sample of 5 m^3.
[c]DL = 0.005 μg 3-nitrotoluene/m^3.
[d]ND = not detected due to interferences.

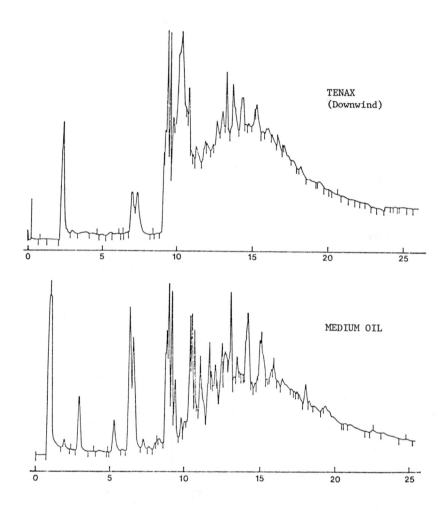

Figure 5. GC-HECD sulfur compound profiles for a downwind Tenax vapor trap extract and for Kosovo medium oil.

REFERENCE

1. Bombaugh, K. J., et al. "Aerosol Characterization of Ambient Air Near a Commercial Lurgi Coal Gasification Plant, Kosovo Region, Yugoslavia," Final Report, U.S. EPA–600/7–80–77 (1980).

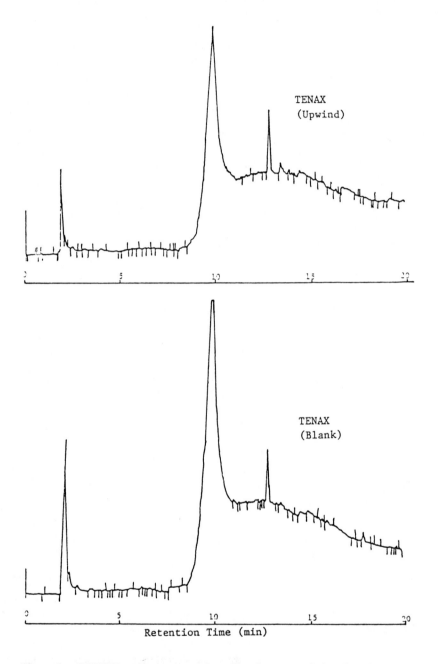

Figure 6. GC-HECD sulfur compound profiles for an upwind and a blank Tenax vapor trap extract.

Table 4. Correlation of Component Concentration[a] with Percent Downwind for Several Tentatively Identified Sulfur Containing Organic Compounds in Kosovo Air[b,c]

Compound(s)	Blank ($\mu g/m^3$)	0% Upwind ($\mu g/m^3$)	18.7% Downwind ($\mu g/m^3$)	40% Downwind ($\mu g/m^3$)	45.2% Downwind ($\mu g/m^3$)	Correlation Coefficient
Thiophene	<DL	<DL	<DL	<DL	<DL	–
2-Methylthiophene	<DL	<DL	0.01	0.13	0.04	0.68
3-Methylthiophene	<DL	<DL	0.01	0.16	0.05	0.69
2,5-Dimethylthiophene	<DL	<DL	<DL	0.18 (merged)	0.03	–
Dimethyl- or Ethylthiophene(s)	ND	ND	0.007	0.07	0.10	0.94
Dimethyl- or Ethylthiophene(s)	ND	ND	0.007	0.07	0.09	0.95
Trimethyl- or Ethylmethyl-thiophene(s)	<DL	<DL	ND	0.02	0.04	0.91
Trimethyl- or Ethylmethyl-thiophene(s)	<DL	<DL	0.01	0.02	0.08	0.78
Benzothiophene	ND	ND	<DL	0.02	ND (0.10)	–
Methylbenzothiophene	<DL	0.005	<DL	<DL	0.12	–
					mean	0.83
					σ	±0.12

[a]Concentrations of all compounds based on the response factor for thiophene.
[b]DL = 0.005 μg of thiophene /m^3.
[c]ND = not detected due to interferences.

CHAPTER 20

GC/MS CHARACTERIZATION OF TRACE ORGANIC COMPOUNDS IN THE AMBIENT AEROSOL ASSOCIATED WITH A COAL GASIFICATION PLANT AT KOSOVO, YUGOSLAVIA

C. H. Williams, Jr., K. J. Bombaugh, P. H. Lin,
K. W. Lee and C. L. Prescott

Radian Corporation
Austin, Texas 78766

INTRODUCTION

The ambient aerosol (particulate plus vapor) on the perimeter of the full scale, Lurgi-process coal gasification complex in Kosovo, Yugoslavia, was sampled for successive 24-hr periods over 16 days in May 1979. These samples were collected by Yugoslavian personnel, directed by an onsite management team of personnel from Radian Corporation and the U.S. Environmental Protection Agency (EPA).

The sampling media for the collection of organic pollutants was prepared by the Oak Ridge National Laboratory. Analysis of this media was divided between Radian Corporation, Oak Ridge National Laboratory (cf. Chapter 21) and the Institut za Primenu Nuklearne Energye (INEP) in Yugoslavia.

This chapter presents a summary of the analytical approach followed by Radian in the characterization of the ambient aerosol and the results for aerosol samples collected on 4 days (days 5 through 8).

EXPERIMENTAL APPROACH

Sampling

Five sites on the perimeter of the coal gasification complex at Kosovo were selected to represent a range of upwind-downwind conditions. A map of the gasification complex showing the sampling sites is shown in Figure 1.

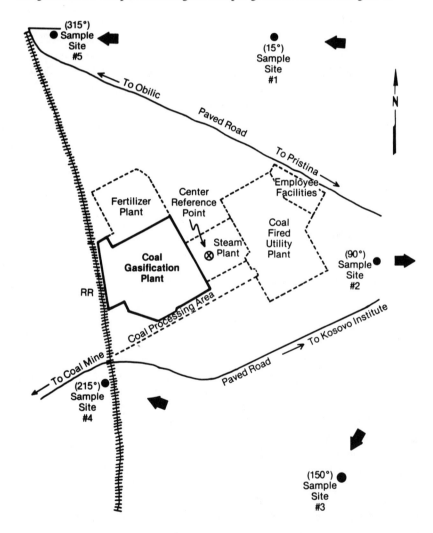

Figure 1. Map of the Kosovo gasification complex showing the five aerosol sampling sites.

Aerosol samples were collected for 24 hours each day at each site over 16 consecutive days. The organic aerosol components were collected on the sampler shown in Figure 2: a Hi-Vol filter sampler equipped with a downstream two-section Tenax resin vapor trap which sampled ~0.5% of the air flow that had passed through the Hi-Vol filter.

About 1400 m³ of ambient air was drawn through the Hi-Vol filter in a 24-hr sampling period and ~6 m³ of this passed through both the front and rear Tenax traps. Radian received the 20 Hi-Vol filters (5 sites x 4 days = 20 samples) and 40 Tenax traps for days 5, 6, 7 and 8.

In addition to the environmental samples, some examples of the coal gasifier's by-products were received by Radian. These by-products included gasoline and medium oil, and the solids, tar and heavy tar. Some of these by-products were analyzed by gas chromatography/mass spectrometry (GC/MS) and their components were identified as an aid in the characterization of the aerosol samples.

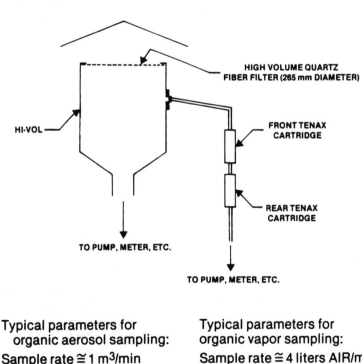

KOSOVO AEROSOL

HIGH VOLUME QUARTZ FIBER FILTER (265 mm DIAMETER)

HI-VOL

FRONT TENAX CARTRIDGE

REAR TENAX CARTRIDGE

TO PUMP, METER, ETC.

TO PUMP, METER, ETC.

Typical parameters for organic aerosol sampling:
Sample rate ≅ 1 m³/min
Sample volume ≅ 1400 m³/24 hr

Typical parameters for organic vapor sampling:
Sample rate ≅ 4 liters AIR/min
Sample volume ≅ 5.8 m³/24 hr

Figure 2. Configuration and parameters for collecting aerosol samples for organic analysis.

The overall approach for aerosol sample and gasifier by-product analysis is shown in Figure 3.

Separations

Two methods were used to separate the organic aerosol components from the sampling media: the Hi-Vol filter and Tenax resin. These are shown in Table 1 and Figure 3.

The glass fiber Hi-Vol filters were only treated by Soxhlet extraction using methylene chloride as a solvent and an 8-hr extraction cycle. Either a whole filter or a weighed half filter were extracted.

The Tenax resin used in the vapor traps was homogenized by thorough mixing in a tightly closed container, and aliquots were taken for separation by Soxhlet extraction or thermal desorption. The Soxhlet extraction was carried out on 1- to 1.5-g portions (40-60% aliquots) using a cyclopentane as the solvent in an 8-hr extraction. Thermal desorption was performed on 40- to 60-mg portions (1.6-2.4% aliquots) heated to 250°C for 10 minutes under a

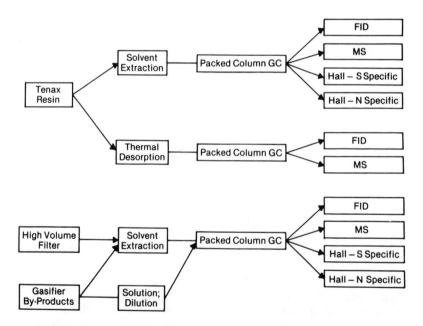

Figure 3. Schematic diagram of the multiple separation and analysis methods used for ambient aerosol samples and gasifier by-product characterization. GC = gas chromatography, FID = flame ionization detector; MS = mass spectrometer; Hall = Hall electroconductivity detector in the sulfur (S) or the nitrogen (N) mode.

Table 1. Separation Methods Used for Each Sampling Medium

Sampling Medium	Soxhlet Extraction	Thermal Desorption
Tenax Resin	Yes	Yes
Hi-Vol Filter	Yes	No

helium purge at 30 mL/min. The desorbed vapors were collected in a trap at liquid nitrogen temperature placed in the GC oven. This trap was subsequently heated by the GC column oven during temperature programming to 250°C to flash vaporize the sample vapor species.

Soxhlet extracts were concentrated to a minimum final volume that ranged from 0.5 mL for the Tenax resin extracts to 6.0 mL for several of the heavily loaded Hi-Vol filters.

Sample Spiking for QC and Methods Validation

A series of deuterated aromatic compounds and PNA were spiked into the blank sampling media or into the actual aerosol samples, as shown in Table 2. The purpose of this spiking program was twofold: to validate the analytical methodology for recovery, precision, specificity and sensitivity for these representative compounds; and to validate the analysis of the actual aerosol samples by using the deuterated isotopes as internal standards for recovery, (d_8-toluene and d_8-naphthalene) and quantitation (d_{10}-anthracene).

Gas Chromatography

All of the gas chromatographic separations were carried out in packed Pyrex® columns. Screening analyses by gas chromatography/flame ionization detection (GC/FID) and speciation analyses by GC/mass spectrometry (GC/MS) were both performed using the same set of GC parameters:

- 10% OV-101 on 80/100 mesh Supelcoport,
- 3 meters x 2 mm i.d. glass column,
- 50°C hold for 5 min; programmed at 20°C/min to 250°C, and
- helium or nitrogen carrier gas at 30 mL/min.

This set of GC parameters was established as providing convenient elution times and adequate resolution for the most significant organic species found in the Kosovo aerosol samples. The range of compounds included was:

C-H (aromatic)	benzene to benzo(a)pyrene [B(a)P]
C-H (aliphatic)	C_7 to C_{30} alkanes

C-H-S	methyl thiophene to benzothiophene
C-H-N	pyridine to alkylquinoline
C-H-O	phenol to diphenylquinone

A packed liquid crystal column was used in an isothermal mode for high-resolution separation of the PNA isomer group of molecular weight 252. The purpose was to confirm the presence of benzo(a)pyrene in the isomeric species detected by GC/MS in the aerosol sample. The GC parameters of the liquid crystal columns were:

- 1.5% SP-301 on 100/120 mesh Supelcoport,
- 2 meters x 2 mm i.d. glass column,
- 290°C isothermal operation and
- helium carrier gas at 30 mL/min.

Species Profiling, Identification and Quantification

GC/FID was used for the methods development and validation where media blanks were spiked with the compounds listed in Table 2. In addition, GC/FID was used for the screening or profiling of all aerosol samples (Hi-Vol filters or Tenax resin) analyzed. This screening was useful for the prioritization of samples for detailed characterization.

GC/MS was used for the identification and quantification of the organic species detected in the aerosol samples. Electron-impact ionization on a Hewlett Packard Model 5982 quadrupole GC/MS was combined with both the scanning (50–350 amu) and selected ion monitoring (SIM) modes of operation.

Species identifications were made on the basis of their mass spectra generated in scanning GC/MS operation. All identified compounds displayed a prominent molecular ion. Retention times were employed to aid in the identification of some isomeric species, e.g., BAP in the 252 group.

Quantification was exact for those PNA species spiked as internal standards into the aerosol samples and estimated from peak area ratios for all other species.

EXPERIMENTAL RESULTS

Methods Validation

The thermal desorption procedure and the GC conditions developed in the course of this work were effective in achieving good chromatographic resolution and quantitative recovery for PNA vapors sorbed on Tenax.

Table 2. Strategy of Aerosol Sample Spiking with Deuterated Internal Standards for GC/FID and GC/MS Analyses

Sample and Sequence	Spiking Compounds	Equivalent Air Concentration Level	Purpose[a] RT	REC	IR
Blank Tenax resin aliquots, before thermal desorption or before Soxhlet extraction	d_6-Benzene[b] d_8-Toluene d_8-Naphthalene d_{10}-Anthracene	$1.7 \ \mu g/m^3$	X	X	
Blank Hi-Vol filter aliquots, before Soxhlet extraction	1,2-Benzanthracene Benzo(a)pyrene[c]	$0.07 \ \mu g/m^3$	X	X	
Tenax aerosol aliquots, before thermal desorption or before Soxhlet extraction	d_8-Toluene d_8-Naphthalene	$1.7 \ \mu g/m^3$	X	X	
Hi-Vol filter aliquots, before Soxhlet extraction	d_8-Toluene d_8-Naphthalene	$0.07 \ \mu g/m^3$	X	X	
Concentrated aerosol (Tenax and Hi-Vol) extracts, before GC/MS analysis	d_{10}-Anthracene	(20 ng/μL injected)	X		X

[a]RT = retention time; REC = recovery; IR = instrument response.
[b]d_6-Benzene spiked for thermal desorption only.
[c]Benzo(a)pyrene spiked for Soxhlet extraction only.

Table 3. Mean Recoveries by Thermal Desorption for Five Aromatic Compounds
Spiked on Blank Tenax Resin

Compound Identification	Mean Recovery (%)	Precision (%)
1. d_6-Benzene	90	±10
2. d_8-Toluene	91	±10
3. d_8-Naphthalene	92	±10
4. d_{10}-Anthracene	94	±10
5. 1,2-Benzanthracene	113	±20

Table 4. Mean Recoveries of Five Aromatic Compounds Soxhlet Extracted from
Spiked Blank Tenax with Cyclopentane

Compound Identification	Mean Recovery (%)	Precision (%)
1. d_8-Toluene	64	±30
2. d_8-Naphthalene	99	±10
3. d_{10}-Anthracene	97	±10
4. 1,2-Benzanthracene	108	±15
5. Benzo(a)pyrene	110	±20

Table 5. Sample Vapor Breakthrough: Quantitation of Sample Species
on Front and Rear Sections of the Tenax Vapor Trap

Tentative Peak Identification	Retention Time (min)	Percent of Total Peak on Rear Trap
1. Benzene	3.9	48
2. C_7-Alkane	5.1	40
3. Toluene	6.5	48
4. C_8-Alkane	7.3	25
5. Ethylbenzene	8.1	16
6. p- and m-Xylenes	8.3	11
7. o-Xylene	8.7	15
8. Naphthalene	12.4	2
9. Methylnaphthalene isomer	13.5	2
10. Anthracene	18.0	4
11. Methylanthracene isomer	18.9	5
12. Unidentified	21.6	6

Table 3 gives the measured recoveries of five PNA, from benzene to 1,2-benzanthracene spiked onto blank Tenax resin and eluted by thermal desorption. Table 4 shows that Soxhlet extraction of the Tenax resin with cyclopentane was equally as effective for PNA compounds from naphthalene through benzo(a)pyrene, although recoveries for toluene were affected by losses in the sample concentration step.

The breakthrough of the aerosol sample vapor was determined by quantitating the same compounds on both the front and rear sections of the Tenax vapor trap. Table 5 shows that the air sample volume of 6 m³ was sufficient to cause both benzene and toluene to be completely broken through. However, the xylenes and the less-volatile species were quantitatively collected on the front Tenax trap.

Aerosol Species Identification

A GC/MS total ion current (TIC) chromatogram of an unspiked Tenax aerosol sample collected at a site (Site 4) predominantly downwind of the Kosovo gasification complex is shown in Figure 4. This chromatogram shows a complex series of peaks, the majority of which were subsequently identified as aromatic hydrocarbons. Aromatic species from benzene to pyrene were identified in the vapors collected on the Tenax, with extensive alkyl substitution in the range C_1 to C_4.

Tenax aerosol samples were also spiked with d_8-toluene and d_8-naphthalene as internal standards before analysis. Figure 5 shows the TIC GC/MS chromatogram of a spiked aerosol sample with extracted ion current profiles (EICP) at: mass 78 for benzene, mass 128 for (unlabeled) naphthalene and mass 202 for fluoranthene and pyrene.

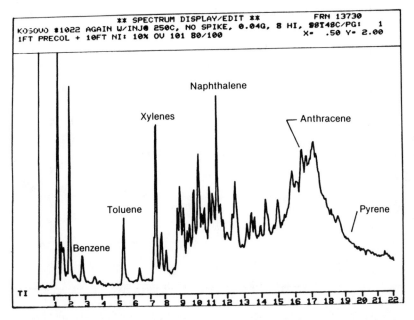

Figure 4. Thermal desorption GC/MS chromatogram of unspiked Tenax aerosol sample.

A GC/MS chromatograph of the Soxhlet extract of a Hi-Vol filter sample collected at the same downwind sample site (Site 4) is shown in Figure 6. The EICP at mass 202, 228 and 252 indicate the presence of the PNA isomer groups of these molecular weights.

The presence of benzo(a)pyrene in the mass 252 isomer group was subsequently demonstrated by SIM-GC/MS analyses of the downwind Hi-Vol filter extract and the gasifier by-product, middle oil. Figure 7 compares three SIM traces at mass 252: the top trace is the Hi-Vol filter extract on the packed 10% OV-101 GC column and indicates the presence of at least 3 isomers. The center trace is of a sample of middle oil on the same column and exhibits a profile very similar to the Hi-Vol filter extract, i.e., composed of at least 3 isomers.The bottom trace shows the SIM-GC/MS chromatogram of the middle oil on the liquid crystal GC column. At the higher GC resolution, the middle oil is seen to contain at least 5 isomers of mass 252. The latest eluting is benzo(a)pyrene, which comprises about 20% by weight of all the 252 isomers. The Hi-Vol filter extract was not analyzed on the liquid crystal column, but the implication is clear that the isomers detected probably include benzo(a)pyrene.

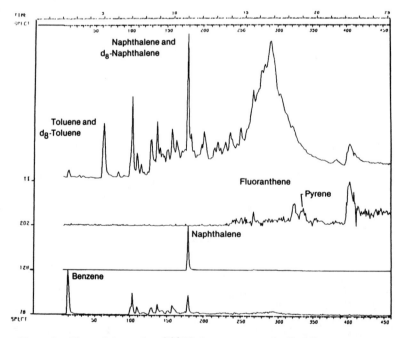

Figure 5.　Thermal desorption GC/MS chromatogram of spiked Tenax sample.

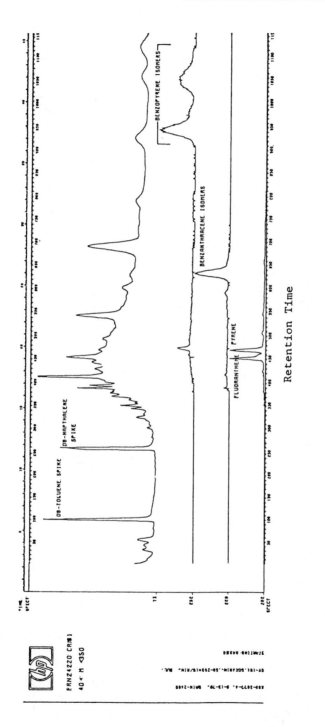

Figure 6.   GC/MS chromatograph of downwind Hi-Vol filter.

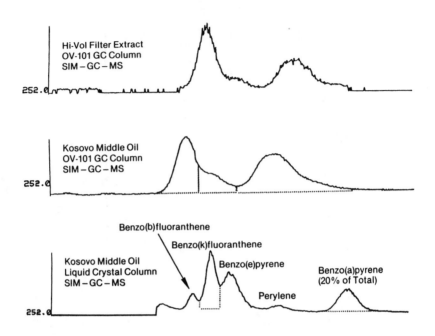

Figure 7. SIM-GC/MS chromatograms of mass 252 isomer group in Kosovo aerosol Hi-Vol extract and middle oil by-product.

A summary of all the organic compounds identified as being present in the ambient aerosol downwind of the gasifier complex is given in Table 6, along with the number of isomers. It can be seen that the ambient aerosol contains a very complex and potentially hazardous mixture of organic pollutants.

Aerosol Species Quantification

Fourteen aromatic compounds or isomeric groups, including PNA were selected to constitute a chemical profile of the gasification plant's major emissions for the purpose of comparing the ambient air samples from different sites and different days. Table 7 presents the profile of these fourteen compounds collected on Tenax vapor traps and Hi-Vol filters at Sites, 1, 3 and 4 on day 5. The mean percentage of time each site was downwind of the gasification complex on day 5 is also given. It can be seen that Site 1 at 0.6% downwind (DW) represents an upwind sample site, Site 4 at 40.0% DW is the most downwind and Site 3 at 18.7% DW is intermediate. Table 8 presents the same profile for samples collected at Site 4 (the predominantly downwind site) for sampling days 5, 6, 7 and 8.

Table 6. Listing of Organic Compounds and Isomer Groups Identified by GC/MS Both in the Kosovo Aerosol and Gasification By-Products

Compound or Isomer Group Identification	Molecular Weight	Number of Isomers Detected
Hydrocarbons		
Benzene	78	1
Toluene	92	1
C_2-Alkyl benzenes	106	$\geqslant 3$
C_3-Aklyl benzenes	120	$\geqslant 6$
C_4-Alkyl benzenes	134	$\geqslant 9$
Naphthalene	128	1
Methyl naphthalenes	142	$\geqslant 2$
Acenaphthene	154	1
Fluorene	166	1
Anthracene/Phenanthrene	178	$\geqslant 1$
Methylanthracene	192	$\geqslant 2$
Fluoranthene	202	1
Pyrene	202	1
Chrysene		
Benzanthraacene	228	$\geqslant 1$
Triphenylene		
Benz(a and e)pyrenes		
Benzo(b and k)fluoranthenes	252	$\geqslant 3$
Perylene		
Terphenyl	230	$\geqslant 3$
Trimethylphenylindane	236	1
C-H-O Compounds		
Phenol	94	1
Methyl phenols	108	$\geqslant 2$
C_2-Alkyl phenols	122	$\geqslant 5$
Acetophenone	120	1
Dibenzofuran	168	1
Diphenylquinone	260	1
Phthalate esters		2
C-H-S Compounds		
Methyl thiophenes	98	2
C_2-Alkyl thiophenes	112	$\geqslant 4$
C_3-Aklyl thiophenes	126	$\geqslant 3$
C-H-N Compounds		
Pyridine	79	1
Methyl pyridines	93	$\geqslant 4$
C_3-Alkyl pyridines	121	$\geqslant 5$
Quinoline	129	1
Isoquinoline	129	1
Methyl quinolines	143	$\geqslant 5$
C_2-Alkyl quinolines	157	$\geqslant 10$

Table 7. Comparative Chemical Profile of Kosovo Aerosol Samples Collected on Day 5 at Sites 1, 3 and 4

Compound Name	Molecular Weight	Site 1 (0.6% DW)		Site 3 (18.7% DW)		Site 4 (40.0% DW)	
		Tenax	Hi-Vol	Tenax	Hi-Vol	Tenax	Hi-Vol
Benzene	78	303	ND	20	ND	89	ND
Toluene	92	565	ND	76	ND	101	ND
C_2-Alkyl benzene	106	442	ND	107	ND	131	ND
C_3-Alkyl benzene	120	260	ND	123	ND	141	ND
C_4-Alkyl benzene	134	40	ND	30	ND	40	ND
Naphthalene	128	100%[a]	ND	100%[a]	7%	100%[a]	ND
Methylnaphthalene	142	35	ND	26	ND	50	6
Acenaphthene	154	15	5	11	ND	17	ND
Fluorene	166	43	2	7	ND	8	5
Anthracene Phenanthrene	178	33	100%[a]	15	100%[a]	19	100%[a]
Fluoranthene	202	ND	1.4	3.5	76	1.9	230
Pyrene	202	ND	1.6	3.1	97	1.6	240
Chrysene Benzanthracene Triphenylene	228	ND	ND	ND	ND	ND	360
Benz( )pyrenes Benz( )fluoranthenes Perylene	252	ND	ND	ND	ND	ND	1100
Ambient air concentrations of naphthalene or anthracene, $\mu g/m^3$ Air (= 100% normalized value)		0.2[a]	0.001[a]	1[a]	0.001[a]	3[a]	0.003[a]

[a]Relative concentrations in upper table are normalized to naphthalene for vapor and anthracene for particulate; lower table presents absolute ambient concentrations calculated for these two species only.

Table 8. Comparative Chemical Profile of Kosovo Aerosol Samples Collected at Site 4 on Days 5, 6, 7 and 8

Compound Name	Molecular Weight	Day 5 (40.0% DW)		Day 6 (45.2% DW)		Day 7 (39.4%)		Day 8 (18.4% DW)	
		Tenax	Hi-Vol	Tenax	Hi-Vol[a]	Tenax	Hi-Vol	Tenax	Hi-Vol
Benzene	78	89	ND	31%		11	ND	20	
Toluene	92	101	ND	41		43	ND	44	
C_2-Alkyl benzene	106	131	ND	73		72	ND	76	
C_3-Alkyl benzene	120	141	ND	56		60	ND	76	
C_4-Alkyl benzene	134	40	ND	26		25	ND	38	
Naphthalene	128	100%[a]	ND	100%[a]		100%[a]	23%	100%[a]	
Methylnaphthalene	142	50	6	43		35	ND	24	
Biphenyl	154								
Acenaphthene	154	17	ND	9.0		10	ND	8.3	
Fluorene	166	8	5	9.3		7.3	9	4.7	
Anthracene Phenanthrene	178	19	100%[a]	20		14	100%[a]	11.2	
Fluoranthene	202	1.9	230	2.7		3.6	230	2.5	
Pyrene	202	1.6	240	2.2		2.3	230	1.6	
Chrysene Benzanthracene Triphenylene	228	ND	360	ND		ND	370	ND	
Benz( )pyrenes Benz( )fluoranthenes Perylene	252	ND	1100	ND		ND	940	ND	
Ambient air concentrations of naphthalene or anthracene, $\mu g/m^3$ Air (= 100% normalized value)		3[b]	0.003[b]	5[b]		2[b]	0.003[b]	2[b]	

[a] Soxhlet malfunction.
[b] Relative concentrations in upper table are normalized to naphthalene for vapor and anthracene for particulate; lower table presents absolute ambient concentrations calculated for these two species only.

The profile presented in these tables has two principal elements:

- a relative quantitation of all detected peaks based on a 100% normalization to naphthalene in the Tenax samples of organic vapor and to anthracene in the Hi-Vol samples of particulate matter, and
- an absolute quantitation (accurate to about a factor of 2 in $\mu g/m^3$ of air) of the measured ambient level of naphthalene in the vapor and anthracene on the particulate matter.

The relative quantitation is based on the ratio of peak areas of the appropriate molecular ions listed in the table, as measured by GC/MS. Absolute quantitation involved the use of all of the following factors: instrumental response, extraction recovery, aliquot fractions, the volume of air sample, the final volume of concentrated extracts, the concentration of deuterated internal standards and GC/MS peak area ratios.

The approximate detection limit for aromatic organic species collected as vapor on the Tenax resin traps for approximately 6 m^3 of air was found to be about 10 ng/m^3. The approximate detection limit for aromatic organic species detected on the particulate matter collected for 1400 m^3 of air was 0.01 ng/m^3.

DISCUSSION AND CONCLUSIONS

Interpretation of the data in Table 7 leads to the following conclusions:

- the measured ambient concentrations of naphthalene and anthracene are a strong function of sample site location and correlate positively with the percent downwind parameter, suggesting that they are indeed compounds emitted by the Kosovo coal gasification plant.
- there is a clear distinction in chemical speciation and pollutant concentration between the vapor samples on the Tenax and the particulate samples on the Hi-Vol filters: Tenax samples are found to contain species from benzene to pyrene at the $\mu g/m^3$ level, and Hi-Vol samples contain PNA mostly heavier than anthracene but at levels in the ng/m^3 range ($\sim$1000 x lower).

It should be pointed out that the relative concentrations of benzene and toluene are minimum values due to the demonstrated vapor breakthrough of these species in the sample collection on Tenax.

Examination of the data in Table 8 indicates that the concentrations of selected pollutants in the air samples showed only small variations on a day-to-day basis at the downwind Site 4 during the sampling period days 5 through 8 (May 18 through 21, 1979).

Since the day-to-day variations in pollutant concentrations seem small, all measured concentrations of naphthalene can be plotted against the

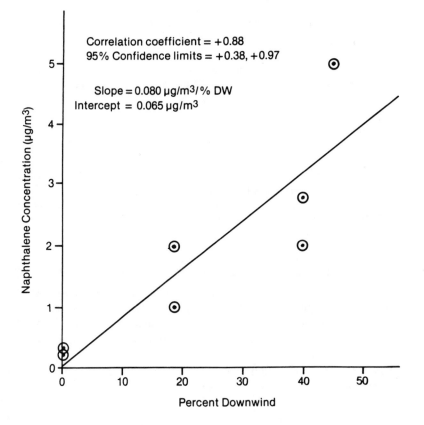

Figure 8. Correlation of ambient naphthalene vapor concentration with the measured time percentage that the sample site was downwind of the coal gasification plant.

measured percent downwind value for the particular day and site. The data in Tables 7 and 8 are plotted in Figure 8 to determine the degree of correlation between organic vapor loading and percent downwind. A linear regression analysis gave the following results for this plot:

$$\begin{aligned} \text{Slope} &= 0.080 \; \mu g/m^3/\text{percent downwind} \\ \text{Intercept} &= 0.065 \; \mu g/m^3 \\ \text{Coefficient Correlation} &= 0.88 \end{aligned}$$

Thus, the ambient air concentration of naphthalene as collected on the Tenax cartridges correlates positively with the percent of time that the sampling site was downwind of the Kosovo gasification plant. The extrapolated concentration of naphthalene at 100% DW is $8.0 \; \mu g/m^3$.

Table 9. Comparison of Selected Compounds in the Kosovo Aerosol
to Ambient Air Goals [1]

Chemical Species	Ambient Levels at 100% DW ($\mu g/m^3$)	A-MEG Goals ($\mu g/m^3$)
1. Benzene	6.8[a]	7.1
2. Naphthalene	8.0	100
3. Benzo(a)pyrene	0.08[b]	0.00005
4. Phenol	6.0	3.8
5. Methyl thiophene	0.3	41
6. Pyridine	0.08	36

[a]Minimum benzene concentrations because of breakthrough.
[b]Quantitation of all mass 252 isomers; if B(a)P is 20% of the total, then the ambient level of B(a)P is ~0.016 $\mu g/m^3$, ~300 x the A-MEG value.

In Table 9 several of the measured compound concentrations in the ambient air are compared with the corresponding Multimedia Environmental Goals (MEG) Ambient Level Goals (A-MEG) based on health effects. The ambient air pollutant concentrations are derived from the measured values, extrapolated to 100% DW.

The A-MEG concentration goals are probably exceeded in the Kosovo aerosol for benzene, phenol and benzo(a)pyrene.

REFERENCE

1. "Multimedia Environmental Goals for Environmental Assessment, Volumes III and IV," EPA-600/7-79-176a and EPA-600/7-79-176b (1979).

CHAPTER 21

CHARACTERIZATION OF VAPOR- AND PARTICULATE-PHASE ORGANICS FROM AMBIENT AIR SAMPLING AT THE KOSOVO GASIFIER

W. H. Griest, C. E. Higgins, R. W. Holmberg,
J. H. Moneyhun, J. E. Caton, J. S. Wike and R. R. Reagan
    Analytical Chemistry Division
    Oak Ridge National Laboratory
    Oak Ridge, Tennessee 37830

INTRODUCTION

The ambient air sampling and analysis of vapor-phase and particulate-phase organics is necessary to detect organic emissions from heavy industry. Although there are no universally validated and accepted methods for achieving this goal, substantial progress has been made in improving the state-of-the-art, and at the same time in providing data useful for environmental assessment.

The role of Oak Ridge National Laboratory (ORNL) in this study sponsored by the U.S. Environmental Protection Agency (EPA) was to design and provide the vapor-phase organics sampling system and also to conduct the particulate-phase organics and vapor-phase organics analyses on samples collected near a Lurgi coal gasifier in Kosovo, Yugoslavia. This chapter describes the methodology and results from the ORNL part of this study.

EXPERIMENTAL

Samples

The samples analyzed in this study were collected in Kosovo, Yugoslavia, under the direction of personnel from the EPA and the Radian Corporation, as described elsewhere in this volume. Twenty-two air particulate filter pads and forty-two Tenax vapor-phase organic traps (the latter supplied by this laboratory for the sampling) from sampling days 9 through 12 of the collection campaign in Kosovo were shipped to Oak Ridge for analysis.

Analysis of Vapor-Phase Organics

Each Tenax vapor-phase organics trap was unloaded in a clean-air box into a tared vial, and the Tenax was well mixed before analysis. A weighed aliquot ($\sim$100 mg) of the homogenized sample was placed in an 11-cm x 9-mm-o.d. x 5-mm-i.d. Pyrex$^{\circledR}$ desorption tube having a tapered end [1, 2]. Solvent washed and thermally conditioned glass wool held the Tenax sample in the tube and separated it from an additional 100^+ mg of Tenax used as a back-up trap to catch sample components which might elute during the addition of the internal standard.

A known weight of the internal standard, 1,2,4-trichlorobenzene, was added in carbon disulfide solution (1 to 2 μL) pipetted onto the glass wool in front of the sample while pulling clean air through from the opposite end at 100–160 mL/min for 10 minutes. This transferred the standard to the Tenax and removed most of the carbon disulfide without loss of any except the most volatile sample components.

Thermal desorption [1, 2] without split was performed in the injector port of a Perkin-Elmer Model 900 GC. The tapered end of the desorption tube was held tightly against the tapered end of the injector port barrel by a spring between the tube and the Teflon$^{\circledR}$*-lined injector port cap. The desorbed vapor phase organics were cryogenically ($-70°$C) trapped at the head of a glass capillary column. After a 15-min desorption at $250°$C the trap was removed. The port was recapped, and the components were chromatographed. The chromatographic conditions were as follows: the column was 66 m x 0.25 mm i.d. HCl-etched [3] soft glass, coated statically [4] with 0.21% UCON 50 HB 660; the injection and manifold temperatures were $250°$C; the carrier was helium at 24 psig to give a flow of 1.0 mL/min at $25°$C; the column oven temperature program was 16 min at $-70°$C, then $24°$C/min to $20°$C, then $2°$C/min to $160°$C with a 30-min hold at the latter temperature.

---

*Registered trademark of E. I. du Pont de Nemours & Company, Inc., Wilmington, Delaware.

Components eluting from the column were measured by flame ionization. A PDP 8/e computer calculated peak areas and directed a printout on a teletypewriter.

Peak identification was made by comparing retention times with known standards and by gas chromatography/mass spectrometry (GC/MS). Known organic compounds were desorbed and chromatographed under the same conditions as the sample. GC/MS was performed using the same capillary column but under necessarily slightly different run conditions in a Hewlett-Packard model 5985 GC/MS. Quantitation was made by the method of internal standards. Authentic standard mixtures were run every three days between sample runs.

Analysis of Particulate-Phase Organics

The air particulate sampling pads were stored at $-30°C$ after receipt from the sampling site. Extraction was performed by spiking with 80,000 dpm of carbon-14 benzo(a)pyrene (^{14}C-BAP) and performing three ultrasonic extractions, each with 100 mL of benzene, using a waterjacketed sintered glass funnel held at $20°C$ and a 350-watt Branson ultrasonic cell disruptor. The volume of the combined solvent extracts was reduced to 0.5 mL by evaporative concentration with dry, flowing nitrogen under reduced temperature and pressure, and the concentrate was analyzed by gas chromatography (GC), utilizing a 3 m x 2 mm i.d. glass column packed with 3% Dexsil 400 on 100/120 mesh Supelcoport, temperature programmed from 100 to $320°C$ at $2°C/min$, and flame ionization detection. The extraction recovery was estimated by liquid scintillation spectrometric measurements [5] of the ^{14}C-BAP tracer in a portion of the extract. Quantitation was by the method of external standards.

The solvent extracts were fractionated by semipreparative-scale high-performance liquid chromatography (HPLC) on a 25-cm x 1-cm Whatman PAC column (cyano/amino silane bonded to 10-μm silica) using the following solvent step-gradient: fractions 1 and 2, hexane (30 mL each); fraction 3, hexane/methylene chloride (50/50 by volume, 60 mL total); fraction 4, hexane/methylene chloride (10/90, 60 mL); and fraction 5, isopropanol/hexane/methylene chloride (80/10/10, 60 mL). Each fraction was concentrated to 0.2 mL in CH_2Cl_2 along with a known amount of n-$C_{34}H_{70}$ internal standard and was analyzed by GC on a 50-m OV-101 fused-silica capillary column, using a column temperature program of $50°C$ (5 min) to $100°C$ at $25°C/min$, and then to $280°C$ at $2°C/min$, splitless injection and flame ionization detection. GC/MS was conducted under similar conditions on a Hewlett-Packard model 5985. Fraction 2 was analyzed by analytical HPLC using a Du Pont Zorbax ODS C_{18} reversed-phase column, an acetonitrile/water (70/30) mobile phase, and fluorescent detection (λ_{ex} = 360 nm, λ_{em} = 425 nm). Recovery of the second fraction was calculated from liquid scintillation counting the ^{14}C-BAP tracer in an aliquot.

RESULTS AND DISCUSSION

Existing vapor-phase organics sampling systems were found unsuitable for the Yugoslavian sampling requirements because of their large size, the need to ship the entire sampling system back to the laboratory for sample analysis, and also because of difficulties in their disassembly and reassembly and their tendency to leak. Because of these deficiencies, an improved sampling system was designed at Oak Ridge, and the necessary number of systems and resin traps were constructed for the sampling in Yugoslavia.

The vapor-phase sampling system diagrammed in Figure 1 is designed to mount onto the side of a high volume (Hi-Vol) air sampler and to collect vapor-phase organics passing through or subliming from the Hi-Vol filter pad into the sampler head. The manifold of the vapor-phase sampling system connects to the Hi-Vol head through a flexible metallic braided hose, and is driven with a separate pump at 3-4 L/min. The system features glass, Teflon and stainless steel construction, and dual parallel sampling channels with independent flow control for each channel to allow simultaneous sampling in duplicate or individual samples in time sequence without changing traps. The system also has a ball-type rotameter to allow monitoring of sampling flowrates to 12 L/min. Each sampling channel has two adsorbent traps in series constructed of 15-cm x 1.5-cm-i.d. glass pipe. In this study, each trap was filled with 17 mL of 35/60 Tenax resin, but different resins could be employed in the front and backup traps.

This sampling system is lightweight, easily assembled and leak-checked, and easily used. It offers the advantages over most previous systems in that only the sampling tubes themselves (sealed with Teflon end caps) need to be sent to and from the sampling site. In addition, any flowrate to 12 L/min can easily be set and controlled onsite.

The vapor-phase organics collected by the Tenax traps can be conveniently analyzed by thermal desorption GC. Our new procedure provides an accurate and precise means of measuring these species. As opposed to previous methods (e.g., 1, 2) which allow only a "one shot" analysis of the vapor-phase traps, this new procedure allows repetitive analyses (up to 20–30) or multiple different analyses (e.g., with different element-selective detectors) to be performed on each Tenax trap.

The entire analytical procedure for vapor-phase organics, as described in the Experimental Section, was evaluated in two separate experiments. In the first, the accuracy of the procedure for measuring volatile organics collected

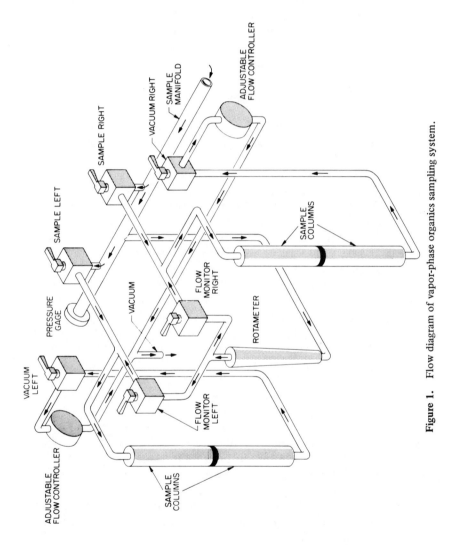

Figure 1. Flow diagram of vapor-phase organics sampling system.

on Tenax was determined by applying in separate runs, 2 μg, 8 μg, 40 μg, and 2 mg each of benzene, toluene, m-xylene and naphthalene to a Tenax trap. These compounds are representative of those found in actual field sampling, and the 2–40 μg amounts of each span the range which has been collected in the field during 24-hr sampling. The 2-mg loading represents a hypothetical "worst case" in air sampling. Recoveries by the analytical procedure ranged from 78 to 123%, with relative standard deviations usually <10%, demonstrating the accuracy and precision of this multi-component analytical procedure.

Repetitive analyses of standards and actual samples collected in the field also provide an evaluation of precision. The relative standard deviation of replicate internal standard runs of an authentic standard over a period of three weeks averaged 3%; however, a more realistic evaluation is obtained with "real world" samples. Repetitive external standard analyses of a sample collected in the field produced relative standard deviations of 8-15% for major species, and with internal standardization, precision averaged 3-8%. However, for minor peaks in the chromatogram where reproducible electronic integration is difficult, precision may degrade considerably.

The application of this analytical procedure to the compounds collected on two of the Tenax traps in the Yugoslavian sampling campaign is illustrated by the chromatograms in Figure 2. The upper trace is the chromatogram from the analysis of the front trap at a predominantly upwind sampling site, the middle chromatogram corresponds to a predominantly downwind sampling site, and the bottom trace is from a blank trap carried to the sampling site and back unopened. The large peak at 80 minutes is the added internal standard. More than 100 vapor-phase organic components are visualized by the procedure; the major components identified by GC retention and/or mass spectrum are identified in Table 1 along with their air concentrations. Many of these same compounds have been observed [6, 7] in other urban air studies, and their concentrations fall within the ranges reported [6] for urban air. Toluene is seen to be one of the major vapor-phase constituents. Use of a more efficient resin in the trapping system (see below) for the very volatile constituents probably also would indicate benzene as another major constituent. Phenol and the cresols appear to stand out as unusual in these samples; a recent summary [8] of organic particulate pollutants observed that such simple phenols are rarely observed in urban atmospheres. Although they are volatile and thus probably are lost from particulates by sublimation, they could be characteristic of this coal gasifier source.

The upwind and downwind samples differ quantitatively mainly in the less volatile components such as naphthalene, phenol and the cresols, which elute toward the end of the chromatogram. There are lesser (and more variable from sampling day to day) differences for the monoaromatics. Although the

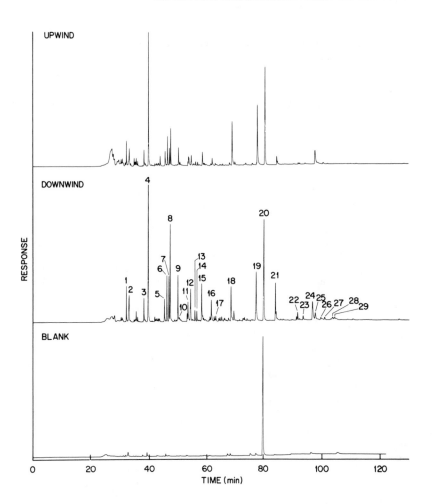

Figure 2. Capillary column GC profiles of vapor-phase organics collected on Tenax traps (see Table 1 for peak identification).

heavier species are more consistently concentrated in the downwind air samples, their air concentrations are lower than those of the more volatile alkyl benzenes. The average downwind/upwind enhancement is listed in Table 1 also. It would appear from these data that the gasifier at Kosovo is increasing the ambient air concentrations of these species by a factor of 2 to 4 between the upwind and downwind sampling points.

The efficacy of the Tenax vapor-phase traps is illustrated in the chromatograms in Figure 3, showing the vapor-phase organics collected in 24 hours from 5 m³ of air by a front trap (upper trace) and its backup trap (lower

Table 1. Vapor-Phase Organics Determined by Thermal Desorption GC

Peak No.[a]	Tentative Identification	Range of Concentrations[b] ($\mu g/m^3$)	Downwind Enrichment[c]
1	Methylene chloride[d]	–	–
2	Benzene	0.33–1.8	2.0
3	$n\text{-}C_9H_{20}$	0.16–1.0	1.1
4	Toluene	0.74–9.0	1.3
5	$n\text{-}C_{10}H_{22}$	0.16–0.60	1.2
6	Ethylbenzene	0.46–1.3	1.2
7	p-Xylene	0.20–1.3	1.7
8	m-Xylene	0.38–3.2	1.8
9	o-Xylene	0.24–1.6	1.7
10	Cumene	0.02–0.38	2.5
11	C_3-Benzene isomer	0.11–0.52	2.1
12	C_3-Benzene isomer	0.25–2.0	1.9
13	Mesitylene	0.06–0.58	2.0
14	C_3-Benzene isomer	ND–0.51	2.1
15	C_3-Benzene isomer	0.21–2.2	1.7
16	C_3-Benzene isomer	0.10–0.81	2.2
17	o-Methylstyrene	ND–0.11	$\gg 1.2^e$
18	Benzaldehyde	1.1 –2.8	0.9
19	Acetophenone	1.3 –3.0	0.9
20	1,2,4-Trichlorobenzene[f]	–	–
21	Naphthalene	0.02–1.5	2.4
22	2-Methylnaphthalene	0.03–0.25	4.5
23	1-Methylnaphthalene	0.01–0.15	4.9
24	Phenol	0.16–2.3	1.8
25	o-Cresol	ND–1.0	4.4
26	Biphenyl	0.04–0.09	1.3
27	Indole	0.02–0.13	2.7
28	p-Cresol	ND–0.24	$\gg 1.6^e$
29	m-Cresol	ND–0.36	$\gg 4.3^e$
30	p-Ethylphenol[g]	ND–0.16	$\gg 3.9$

[a]Figure 2.
[b]Range of air concentrations for all sampling sites. ND = not detected.
[c]Average downwind/upwind concentration ratio.
[d]Contaminant.
[e]Not detected at the upwind sampling site on some sampling days.
[f]Internal standard.
[g]Not shown in Figure 2.

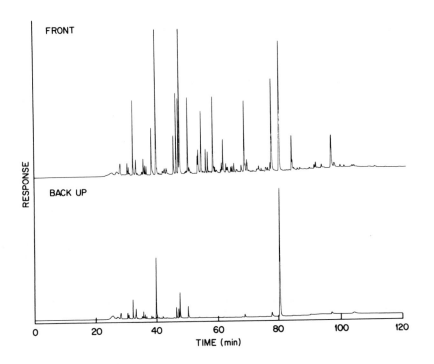

Figure 3. Capillary column GC profiles of vapor-phase organics collected on front and backup Tenax traps (same time scale and peak identification as in Figure 2).

trace). Considerable breakthrough from the front trap is evident for benzene and toluene (to 100% breakthrough in some samples), but breakthrough decreases to well below 10% for *p*-xylene, and is essentially insignificant for less volatile constituents. It is most likely that benzene and toluene also break through the backup trap. The collection efficiency for benzene and toluene possibly could be improved by employing a carbonaceous adsorbent, such as Ambersorb®, in the backup trap.

The particulate-phase organics are extracted from the Hi-Vol filter pads by ultrasonication in benzene. This solvent yields good extraction recoveries of many organics [9, 10] and the ultrasonic procedure is much more rapid and efficient than the classical Soxhlet procedure [11, 12]. Most of the ^{14}C-BAP extraction recoveries from the filter samples collected in Yugoslavia were quantitative.

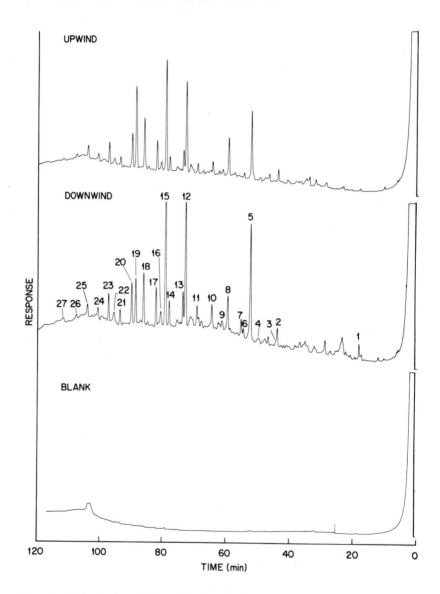

Figure 4. Packed column GC profiles of particulate-phase organics (peak identifications in Table 2).

Direct packed column GC profiling of the concentrated extracts provides a ready means of determining the major gas chromatographable organics collected by the Hi-Vol filters. Figure 4 shows the GC profiles of an upwind particulate extract (top trace), a downwind particulate extract (middle trace) and a blank filter pad extract (bottom trace). As indicated in Table 2, a rather wide range of paraffins ranging from n-C_{19} to at least n-C_{35}, some PAH, n-heterocyclics, and a phthalate are visualized, as others have reported [8, 13, 14] for particulates collected elsewhere. Air concentrations of these particulate-phase organics are 2–3 orders of magnitude lower than those of the vapor-phase organics.

The direct GC profiles of the particulate-phase organics do not show as great differences between upwind and downwind sampling sites as do the vapor-phase organics profiles. Although the average downwind/upwind ratio is more than 1.0, there was considerable variation among sampling days, and the ratio probably is not significantly different from 1.0. This observation may reflect a more gaseous nature of the gasifier emissions, or possibly the appropriate classes of particulate-phase organics are not being visualized in the GC of the crude extract. Indeed, some classes of organics (e.g., fatty acids) must be derivatized before GC and thus are not detected by this procedure. These compounds might be visualized if an additional derivatization step were included in the protocol. Some workers have found [15] only PAH concentrations were different between particulates collected at some urban and rural background sites.

Fractionation and analysis of the separated fractions does provide a powerful tool for visualizing and determining industrial emissions in the complex mixture of ambient air particulates. HPLC is a rapid and convenient means [16] of fractionating complex mixtures for further analysis. The benzene extracts of the filter pads were separated in this study by semipreparative scale HPLC into five fractions on the basis of polarity, quite similarly to the procedure used by others [16] for fractionating shale oil. Although all five fractions from the Yugoslavian particulate extracts were profiled by capillary column GC, the second and third fractions showed identifiable differences worthy of note.

The second fraction contains PAH and some phthalates which are best analyzed by HPLC because of their low concentrations in the ambient air samples. Figure 5 shows an isocratic reversed-phase HPLC separation of the PAH in one filter pad fraction. The fluorescence detector easily provides the necessary sensitivity, while the reversed-phase column allows the important separation of the benzofluoranthenes and benzopyrenes. Quantitative data for some PAH are included in Table 2, and the results indicate an $\sim$ 1.5- to 3-fold increase in PAH air concentrations between the upwind, off-wind and downwind sampling site.

These quantitative data for PAH and n-paraffin air concentrations fall within the ranges summarized [8] for the "average American urban atmosphere." To place the results in context with other sites or emission

Table 2. Particulate-Phase Organics Determined in the Crude Filter Extract
and in the PAH Isolate of the Extract

Peak No.[a]	Tentative Identification	Range of Concentrations[b] (ng/m^3)	Downwind Enrichment[c]
	Crude Extract:		
1	Biphenyl	0.3–4.2	1.0
2	n-$C_{19}H_{40}$	1.8–11	1.1
3	Phenanthrene	_d	–
4	n-$C_{20}H_{42}$	0.44–2.0	1.2
5	C_{14}-Benzene isomer	_e	–
6	n-$C_{21}H_{44}$	1.0–4.7	1.5
7	C_{14}-Benzene isomer	_e	–
8	n-$C_{22}H_{46}$	8.5–28	1.0
9	Fluoranthene (+ hydrocarbon)	_d	–
10	n-$C_{23}H_{48}$	5.4–13	1.3
11	n-$C_{24}H_{50}$	1.6–8.8	1.6
12	MW 256 + 274	–	–
13	n-$C_{25}H_{52}$	6.2–18	1.2
14	n-$C_{26}H_{54}$	3.9–16	1.4
15	bis-(2-Ethylhexyl) phthalate	43–120	1.0
16	MW 226	–	–
17	n-$C_{27}H_{56}$	19–40	1.2
18	n-$C_{28}H_{58}$	13–42	1.1
19	C_4-Quinoline	_e	–
20	n-$C_{29}H_{60}$	11–21	1.1
21	n-$C_{30}H_{62}$	2.2–7.9	1.2
22	Benzo(b,j, or k)fluoranthene (+ hydrocarbon)	_d	–
23	n-$C_{31}H_{64}$	7.4–13	1.1
24	n-$C_{32}H_{66}$	1.4–7.2	1.6
25	n-$C_{33}H_{68}$	2.2–6.5	1.1
26	n-$C_{34}H_{70}$	1.1–3.6	0.9
27	n-$C_{35}H_{72}$	0.8–2.9	1.1
	n-$C_{36}H_{74}$	ND-2.7	1.1
	PAH Isolate:		
_f	Fluorene	0.05–0.21	_g
–	Fluoranthene	0.05–1.2	>1.4
–	Benzo(b)fluoranthene	0.5–2.7	1.5–2.7
–	Benzo(k)fluoranthene	0.3–1.5	1.0–2.4
	Benzo(a)pyrene	0.3–0.9	1.0–1.5
	Benzo(ghi)perylene	0.4–2.1	_g

[a]Figure 4.
[b]Range of air concentrations for all sites. ND = not detected.
[c]Average downwind/upwind concentration ratio, except for PAH (range of all sites).
[d]Incomplete resolution prevents quantitation.
[e]No authentic standard available for quantitation.
[f]See Figure 5.
[g]Variable.

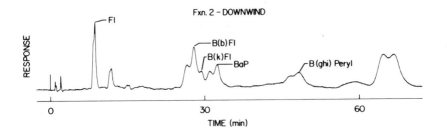

Figure 5. HPLC separation of fraction 2 from a downwind air particulate sample. (Fl = fluoranthene, B(b or k)Fl = benzo(b or k)fluoranthenes, BAP = benzo(a)-pyrene, B(ghi)Peryl = benzo(ghi)perylene).

sources, the PAH concentrations are lower than those found in coke oven emissions [17] or in automotive tunnels [18]. Because the highly specific HPLC analytical procedure does not provide as detailed information on PAH content as does GC, the second fractions were composited and the PAH from eight samples were reisolated by the semipreparative scale HPLC procedure to provide sufficient sample for analysis by capillary column GC. The PAH that were identified by GC retention and/or GC/MS are listed in Table 3 along with their relative concentrations in the combined sample, which included both upwind, downwind and off-wind sampling sites. The major PAH are seen to consist of the parent PAH; some simply alkylated (predominantly methyl) PAH also are present. Most of these PAH have been reported previously (e.g., [8, 14]). The observation of phenyl naphthalenes is of interest, but must be considered tentative at the present.

The third fraction is important from its apparent n-heterocyclic content. Figure 6 shows that a downwind particulate sample is enriched in these species relative to an upwind sample. GC/MS examination of the isolates suggested the presence of C_7- through C_9-, C_{12}-, and C_{13}-pyridines and C_3- (2 isomers), C_4- (2 isomers) and C_8-quinolines. These identifications must be considered tentative in the absence of confirming data. The apparent lack of intermediate members (e.g., C_{10}- and C_{11}-pyridines) in the homologous series of alkyl derivatives is surprising. Also, most investigations of n-heterocyclics in air particulate matter (albeit, urban sources) have reported (e.g., [8]) only simply alkylated quinolines, acridines and their benzo-derivatives. Because no radio-tracer was included for these species in the protocol, and because authentic standards are not available, only relative air concentrations were calculated. These showed considerable variation in the downwind/upwind enrichment ratio (from 0.3 to >20), probably because equivalent isolation recoveries were not achieved for each sample. However, the approximate "average" increase was a factor in the vicinity of three.

Table 3. GC/MS Analysis of PAH Fractions Combined from Eight
Yugoslavian Gasifier Particulate-Phase Air Samples

Tentative Identification	Criteria		Relative Concentration
	GC/RT	GC/MS	
Naphthalene	†		<0.01
2-Methylnaphthalene	†		<0.01
1-Methylnaphthalene	†		0.01
Biphenyl	†		0.01
Phenanthrene	†	†	0.1
Anthracene	†		0.02
1-Phenylnaphthalene	†	†	0.04
1-Methylphenanthrene	†	†	0.04
2-Methylanthracene	†	†	0.04
9-Methylphenanthrene	†		0.04
2-Phenylnaphthalene	†	†	0.05
C_2-Phenanthrene		†	0.04
Fluoranthene	†	†	1.0
Pyrene	†	†	0.9
C_1-Pyrene isomer		†	0.07
C_2-Pyrene isomer		†	0.08
Benzo(a)fluorene	†	†	0.4
Benzo(b)fluorene	†	†	0.3
3-Methylpyrene	†	†	0.1
Benzo(ghi)fluoranthene	†	†	0.4
Benz(a)anthracene	†	†	0.8
Chrysene	†	†	0.9
Benzo(b)fluoranthene	†	†	2.1
Benzo(e)pyrene	†	†	0.9
Benzo(ghi)perylene	†	†	0.9

[a]Concentration relative to fluoranthene.
[b]Possibly b, j and k isomers.

SUMMARY

The results of this study suggest that the coal gasifier located near Kosovo, Yugoslavia, is emitting alkyl benzenes, diaromatics, PAH, phenols and n-heterocyclics to the atmosphere. However, their air concentrations appear to be similar to those found in a typical urban American atmosphere, with the possible exception of simple phenols.

The methodology developed in this study should be useful in the examination of atmospheric samples from other sources, and other complex mixtures such as coal- or shale-derived oils.

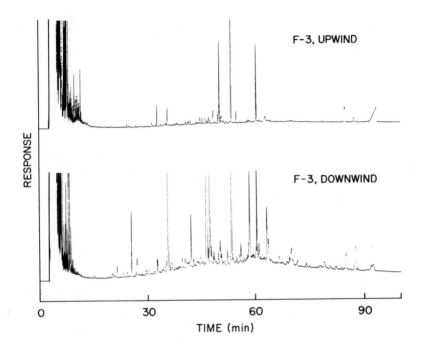

Figure 6. Capillary column GC separation of fraction 3 from upwind (upper trace) and downwind (lower trace) air particulate samples.

ACKNOWLEDGMENT

This work was sponsored by the Aerosol Research Branch, Division of Atmospheric Chemistry and Physics, Environmental Science Research Laboratory, U.S. Environmental Protection Agency, R. K. Patterson, Project Officer, under Union Carbide Corporation contract W-7405-eng-26 with the U.S. Department of Energy.

REFERENCES

1. Zlatkis, A., H. A. Lichtenstein and A. Tishbee. "Concentration and Analysis of Trace Volatile Organics in Gases and Biological Fluids with a New Solid Adsorbent," *Chromatographia* 6:67–70 (1973).
2. Zeldes, S. G. and A. D. Horton. "Trapping and Determination of Labile Compounds in the Gas Phase of Cigarette Smoke," *Anal. Chem.* 50: 779–782 (1978).

3. Schomburg, G., et al. "Gas Chromatographic Analysis with Glass Capillary Columns," *J. Chromatog.* 122:55–72 (1976).

4. Bouche, J., and M. Verzele. "A Static Coating Procedure for Glass Capillary Columns," *J. Gas Chromatog.* 6:501–505 (1968).

5. Griest, W. H., L. B. Yeatts, Jr., and J. E. Caton. "Recovery of Polycyclic Aromatic Hydrocarbons Sorbed on Fly Ash for Quantitative Determination," *Anal. Chem.* 52:199–201 (1980).

6. Louw, C. W., J. F. Richards and P. K. Faure. "The Determination of Volatile Organic Compounds in City Air by Gas Chromatography Combined with Standard Addition, Selective Subtraction, Infrared Spectrometry and Mass Spectrometry," *Atmos. Env.* 11:703–717 (1977).

7. Bunn, W. W., E. R. Deane, D. W. Klein and R. D. Kleopfer, "Sampling and Characterization of Air for Organic Compounds," *Water Air Soil Poll.* 4:367–380 (1975).

8. Hoffman, D. and E. L. Wynder. In: *Air Pollution, Vol. II,* 3rd ed., A. C. Stern, Ed., (New York: Academic Press, 1977), pp. 361–455.

9. Cautreels, W., and K. Van Cauwenberghe. "Extraction of Organic Compounds from Airborne Particulate Matter," *Water Air Soil Poll.* 6: 103–110 (1976).

10. Stanley, T. W., J. E. Meeker and M. J. Morgan. "Extraction of Organics from Airborne Particulates. Effects of Various Solvents and Conditions on the Recovery of Benzo(a)pyrene, Benz(c)acridine, and 7H-Benz(de)-anthracen-7-one," *Environ. Sci. Technol.* 1:927–931 (1967).

11. Chatot, G., et al. "Etude Comparee Des Ultra-Sons Et Du Soxhlet Dan L'Extraction Des Hydrocarbons Polycycliques Atmospheriques," *Anal. Chim. Acta* 53:259–265 (1971).

12. Golden, G., and E. Sawicki. "Ultrasonic Extraction of Total Particulate Aromatic Hydrocarbons (TpAH) from Airborne Particles at Room Temperature," *Int. J. Environ. Anal. Chem.* 4:9–23 (1975).

13. Cautreels, W., and K. Van Cauwenberghe. "Fast Quantitative Analysis of Organic Compounds in Airborne Particulate Matter by Gas Chromatography with Selective Mass Spectrometric Detection," *J. Chrom.* 131: 253–264 (1977).

14. Lao, R. C., et al. "Application of a Gas Chromatograph-Mass Spectrometer-Data Processor Combination to the Analysis of the Polycyclic Aromatic Hydrocarbon Content of Airborne Pollutants," *Anal. Chem.* 45:908–915 (1973).

15. Cautreels, W., K. Van Cauwenberghe and L. A. Guzman. "Comparison Between the Organic Fraction of Suspended Matter at a Background and an Urban Station," *Sci. Total Environ.* 8:79–88 (1977).

16. Hertz, H. S., et al. "Determination of Individual Organic Compounds in Shale Oil," *Anal. Chem.* 52:1650–1657 (1980).

17. Broddin, G., L. Van Vaeck and K. Van Cauwenberghe. "On the Size Distribution of Polycyclic Aromatic Hydrocarbon Containing Particles from a Coke Oven Emission Source," *Atmos. Environ.* 11:1061–1064 (1977).

18. Fox, M. A., and S. W. Staley. "Determination of Polycyclic Aromatic Hydrocarbons in Atmospheric Particulate Matter by High Pressure Liquid Chromatography Coupled with Fluorescence Techniques," *Anal. Chem.* 48:992–998 (1976).

CHAPTER 22

CARBONACEOUS AEROSOL
IN THE VICINITY OF A LURGI GASIFIER

James J. Huntzicker, Richard L. Johnson,
Jitendra J. Shah and Emily K. Heyerdahl
    Department of Environmental Science
    Oregon Graduate Center
    Beaverton, Oregon 97006

INTRODUCTION

The use of synthetic fuels as substitutes for oil is a potential solution to the energy dilemma facing the United States. Before this can become a practical solution, however, the environmental impacts of synthetic fuel production must be evaluated. One process currently in use is the production of substitute natural gas from lignite or coal by the Lurgi process. This chapter reports on measurements of carbonaceous aerosol in the vicinity of a Lurgi gasifier located at the Kosovo industrial complex near Pristina, Yugoslavia. In this plant Kosovo lignite is converted into fuel gas and fertilizer plant feedstocks. A schematic of the complex is shown in Figure 1.

ANALYTICAL

During May 1979, samples on quartz fiber filters were collected by U.S. Environmental Protection Agency (EPA) and Radian Corporation personnel at ~6-hr intervals at the five sites shown in Figure 1. Each site was equipped

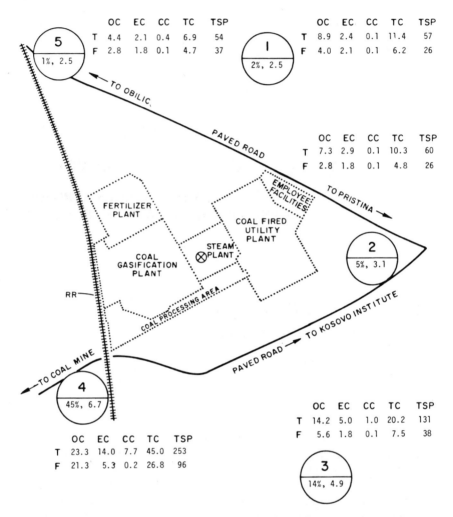

Figure 1. Carbonaceous aerosol concentrations at the Kosovo industrial complex. T = total; F = fine; OC = organic carbon; EC = elemental carbon; TC = total carbon (all carbon concentrations as μg C/m^3); TSP = aerosol concentration in μg/m^3. The upper number in the circles is the site number, and the lower numbers are the per-cent of time the site was downwind from the x (steam plant) and the average wind speed (km/hr). These data are for the period 18–24:00, May 19, 1979.

with two samplers: one that sampled total suspended aerosol (i.e., for $d_p \lesssim$ 15 μm) and one that sampled fine aerosol (i.e., $d_p \leqslant 2$ μm). The 2-μm cut was achieved with a cyclone separator.

Samples provided by Radian Corporation were analyzed at the Oregon Graduate Center for organic, elemental and carbonate carbon. The carbon analysis method combines both thermal and optical processes and has been described in detail elsewhere [1–3]. A filter sample (as four 0.25 cm^2 disks) is inserted into a 350°C oven in a 2% O_2-98% He atmosphere to volatilize organic carbon. The volatilized carbon is oxidized to CO_2 in a MnO_2 bed, reduced to CH_4 in a Ni/firebrick (450°C) bed, and measured in a flame ionization detector. The oven is next purged with He to remove all O_2, and the temperature is raised to 600°C to volatilize any remaining organic carbon. This is measured as above. The carrier gas is then changed to 2% O_2-98% He, and elemental carbon is combusted to CO_2 and measured. Finally, the response of the system to a known amount of CH_4 gas is measured as a calibration at the end of each run. All valve switching operations, temperature control and timing are under the control of a microcomputer built around a Motorola 6802 microprocessor.

During the 600°C volatilization in He some of the organic carbon on the filter is pyrolytically converted to elemental carbon. This effect, which is manifested by a darkening of the filter, is continuously monitored during the thermal analysis with a laser reflectance system. For the concentrations of elemental carbon encountered in this experiment the logarithm of the relative reflectance (i.e., the filter reflectance normalized to the reflectance after elemental carbon combustion) varied approximately linearly with elemental carbon concentration (μg C/cm^2) on the filter. This relationship and the measured values of relative reflectance initially and after the 600°C/He step (i.e., after the pyrolytic conversion had occurred) were used to determine the correction for the pyrolytic production of elemental carbon. Uncertainties in the resultant organic and elemental carbon concentrations are estimated to be ±25%. (In the current version of the carbon analyzer [2, 3] the pyrolysis correction is determined by measuring the amount of elemental carbon oxidation necessary to return the filter reflectance to its initial value.)

Carbonate carbon was determined by acidification of the filter with 20 μL of 1% H_3PO_4 and measurement of the evolved CO_2 after reduction to CH_4. Because carbonate carbon is also measured in the organic mode in the thermal-optical procedure, the organic carbon concentration must be corrected by subtracting the carbonate carbon concentration.

RESULTS

From the five sites shown in Figure 1, 320 samples were analyzed for organic, elemental and carbonate carbon. Of these sites, Site 5 was most remote relative to the gasifier. Site 1 was also relatively remote although it was located near a home and a dirt road. Site 2 was on the grounds of the Kosovo Institute, and Site 3 was in a village. Site 4, which was the principal downwind site, was located near a railroad track and train station. The track carried 48 trains per day—27 diesel and 21 coal-fired steam locomotives. Two lignite conveyors were located just to the north of Site 4. A typical set of results is shown in Figure 1.

Total carbon concentrations at Sites 4 and 5 are compared in Figure 2. Site 4, the principal downwind site, is a very dirty site with carbon concentrations rising to almost 90 μg C/m^3. The origin of the large peaks at Site 5, however, is not clear. The numbers listed by the data points correspond to

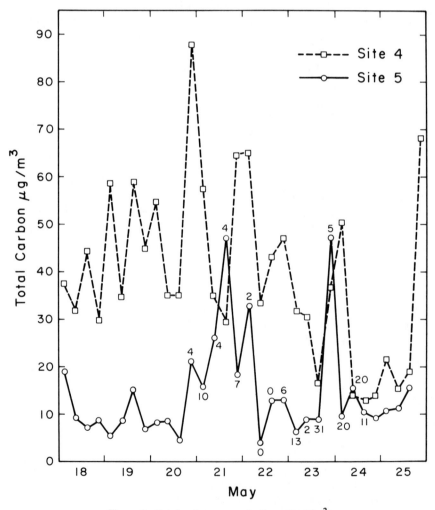

Figure 2. Total carbon concentrations (μg C/m^3).

the percentages of time Site 5 was downwind. No correlation between carbon maxima and these numbers is apparent. It is therefore likely that the peaks are due to local influences.

The diurnal behaviors of coarse particle (i.e., total–fine) organic, elemental and carbonate carbon at Site 4 are plotted in Figure 3. The most prominent

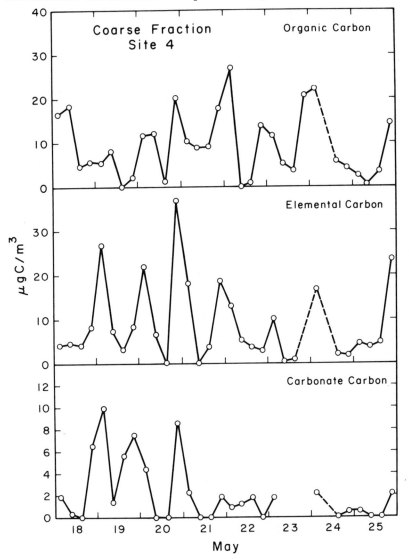

Figure 3. Organic, elemental and carbonate carbon concentrations in coarse aerosol at Site 4.

feature is the nighttime peaking of elemental carbon. This peaking is also observed to some extent for organic and carbonate carbon. Because the mineral matter associated with Kosovo lignite is rich in carbonate [4], high carbonate concentrations are indicative of blowing lignite dust. Thus, the first three

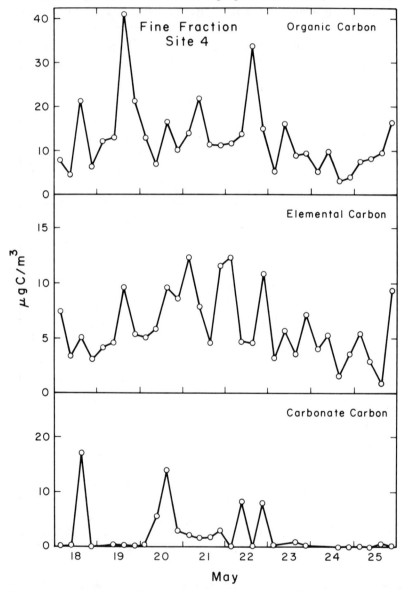

Figure 4. Organic, elemental and carbonate carbon concentrations in fine aerosol at Site 4.

peaks probably are associated with fugitive lignite dust although the reason for the nocturnal peaking is not known.

In Figure 4 the carbon concentrations in the fine particle fraction of the aerosol at Site 4 are plotted. In contrast to the coarse particle situation the most prominent periodicity is a midday peaking of the organic carbon. Again, the cause of this peaking is not known but is probably related to activities within the industrial complex. A more detailed chemical analysis of the compounds present during these periods would undoubtedly yield pertinent information concerning the sources of the carbon.

The peaking in the fine particle carbonate is not expected because the mineral origin of the carbonate implies large particles. Thus, contamination must be suspected for several of the samples.

Figure 5 compares the diurnal behaviors of coarse and fine organic carbon at Site 4. Inspection indicates that the two fractions are anticorrelated, and,

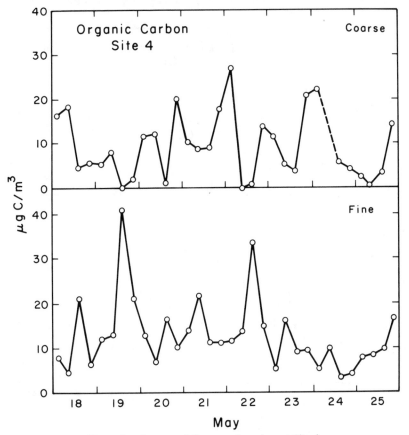

Figure 5. Coarse and fine organic carbon at Site 4.

in fact, the linear regression correlation coefficient between the two is –0.28. This is indicative of different sources for the fine and coarse organic carbon. A small positive correlation ($r = 0.37$) is observed for the coarse and fine elemental carbon (Figure 6). Such a correlation coefficient, however, implies little commonality between the sources of coarse and fine elemental carbon.

Likely sources for elemental carbon at Site 4 include fugitive lignite dust, which should be primarily in the coarse fraction, and combustion emissions. (In the carbon analysis system used for these measurements the noncarbonate carbon in one Kosovo lignite sample analyzed as 40% organic and 60% elemental.) Linear regression analysis between coarse elemental carbon and coarse carbonate carbon, a tracer for lignite dust, gave a correlation coefficient of 0.48. This implies that only 23% of the coarse elemental carbon variance was associated with carbonate carbon. Thus, the only firm conclusions which can be drawn are that fine and coarse carbonaceous particles at Site 4 result from different sources and that sources in addition to fugitive

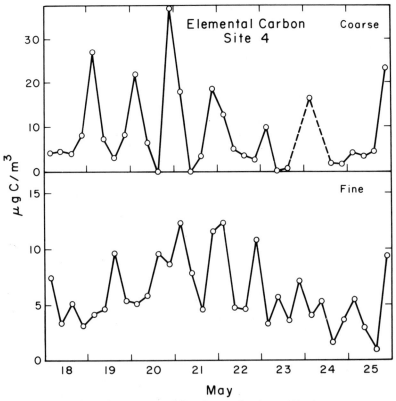

Figure 6. Coarse and fine elemental carbon at Site 4.

Table 1. Correlation Between Coarse and Fine Fractions

| Sites | Linear Regression Correlation Coefficient, r | |
	Organic	Elemental
Site 4	-0.28	0.37
Sites 1, 2, 3, 5 ≥10% Downwind	0.06	0.12
Site 5	0.70	0.40

Table 2. Correlation Between Organic and Elemental Carbon

| Sites | Linear Regression Correlation Coefficient, r | |
	Fine	Coarse
Site 4	0.43	0.55
Sites 1, 2, 3, 5 ≥10% Downwind	0.56	0.56 (0.35)[a]
Sites 1, 2, 3, 5 <5% Downwind	0.48	0.70
Site 5	0.73	0.77

[a]The number in parentheses is the correlation coefficient with one extreme point removed.

dust also contribute to the coarse carbon concentrations. The complexity of Site 4 is undoubtedly a complicating factor in the interpretation of the data.

The results of further correlation analysis between the fine and coarse fractions are presented in Table 1. The low correlation coefficients observed at Sites 1, 2, 3 and 5 when the percent of time downwind was at least 10% again show that fine and coarse carbonaceous aerosols coming from the Kosovo complex originate from different sources. The only large correlation coefficient is between fine and coarse organic carbon at Site 5, the "background" site.

Linear regression analysis between organic and elemental carbon concentrations at the different sites and in the different size fractions was also performed. The resultant correlation coefficients are given in Table 2. All categories show positive correlation coefficients—indicating some degree of commonality between the sources of organic and elemental carbon, but only

Table 3. Average Organic Carbon Concentrations at Downwind and Upwind Sites

| Sites | Organic Carbon Concentrations (μg C/m^3) | | |
	Coarse[a]	Fine[a]	Total
A. Site 4	10.8 (47)	12.2 (53)	23.0
B. Sites 1, 2, 3, 5 ≥10% Downwind	5.8 (45)	7.1 (55)	12.9
C. Sites 1, 2, 3, 5 <5% Downwind	3.5 (44)	4.5 (56)	8.0
Downwind–Upwind (A–C)	7.3 (49)	7.7 (51)	15.0
Downwind–Upwind (B–C)	2.3 (47)	2.6 (53)	4.9

[a]Numbers in parentheses represent the percentages of the respective concentrations in the coarse and fine fractions.

at Site 5 is the organic-elemental correlation strong in both size fractions. If elemental carbon in the fine fraction is taken as a tracer for combustion generated aerosol, 20–50% of the variance associated with fine organic carbon can be attributed to combustion. The value of the correlation coefficient at Site 5 for fine aerosol is similar to that observed at urban sites in the United States [5]. The smaller correlation coefficients in the fine aerosol at the other sites suggest a multiplicity of sources (including local sources) and probably also poor mixing. The correlation coefficients in the coarse fraction are more difficult to interpret because elemental carbon can arise both from combustion as soot and from blowing lignite dust.

In Tables 3–5 the average concentrations of organic, elemental and carbonate carbon for the May 18–25, 1979 period are tabulated for coarse, fine and total aerosol at downwind and upwind sites. The numbers in parentheses represent the percentages of the respective concentrations in the fine and coarse fractions. For the case of organic carbon (Table 3) the split between coarse and fine was approximately the same for the three site categories with slightly more than half of the mass in the fine fraction. For elemental carbon (Table 4), however, 60% of the mass at Site 4 was in the coarse fraction whereas the reverse occurred for the other site categories. This is most likely a consequence of fugitive lignite dust at Site 4. For carbonate carbon (Table 5), the coarse fraction predominated for all site categories. This result is consistent with the large particle, mineral origin of the carbonate. Tables 3–5 also indicate that the highest average concentrations were observed at Site 4, the site most often downwind of the gasifier. As noted above, however, the complexity of both the sampling site itself and the industrial area do

Table 4. Average Elemental Carbon Concentrations at Downwind and Upwind Sites

Sites	Elemental Carbon Concentrations (μg C/m^3)		
	Coarse[a]	Fine[a]	Total
A. Site 4	9.4 (60)	6.2 (40)	15.6
B. Sites 1, 2, 3, 5 $\geqslant$10% Downwind	2.6 (44)	3.3 (56)	5.9
C. Sites 1, 2, 3, 5 $<$5% Downwind	1.3 (39)	2.0 (61)	3.3
Downwind–Upwind (A–C)	8.1 (66)	4.2 (34)	12.3
Downwind–Upwind (B–C)	1.3 (50)	1.3 (50)	2.6

[a]Numbers in parentheses represent the percentages of the respective concentrations in the coarse and fine fractions.

Table 5. Average Carbonate Carbon Concentrations at Downwind and Upwind Sites[a]

Sites	Carbonate Carbon Concentrations (μg C/m^3)		
	Coarse[b]	Fine[b]	Total
A. Site 4	2.0 (59)	1.4 (41)	3.4
B. Sites 1, 2, 3, 5 $\geqslant$10% Downwind	0.70 (77)	0.21 (23)	0.91
C Sites 1, 2, 3, 5 $<$5% Downwind	0.23 (72)	0.09 (28)	0.32
Downwind–Upwind (A–C)	1.8 (58)	1.3 (42)	3.1
Downwind–Upwind (B–C)	0.5 (83)	0.1 (17)	0.6

[a]Note that concentrations as CO_3^{2-} are obtained by multiplying the listed concentrations by 5.

[b]Numbers in parentheses represent the percentages of the respective concentrations in the coarse and fine fractions.

not permit the identification of the specific sources responsible for the aerosol at Site 4. Further insight could undoubtedly be gained by use of other trace element and chemical data.

Also tabulated in Tables 3–5 are the differences between downwind and upwind concentrations. In these tables the average concentration at Sites 1, 2,

Table 6. Increase in Average Carbonaceous Aerosol Concentrations in Air
Passing over the Kosovo Complex

Carbon Type	Concentration Range[a] (μg C/m^3)
Organic Carbon	4.9–15.0
Elemental Carbon	2.6–12.3
Carbonate Carbon	0.6– 3.1
Total Carbon	8.1–30.4

[a]The ranges correspond to A–C and B–C in Tables 3–5.

3 and 5 during periods when the sites were downwind less than 5% of the
time was taken to represent upwind conditions. Downwind concentrations
were represented by two cases: the average concentration at Site 4 and the
average concentration at Sites 1, 2, 3 and 5 when these were downwind at
least 10% of the time. The two sets of differences which result are reasonable
estimates of the maximum and minimum impacts of the industrial complex
on nearby receptor sites. These results are summarized in Table 6. Because
half or more of the incremental concentrations are associated with coarse
particles, significant reductions in the carbonaceous aerosol concentrations at
the receptor sites should be achievable by conventional control procedures.
Such control, however, is complicated by the lack of specific knowledge concerning the identity of the sources. The remaining fine particle mass is more
difficult to control and is a more serious cause for concern because of the
greater potential for respiratory tract penetration of such particles. Specific
knowledge as to both the sources and chemical composition of the fine
organic carbon is highly desirable.

SUMMARY

Filters from the vicinity of a Lurgi lignite gasifier in Kosovo, Yugoslavia,
have been analyzed for organic, elemental and carbonate carbon. Linear regression analysis was performed on the data to identify the sources of the
carbonaceous aerosol. Because of the complexity of the principal downwind
site and of the industrial area in which the gasifier was located, an unequivocal
determination of the impact of the gasifier itself could not be made. The data
did, however, suggest that fugitive lignite dust and combustion emissions were
significant sources of carbonaceous aerosol in the vicinity of the Kosovo
complex. Almost half of the carbonaceous aerosol was associated with coarse

particles, and the coarse and fine carbon were determined to result from different sources. Because of the large fraction of the aerosol in the coarse particles a significant reduction in the total carbon mass concentration should be obtainable by conventional particle control procedures. Further research is needed, however, to identify the *specific* sources and chemical composition of the carbonaceous aerosol emanating from the Kosovo complex. Particular attention should be given to the fine particle aerosol, which is difficult to control and has the greatest potential for retention in the lung.

ACKNOWLEDGMENT

This work was supported in part by U.S. Environmental Protection Agency Grant No. R806274.

REFERENCES

1. Johnson, R. L., J. J. Shah and J. J. Huntzicker. "Analysis of Organic Elemental, and Carbonate Carbon in Ambient Aerosols," in *Sampling and Analysis of Toxic Organics in the Atmosphere* (Philadelphia, PA: American Society for Testing and Materials, 1980), pp. 111–119.
2. Johnson, R. L., et al. "An Automated Thermal-Optical Method for the Analysis of Carbonaceous Aerosol," in *Atmospheric Aerosol: Source/Air Quality Relationships,* E. S. Macias and P. K. Hopke, Eds. (Washington, DC: American Chemical Society, 1981), pp. 223–233.
3. Huntzicker, J. J. "Analysis of Organic and Elemental Carbon in Ambient Aerosols by a Thermal-Optical Method," presented at the International Symposium on Particulate Carbon: Atmospheric Life Cycle, General Motors Research Laboratories, Warren, MI, October 1980 (to be published by Plenum Press, 1981).
4. Mitrovic, M., S. Tomasic and S. Bratuljevic. "Should High-Ash Lignite be Burned at Power Plants (Kolubara and Kosovo Lignite)?" translated from Serbo-Croatian by the Ralph McElroy Co., Custom Division, Austin, TX.
5. Shah, J. J., R. L. Johnson and J. J. Huntzicker. Unpublished data.

CHAPTER 23

PARTICLE SIZE DISTRIBUTIONS OF TRACE ELEMENT EMISSIONS NEAR A COAL GASIFICATION PLANT IN KOSOVO, YUGOSLAVIA

L. C. S. Bouéres and J. W. Winchester

Department of Oceanography
Florida State University
Tallahassee, Florida 32306

J. W. Nelson

Department of Physics
Florida State University
Tallahassee, Florida 32306

INTRODUCTION

An aerosol sampling program near a coal gasification plant at Kosovo, Yugoslavia, was carried out in order to determine the contribution of its particulate emissions to nearby air quality. The results of this study are significant in the United States for predicting the effects of similar technology which may be established in the future. A report on sampling and particle-induced X-ray emission (PIXE) analysis methods together with representative data obtained for variation of elemental composition with time and some particle size distributions has been given elsewhere [1]. In that report the correlations of variability in time of elemental concentrations at Site 4, predominantly downwind of the plant, were shown to be strong for sulfur and iron, and for zinc and lead, but not between the pairs, suggesting distinct

425

pollution sources in the vicinity. In this chapter evidence based on particle size distributions which tend to support these elemental groupings is presented. Moreover, the elemental data at all sites plus the meteorological information indicate that the former pair (sulfur and iron) but not the latter pair may be associated with the coal gasifier emissions.

EXPECTED AEROSOL EMISSIONS FROM THE KOSOVO COMBINE

The Kosovo industrual complex is located in a large valley surrounded by farming communities [2]. This complex comprises five potential sources of aerosols: (1) a coal gasification plant, (2) a coal fired utility plant, (3) a steam generating plant, (4) a fertilizer plant and (5) the coal processing area. To these sources we should add the regional background emissions such as the soil-derived aerosols and those from farming operations.

The industrially derived aerosol may be subdivided into two types: primary and secondary. Among the primary aerosols we expect the major components to be: (1) coal fly ash from both the utility and the steam generating plants, (2) organic aerosols from the coal gasification plant and (3) coal dust from the processing area and the gasifier itself. Among the sources of secondary aerosols we expect the major ones to be: (1) sulfur species (H_2S, COS, mercaptans and thiophenes) from the coal gasification plant, (2) organic species from the coal gasifier, and (3) SO_2 from the utility, steam generating and gasifier plants.

In the particular case of the Kosovo aerosol we expect three important contributions of trace metals: (1) coal fly ash, (2) coal dust, and (3) topsoil particles. However, since the ash from Kosovo lignite is mainly composed of calcium carbonate ($CaCO_3$), the major trace metal contribution from the industrial complex should be in the form of calcium carbonates (detected as Ca in our PIXE analysis).

EXPERIMENTAL

Cascade impactor samples were collected at five sites [1, 2] for approximately 6-hr periods (designated as A, B, C, D, for the periods nearest the times 0000-0600, 0600-1200, 1200-1800, 1800-2400, respectively) and analyzed for elemental composition by particle-induced X-ray emission as described previously [1].

Meteorological measurements made during the sampling interval (18-25 May 1979) indicated that the distinction between upwind and downwind locations of Site 1 and Site 4, respectively, was most pronounced during the afternoon periods, viz., samples 5C, 6C, 7C and 9C at the two sites. In order to determine qualitatively the importance of particulate emissions of the plants to aerosol composition in the region, these four sampling periods at the two sites were selected for close examination.

RESULTS

The particle size distributions of elements calcium, iron, manganese, sulfur, silicon, zinc and lead at Sites 1 and 4 for the above-mentioned periods are given in Figures 1, 2, 3 and 4. Where differences in concentration are found between the two sites, they are most marked for the coarsest particles. Accordingly, Table 1 summarizes the concentration ratios downwind/upwind in the >4 μm aerodynamic diameter (μmad) fraction, impaction stage 5, for the seven elements considered here. These results provide a qualitative estimate of which elements in the industrial particulate emissions are most significant in changing the aerosol composition from its regional character or as affected by other pollution sources in the vicinity.

Calcium, iron, manganese and sulfur in coarse particles are clearly associated with emissions from the industrial complex. Silica and fine particle sulfur are not so clearly associated. Zinc and lead concentrations are actually lower at the downwind site during these four sampling periods, indicating that they are not associated with the source of coarse particle calcium, iron, manganese and sulfur. However, detailed measurements of their variability in time indicate that another pollution source of zinc and lead in the vicinity is likely [1].

DISCUSSION

It is noteworthy that all seven elements considered here exhibit their highest concentrations in the coarse particle fraction. These elements include calcium, iron, manganese and silicon which are often associated with coarse particle dusts of natural or pollution origin. However, sulfur is also mainly in the coarse particle fraction, in contrast to its frequent occurrence elsewhere as a major fine aerosol constituent. As discussed in our previous report [1], the sources of particulate sulfur may be of a complex nature including homogeneous sulfur dioxide and hydrogen sulfide conversion, heterogeneous sulfur dioxide and hydrogen sulfide conversion, and direct emission as inorganic sulfur (sulfates or sulfides) in fly ash and/or coal dust.

Most of the observed increase, at Site 4, of coarse particles containing calcium, iron, manganese and part of sulfur may be attributed to the fly ash

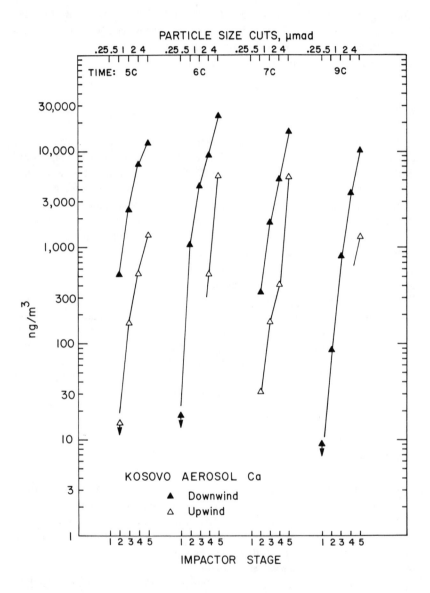

Figure 1. Particle size distributions for Ca measured at upwind Site 1 and downwind Site 4 for the indicated time periods 5C, 6C, 7C and 9C, on the afternoons of May 18, 19, 20 and 22, 1979, respectively. Minimum detection limits (MDL) are marked by arrows.

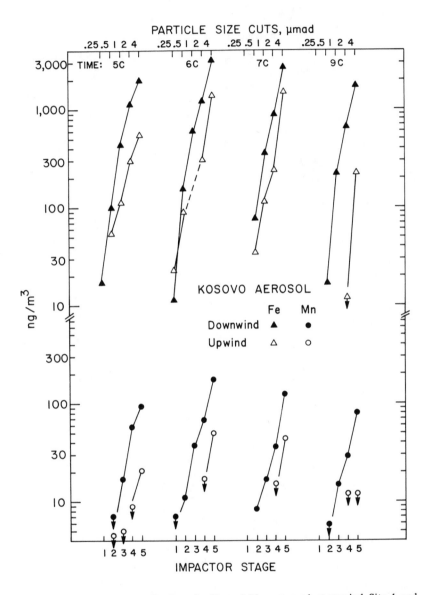

Figure 2. Particle size distributions for Fe and Mn measured at upwind Site 1 and downwind Site 4 for the indicated time periods 5C, 6C, 7C and 9C, on the afternoons of May 18, 19, 20 and 22, 1979, respectively. Minimum detection limits (MDL) are marked by arrows.

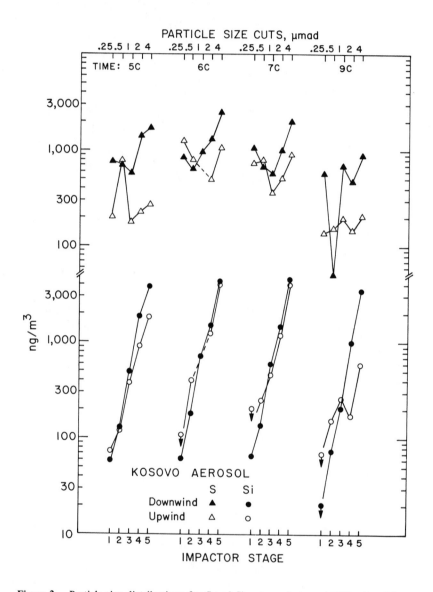

Figure 3. Particle size distributions for S and Si measured at upwind Site 1 and downwind Site 4 for the indicated time periods 5C, 6C, 7C and 9C, on the afternoons of May 18, 19, 20 and 22, 1979, respectively. Minimum detection limits (MDL) are marked by arrows.

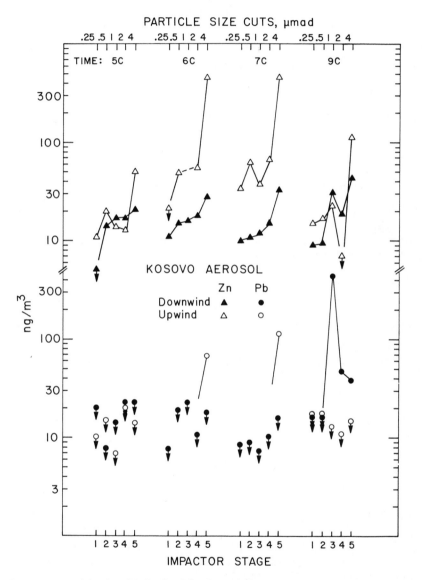

Figure 4. Particle size distributions for Zn and Pb measured at upwind Site 1 and downwind Site 4 for the indicated time periods 5C, 6C, 7C and 9C, on the afternoons of May 18, 19, 20 and 22, 1979, respectively. Minimum detection limits (MDL) are marked by arrows.

Table 1. Site 4/Site 1, Downwind/Upwind, Concentration Ratios for
the Coarsest Particle Fraction ($>$4-μmad) in the Kosovo Aerosol

	Time Period			
Element	5C	6C	7C	9C
Ca	9.2	4.2	2.9	7.8
Fe	3.5	2.3	1.8	7.6
Mn	4.5	3.5	2.8	$>$6.8
S	6.3	2.3	2.2	4.2
Si	2.1	1.0	1.1	6.0
Zn	0.43	0.06	0.07	0.39
Pb	–	$<$0.27	$<$0.14	$>$3.0
% Downwind[a]:				
Site 4	59.0	60.0	47.0	48.0
Site 1	0.0	0.0	0.0	0.0

[a]Percent of time the site was downwind of coal gasification plant according to meteorological observations.

plus coal dust components of a complex source region in the Kosovo basin. More insight into the associations of these and other elements with specific sources within the region may be obtained from detailed statistical interpretation of the entire Kosovo project data set, currently in progress.

ACKNOWLEDGMENTS

This work was supported in part by the U.S. Environmental Protection Agency, grant R806652. The authors are indebted to Ronald K. Patterson, Project Director, for substantial technical assistance.

REFERENCES

1. Bouéres, L. C. S., and R. K. Patterson. "Aerosol Emissions Near a Coal Gasification Plant in the Kosovo Region, Yugoslavia," *Nucl. Instr. Methods* 181: 353–358 (1981).
2. Bombaugh, K. J., et al. "Aerosol Characterization of Ambient Air Near a Commercial Lurgi Coal Gasification Plant: Kosovo Region, Yugoslavia," U.S. EPA Report 600/7-8-177, prepared by Radian Corporation (July 1980).

CHAPTER 24

COLLECTION AND CHARACTERIZATION OF AMBIENT AEROSOLS DOWNWIND FROM A COMMERCIAL LURGI COAL GASIFICATION FACILITY

W. D. Balfour, K. J. Bombaugh and L. O. Edwards

Radian Corporation
Austin, Texas 78766

R. K. Patterson

U.S. Environmental Protection Agency
Research Triangle Park, North Carolina 27711

INTRODUCTION

An ambient aerosol characterization study was performed in the vicinity of a commercial coal gasification plant in Region Kosovo, Yugoslavia. This study represented a unique opportunity to evaluate the potential environmental problems of a full-scale operating plant that was using Lurgi technology. The gasification facility at Kosovo is part of a large industrial complex that includes a lignite mining operation, a coal preparation operation, coal-fired boilers for power and process steam generation and an ammonia-based fertilizer plant. A process overview of the Kosovo gasification facility is described in Chapter 18. The complex is located in a large valley that is surrounded by farming communities. The area has a climate similar to the Midwestern United States.

All of the operations in the Kosovo industrial complex, as well as the local farming activities in the valley, are potential sources of atmospheric emissions.

Table 1. Participating Organizations and Their Responsibilities

United States Participants:

- U.S. Environmental Protection Agency (Aerosol Research Branch)   Sampling equipment, field sampling

- Radian Corporation   Project management, quality assurance, field sampling, soluble species analysis, scanning electron microscopy, elemental analysis, organic analysis

- Northrop Services, Inc.   Field sampling

- Oak Ridge National Laboratory   Organic analysis

- Florida State University   Elemental analysis

- Oregon Graduate Center   Carbon speciation

Yugoslav Participants:

- Rudarski Institut   Project management

- Kosovo Institut   Field sampling

- Institutza Primenu Nuklearne Energije (INEP)   Soluble species analysis, elemental analysis, organic analysis

- Kombinat Kosovo   Field sampling

However, emissions from coal gasification were expected to differ chemically from those attributed to the other sources. The objectives of this study were to provide qualitative and quantitative measurements of ambient air quality as influenced by the specific gasification process emissions using an upwind-downwind sampling strategy, supported by a fingerprinting of the emission sources. This chapter describes the overall sampling program and reports on the results of the physical characterization and of the inorganic chemical analyses of the collected aerosol samples.

SAMPLE COLLECTION

This study was an integrated effort in sampling and analysis by personnel from both the United States and Yugoslavia. Table 1 lists the various participating organizations, and the role(s) each played in the study. Due to the number of participating groups, their language differences and their varying technical backgrounds, an unusually comprehensive management and quality assurance program was implemented.

A total of five sampling stations were deployed within 1-2 km of the gasification facility, as shown in Figure 1. Choice of location was based on

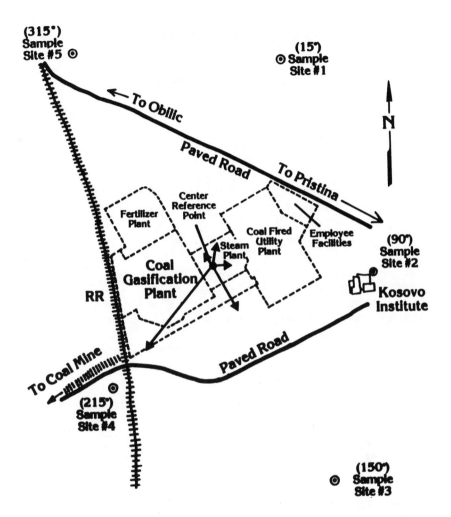

Figure 1. Schematic of the Kosovo complex. Vectors indicate the relative time that each site was downwind.

wind direction and site access. According to historical meteorological data furnished by Kosovo Institute, the wind during May (the test period) is predominantly from the north. The actual selection of the sites reflected the availability of the land (owners' cooperation) and the ability to provide equipment security. Locations of sites, typically, provided upwind, down-wind and offwind data from the gasification emission source. The distances from each site to the gasification facility (reference point) had to be estimated because military security regulations prohibited the use of measuring devices.

Compass headings shown in Figure 1 are in degrees from a reference point within the plant.

Sites 1, 3 and 5 were located near private residences, typically in fields or gardens. The immediate surrounding area was flat and free of obstructions, except for homes 30–40 m away. Sites 1 and 5 were generally upwind. Site 2 was located on the grounds of the Kosovo Institute, which served as the base of operations during the field study. It was located 30 m from the nearest building in an unobstructed area and was relatively free of activity. Site 2 was generally offwind. Site 4 was located in a pasture that was usually downwind from the gasification facility. It was 30 m from a railroad track, that was very active with both diesel (27 daily) and steam (21 daily) locomotives. To the north of the site (~ 50 m) were a paved road, two coal conveyors and a track for an electric train.

Each of the five sampling sites was equipped to collect particles and organic vapors for analyses. Collections were as follows: total and fine particles for gravimetric, inorganic and elemental analysis; size-fractionated particles for elemental analysis; total suspended particles and organic vapors for characterization by gas chromatography (GC) and GC/mass spectrometry (GC/MS). The sampling equipment used for this study is listed in Table 2.

Table 2. Summary of Sampling Equipment

Equipment Item	Collection Medium	Sample Collection	Nominal Volume Sampled (m^3)	Nominal Flow-rate	Sampling Period
High-Volume Sampler (Hi-Vol)	Gelman Microquartz Filters	Particles	1500	1 m^3/min	24 hr
Organic Vapor	Tenax GC Resin	Volatile Organics	6	4 L/min	24 hr
Time Phased Sampler (Streaker)	Nuclepore Polycarbonate Filter	Particles ($<15\,\mu m$)	–	5 L/min	Continuous (4–6 days)
Cascade Impactor (Battelle)	Coated Mylar Surfaces (Nuclepore backup)	Particles (5 stages between 4.0–0.25 μm)	9	25 L/min	6 hr
Low Volume Aerosol Sampler (Lo-Vol)	Gelman Microquartz Filters	Particles ($<15\,\mu m$ and $<2.0\,\mu m$)	11	30 L/min	6 hr

The high-volume sampler (Hi-Vol) was of standard design, except for the head, which was modified to accept round, 265-mm-dia, Gelman Microquartz filters. The Hi-Vol filter medium was selected for its low organic background. The filters were held between stainless plates to prevent contamination from the gasket. A Sierra Model 310 flow control module was used with the Hi-Vol to regulate air flow through the system.

A flexible steel sample line was attached to the plenum upstream of the Hi-Vol blower. This sample line delivered filtered sample air to the organic vapor traps mounted on the Hi-Vol housing. The unit had two identical vapor collection systems with mechanical flow controllers. It also had a rotameter for checking sampling rate and associated valving for directing flow. Each vapor collection system consisted of two glass cartridges filled with Tenax GC adsorbent resin. Stainless steel materials were used upstream of the cartridges to prevent sample contamination.

The low volume sampler (Lo-Vol) served as a dichotomous sampler, providing filter samples of total and fine (<2 μm) particle sizes. A cyclone [1] was used which provided particle classification at 2 μm (nominal). Particles penetrating the cyclone (<2 μm) were collected on 47-mm Gelman Microquartz filters. A second Gelman filter colocated with the cyclone collected total particulate mass. Flowrates were maintained through the Lo-Vol by mechanical flow controllers and were monitored by rotameters.

Additionally, two meteorological stations were operated at Sites 3 and 5. A Climatronics meteorological station was operated at Site 5. It provided a continuous record of wind speed and direction, ambient temperature, barometric pressure and total solar radiation. Data were recorded both on strip chart and magnetic tape (Datel Data Logger II). A Bendix Aerovane was placed at Site 3. This system provided continuous wind speed and wind direction data, which were recorded on strip chart only. Due to the repeated failures of the Climatronic dewpoint sensor, manual wet-bulb/dry-bulb measurements were made at 6-hr intervals to determine ambient relative humidity.

All sites were oriented as shown in Figure 2. Electrical power was provided by portable, gasoline-powered generators (Sears, 3 kW). In each case, the generator was placed approximately 40 m downwind from the sampling station.

Samples were collected over a 16-day period. The sampling period for each of the samplers is shown in Table 2. All samplers were operated continuously during the study.

Hi-Vol Microquartz filters were kept in aluminum foil covers before and after exposure. These filters were not weighed. Handling was kept to a minimum to minimize contamination. Vapor collection and the Hi-Vol sample period coincided, with collection media change at midnight. Flowrates were adjusted and recorded at that time.

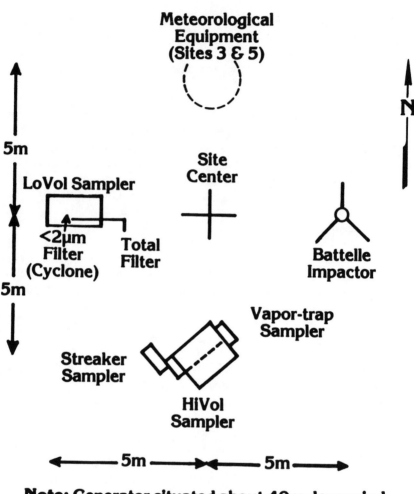

Note: Generator situated about 40m downwind

Figure 2.   Schematic of sampling site geometry.

Impactors were loaded and unloaded in the laboratory. Sampling rates were adjusted, using a calibrated rotameter, at the beginning of each 6-hr sampling period and were recorded at the end of each sampling period. The Lo-Vol Microquartz filters were weighed using a Mettler ME30 microbalance. Prior to being weighed, the filters were desiccated for 48 hr, both before and after field exposure. Filters were stored and transported to and from the field in petri dishes. Flow controllers were adjusted to the set point at the beginning of each 6-hr sampling period. Any deviations from the set point were recorded at the end of each period. The cyclone catch ($>2 \mu m$) was not retained for analysis.

Quality assurance during the field sampling program covered the following areas: sampling media preparation, equipment calibration and operation, initial and final gravimetric measurements, sample storage and transport, and sample documentation. The purpose of the quality assurance program was to detect and correct improper procedures and to provide a means of evaluating the accuracy of the data. In addition to the daily flow adjustments by the field crew, daily spot checks were made as an independent audit procedure. Control-weights were weighed daily by the analyst to guard against instrument drift and to quantitate the effects of changes in humidity, temperature and pressure. Independent audits of filter weights were conducted at random on 10% of the filters. Records were reviewed on a daily basis to ensure proper documentation. Archived samples were checked for proper packaging and storage.

At the end of the sampling period, all samples were given to the U.S. team along with copies of all data sheets; originals were retained at the Kosovo Institute. Data recovery was greater than 98%. The samples were distributed to the various members of the project team as listed in Table 3. The samples from the middle of the sampling period (days 5-12) were transported to the United States where they were further distributed as follows: samples of the Hi-Vol and Tenax vapor traps for days 9-12 to Oak Ridge National Laboratory for organic analysis (extractions, GC and GC/MS); all impactor and time-phased samples to Florida State University for multielement analysis (PIXE);

Table 3. Samples Collected and Respective Recipients

Participating Organizations	Hi-Vol Filters	Organic Vapor Traps	Battelle Impactor Disc Sets	Streaker Sampler Slides	Lo-Vol Filters
INEP[a]	42	83	157	6	316
Radian/OGC[b]	23	42			326
ORNL	22	42			
Florida State University			161	12	
Totals	87	167	318	18	642

[a]INEP–Institut za Primenu Nuklearne Energije, Belgrade, Yugoslavia.
[b]The Oregon Graduate Center (OGC) received two sections of each Lo-Vol filter for carbon speciation; Radian analyzed for specific water-soluble ions and 15 metallic elements.

three grab sample bombs to EPA/RTP for volatile organic compound analysis (GC). Radian retained the remaining samples; days 5-8 of Hi-Vol filters and Tenax vapor traps and days 5-12 Lo-Vol filters. Organic analyses (extractions, GC and GC/MS) were performed on the Hi-Vol filters and vapor traps. The Lo-Vol filters were segmented and portions were used for CHNS, soluble species (ion chromatography) and multielement analysis. A remaining portion was sent to the Oregon Graduate Center for carbon speciation.

SAMPLE ANALYSIS

Elemental analyses were performed on the Lo-Vol aerosol samples using two techniques—inductively coupled argon plasma emission spectroscopy (ICAP) and combustion analysis. ICAP provides simultaneous measurement of multiple elemental concentrations. The filter samples were digested in a mixture of concentrated HF and HNO_3 and injected directly into the ICAP. A similar procedure was followed with blank filters to quantitate background elemental concentrations. Combustion analyses for hydrogen, carbon, nitrogen and sulfur were performed with a graphite furnace. Filter samples were vaporized in the graphite furnace without pretreatment. Blanks were determined using a similar procedure.

Selected filter samples were extracted in deionized water and analyzed for soluble species using ion chromatography. Separate ion exchange columns were used to analyze for cations (Na^+, NH_4^+) and anions (Cl^-, NO_3^-, SO_4^{2-}).

Scanning electron micrographs (SEM) were obtained for specific Lo-Vol filter samples.

RESULTS OF PHYSICAL CHARACTERIZATION

The SEM of both upwind and downwind particulate samples show that the particles are predominantly composite mineral matter. There was little if any ($<1\%$ by number) sintered fly ash present in either the upwind or downwind samples. The downwind filters visibly were more heavily loaded than the upwind filters. A majority of the particulate matter on the downwind samples appeared to be lignite dust.

Particulate mass concentrations in the vicinity of the gasifier were calculated for total and fine particulate (<2 μm dia) mass as measured by the Lo-Vol sampler. Total mass concentrations ranged from 30 $\mu g/m^3$ to >400 $\mu g/m^3$. Fine particulate mass concentrations ranged from 10 $\mu g/m^3$ to >150 $\mu g/m^3$. In order to determine the ambient particulate mass contribution from the gasifier complex, the mass concentrations were correlated with the percentage of time that the collection site was downwind from the gasifier. As shown in Figures 3 and 4, both total and fine particulate mass concentrations are highly correlated with percent downwind. The intercepts for the total and fine particulate mass concentrations represent the ambient backgrounds. The fine fraction accounts for ~28% of the total particulate mass in the ambient background, but only 20% when downwind from the gasifier. These data indicate that the majority of the increase in total mass downwind from the gasifier is due to the coarse particle fraction (>2 μm).

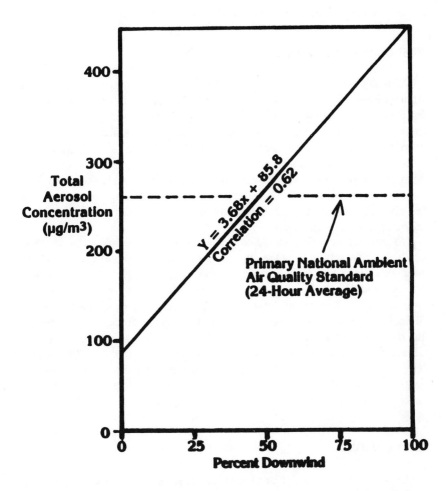

Figure 3. Plot of total (<15 μm) aerosol concentration versus percent downwind.

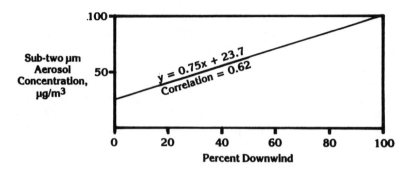

Figure 4. Plot of sub-two micrometer aerosol concentration versus percent downwind.

RESULTS OF CHEMICAL CHARACTERIZATION

A strong correlation was found between total carbon and percent downwind (Figure 5) for both the total particulate mass and the fine-particle fraction. The ratio of carbon associated with the fine-particle fraction to carbon on the total particulate mass decreases significantly as the percent downwind increases. Approximately 80% of the collected carbon is associated with the fine-particle fraction in the ambient background. However, downwind of the plant only 45% of the total particulate carbon is associated with the fine-particle fraction. These data indicate a significant portion of the increase in particulate mass downwind of the plant can be attributed to coarse carbonaceous material.

No correlation was observed between either sulfur (combustion analysis) or the soluble sulfate (analyses by ion chromatography) and percent downwind. Sulfate concentrations were nominally 12 $\mu g/m^3$ for the fine particle fraction and 21 $\mu g/m^3$ for the total particle catch. These sulfate values were from 6 to 30% greater than the sulfur values obtained by combustion analysis. Similar discrepancies have been reported by other researchers [2]. However, the data are consistent within either analytical scheme, and their use here as a basis for identifying trends and differences is considered justifiable. Similarly, no correlations were evident between either the other soluble species analyzed (Na^+, NH_4^+, Cl^-, NO_3^-) or the 15 metallic elements (ICAP) and percent downwind.

CONCLUSIONS

As a result of the chemical and physical characterizations described above, the following conclusions can be presented:

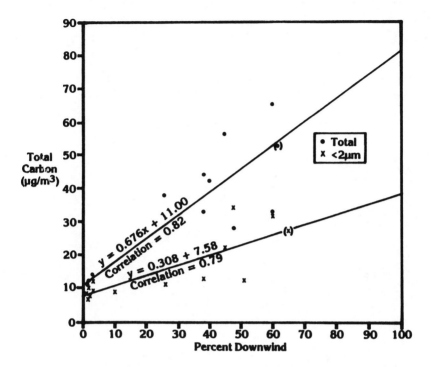

Figure 5. Plot of total carbon in sub-fifteen and sub-two micrometer aerosol fractions versus percent downwind.

1. Increased particulate mass concentrations downwind of the gasification facility appear to be the result of coal dust (lignite) resulting from the coal-handling operation.
2. Any particulate mass emitted directly from the gasification processes is overshadowed by the fugitive emissions from coal handling.
3. Ambient aerosol levels downwind of the gasification plant can be expected to exceed National Ambient Air Quality Standards for total suspended particulate matter. As such, control of coal handling emissions may be of concern in a large U.S. facility.

REFERENCES

1. Smith, W. D., R. R. Wilson, Jr. and D. B. Harris. "A Five Stage Cyclone System for In-Situ Sampling," *Environ. Sci. Technol.* 13:1387–1392 (1979).
2. Husar, R., et al. "Sulfur Budget of a Power Plant Plume," *Atmos. Environ.* 12:549–568 (1978).